Handbook of Mechanical In-Service Inspection

Pressure Systems and Mechanical Plant

Clifford Matthews
BSc, CEng, MBA, FIMechE

**Professional
Engineering
Publishing**

Professional Engineering Publishing Limited,
London and Bury St Edmunds, UK

First published 2004

ISBN 1 86058 416 0

© 2004 Clifford Matthews

Crown Copyright material is reproduced with the permission of the Controller of HMSO and the Queens Printer for Scotland.

A CIP catalogue record for this book is available from the British Library.

About the Author

Initially a Marine Engine Officer in Los Angeles, Cliff Matthews returned to England to act as first mechanical engineer for the Central Electricity Generating Board in 1987. He went on to work as a senior surveyor for Lloyds' Register, before becoming a consultant mechanical engineer in 1991.

Cliff has extensive experience as a consulting/inspection engineer on power/chemical plant projects worldwide: Europe, Asia, Middle East, USA, Central and South America, and Africa. He has been an expert witness in a wide variety of insurance investigations and technical disputes in power plants, ships, paper mills, and glass plants concerning values of $40m. Cliff also performs factory inspections in all parts of the world including China, USA, Western and Eastern Europe. He carries out site engineering in the Caribbean – Jamaica, Bahamas, and the Cayman Islands.

Cliff is also the author of several books and training courses on pressure equipment-related subjects.

Acknowledgements

Special thanks are due to the following:

- Stephanie Evans, for her excellent work in typing the manuscript for this book.
- The staff at Professional Engineering Publishing (PEP) for their editorial skills.
- Neil Haver, for his continuing high-quality advice on power station boilers and superheaters.

If you find any errors in the book or have comments on any of its content (except, perhaps TOTB), I would be pleased to receive them. You can contact me through my website at:

www.plant-inspection.org.uk

Also by the Author

Other Related Titles

How Did that Happen? Engineering Safety and Reliability	Edited by W Wong	1 86058 370 9
Process Machinery – Safety and Reliability	Edited by W Wong	1 86058 046 7
Reliability and Risk Assessment – Second Edition	J D Andrews and T R Moss	1 86058 290 7
The Reliability of Mechanical Systems	Edited by J Davidson	0 85298 81 8
An Engineer's Guide to Pipe Joints	G Thompson	1 86058 081 5
Improving Maintainability and Reliability through Design	G Thompson	1 86058 135 8
Guide to European Pressure Equipment	Edited by S Earland	1 86058 345 8
Guide to European Pressure Equipment	Edited by P Simmons, B Nesbit, and D Searle	1 86058 336 9

For the full range of titles published by Professional Engineering Publishing contact:
Marketing Department
Professional Engineering Publishing Limited
Northgate Avenue
Bury St Edmunds
Suffolk IP32 6BW
UK
Tel: +44 (0) 1284 763277; Fax: +44 (0) 1284 704006
e-mail: marketing@pepublishing.com
website: www. pepublishing.com

Contents

Part 2   Technical Disciplines

Appendices

Introduction – The Handbook of In-Service Inspection

This book is written as a partner to my existing book *The Handbook of Mechanical Works Inspection* published by Professional Engineering Publishing. It starts where that book finishes, at the point where mechanical equipment is released from the manufacturer's works and put into service. The scope of this book is slightly wider but the approach is the same; it covers only items of mechanical plant. Other types of equipment such as electrical, control and instrumentation, and computer hardware/software items are inspected and tested in-service, but this is an entirely different subject.

What is this book about?

It is about the engineering techniques of in-service inspection – what to inspect and how to inspect it. While it is essentially a practical *technical* book it also covers some of the wider aspects of the in-service inspection business such as who does inspections and how they do it. It also looks at how *well* they do it. The book tries to provide information which is applicable to a wide variety of different industries. Power generation, refinery, petrochemical, offshore, and various types of general steam and process plant are subject to in-service inspection so I have tried to provide information that will be useful to the inspection engineer working in any of these fields. Many items such as vessels, pipework systems, etc. are common to many industries, and the basics of their in-service inspection are not *that* different.

How to use this book

The book is intended for use as a technical 'how to do it' guide for inspection engineers involved in in-service inspection. It contains straightforward technical guidance. You can use the methods shown to help you decide what to do during an inspection, and how to interpret the results that you find. The book should point you towards relevant parts of technical codes and published documents, so that you can go and look up any information you need in greater depth.

The book also includes discussion about, and references to, some of the statutory regulations that lie behind the activities of in-service inspection, mainly of lifting equipment and pressure equipment. I have tried to provide usable guidelines to what can be a difficult and confusing area. Treat these as guidelines only – they are not a substitute for, or a full explanation of, statutory requirements. You will need to ask the enforcing authority for that.

Practical or theoretical?

Most (perhaps 90 percent) of this book is pure *practice*. This is also supplemented, however, in selected areas, by some theoretical aspects, mainly design and stress calculations and similar. These go no deeper than absolutely necessary. Their purpose is to help enable the in-service inspection engineer to justify important decisions on future plant operating conditions, fitness for purpose, and service lifetime.

Fact or opinion?

It's both. In-service inspection has its roots in technical fact but is mainly *implemented* through engineering interpretation and judgement, and you cannot have those without opinion. All engineering inspectors carry their own opinions with them and use them to help them form judgements on engineering matters such as what inspection scopes and techniques to use and what to do when defects are found.

Then – thinking outside the box (TOTB)

There is no real reason why inspection engineers *need* to think outside the box – this is the box containing conventional well-proven ideas about in-service inspection. If you do need to do it, then read the TOTB pages included at the end of some of the chapters.

Clifford Matthews BSc, CEng, MBA, FIMechE

Part 1

Chapter 1

In-service inspection – the concept

In-service inspection (ISI) of engineering equipment exists, in some form or other, in most developed countries of the world. It is the regular planned inspection of equipment during its use. In most cases the inspection is *periodic* and continuous, taking place at predetermined time intervals throughout the working life of the equipment.

ISI – why bother?

Opinions vary widely as to the exact purpose of ISI. There are several clear, logical *engineering reasons* why ISI of engineering equipment is necessary and most engineers would agree that these reasons make ISI a good idea. Unfortunately these unambiguous reasons are surrounded by a mass of less well-founded technical opinion. This is where viewpoint and judgement appear, making agreement difficult. The complexity of the situation is compounded by the fact that not all of the viewpoints involved are technical. The spectre of viewpoints based on management and business objectives soon appear to cloud the engineering inputs. The end result is that the real engineering reasons for ISI have to coexist with all the others that surround them – which means that, in practice, some in-service inspections make good sense and some probably do not.

Against this background lie the three main tangible reasons for in-service inspection:

- *The engineering reason.* All engineering equipment has a finite life and deteriorates because of use, elapsed time, or a combination of the two. The concept of ISI fits in with the necessity to assess the condition of, and maintain, critical parts of equipment during its life.
- *Rules and regulations.* Throughout the world, engineering plant is subject to the restrictions of rules, regulations, and legislation of the countries in which it is used. Most of these are related to safety and

apply mainly to pressure equipment, lifting equipment, and a large variety of smaller items that can pose a hazard to people if they are not maintained and operated correctly. The existence of such rules and regulations is the key driver for ISI in some industries. The situation is far from uniform however; some industries are more heavily regulated than others. The culture of compliance also varies – some industries exceed legislative requirements in the interest of good engineering and business practice, while others seem to operate policies of doing the absolute minimum to comply with the law as they themselves interpret it.

- *Insurance requirements.* Organizations that use and operate engineering equipment (ranging from a single machine or item of equipment to the largest plant) usually want to insure themselves against the risks of accidents, breakdowns, business interruption costs, etc. In order to cover such risks most insurers require that plant is subject to regular inspections either by themselves or an independent body of some sort. The extent and quality of such inspections varies hugely *between* industries but tends to be fairly consistent *within* industries. It is fair to say, however, that insurance inspections are limited in scope, perhaps even superficial in some cases, but do nevertheless form one of the driving forces behind in-service inspection.

ISI and integrity management

Integrity management – what is it?

Integrity management is a term which is being used with increasing frequency by petrochemical, power, and process industries, where the issue of plant integrity forms part of their overall strategy of asset management. Interestingly, this does not necessarily mean that these utilities necessarily view plant integrity in a pure engineering sense; it can be the opposite, with technical aspects relegated to a subset of the overall task of management of the asset. This often means that plant managers do not retain full technical (and inspection) capability in-house, instead preferring to subcontract these services from outside. Integrity management can therefore manifest itself either as a wide-ranging set of technical activities, or as a pure management exercise.

What about inspection?

In-service inspection *does* play its part in integrity management – it is obviously necessary to inspect plant to find out its condition. One general effect of the integrity management concept, however, has been to diversify the nature of the inspection activity. Inspection frequencies have become less rigid, and the scope of inspections themselves has changed from rigid scopes of work including full stripdown and non-destructive testing (NDT) to a more risk-based approach concentrating on only the more critical areas. Non-intrusive inspection techniques that can be performed without shutting down the plant are becoming more popular. The concept of integrity management has increased, therefore, the *sharpness of focus* on in-service inspections. The effect has been to question the effectiveness of inspections, rather than to do them 'by rote' to a set of prescribed frequencies and guidelines.

In-service inspection versus construction inspection

Irrespective of the variety of types of in-service inspection that can exist, there is a clear boundary between the activities of *in-service* inspection and those of *construction* inspection. Construction inspection is the inspection of new equipment during manufacture in the works and its pre-use commissioning on site. As for in-service inspection, construction inspection is influenced by technical codes and standards (and for some equipment, statutory requirements) but the main drive tends to be the commercial requirements and preferences of the purchaser. It would be wrong to say that safety and integrity issues do not play a part in the activities of construction inspection but they are often seen as being secondary to the requirements of the purchase order or to the client's technical specifications.

Paradoxically, from a technical viewpoint, the *scope* of construction inspection is actually wider than that of in-service inspection. This is because most items of engineering plant are not covered by in-service legislative requirements. Items such as pressure equipment, lifting equipment, some structural items, vehicles, etc. are subject to in-service inspection legislation in most developed countries but vast amounts of other types of engineering items are not. In-service inspection of these excluded items is therefore an option rather than being mandatory and is left to the owner or user to either do it or not, as they think fit. Figure 1.1 shows the general situation.

Technically, the *activities* of in-service inspection are wider than those of construction inspection. Once a piece of equipment has been put into use it is subject to various degradation mechanisms: corrosion, fatigue, creep, and straightforward wear and tear, etc. that are not an issue with

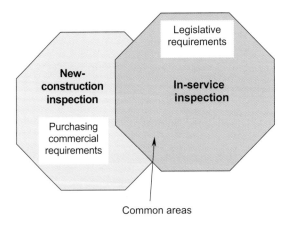

Fig. 1.1   **In-service inspection versus construction inspection**

new equipment. With most equipment, the issues of integrity and fitness for purpose (FFP) are made more complex by the effects of these degradation mechanisms. As a rule, the more complex the design and construction of a piece of equipment, the greater is the complexity of the effects of its degradation in use. This means that, for complex equipment such as turbine pumps, pressure systems, etc. the assessment of FFP and integrity becomes progressively more difficult as time progresses. Worse still, the effect of most degradation mechanisms are not linear and so general levels of uncertainty and risk increase unpredictably during a piece of equipment's operational life.

Does this mean that in-service inspection is *difficult*? Opinions vary on this but there is little doubt that however simple a piece of equipment in an engineering sense, there will always exist a degree of uncertainty about the condition of that equipment during its working life. In-service inspection rarely has the degree of predictability that can (sometimes) exist in new-construction inspection. For this reason in-service inspection rarely involves quantitative aspects alone; qualitative techniques such as risk-based analysis have to be used in order to handle the uncertainty. This is what makes in-service inspection interesting.

In-service inspection – predictive or defensive measure?

Huge rambling debates exist about whether in-service inspection is a true predictive measure, i.e. in which the objective is to incisively discover the true condition of a piece of equipment and then predict the path of its future life, or whether it is *defensive*, i.e. done only to comply with external rules and regulations. Despite the apparent endless nature of the arguments about this, the answer is, fortunately, easy: it depends on the *type of equipment*. This means that it is not really related to whether the equipment operators are conscientious or responsible, or knowledgeable. Neither is it related to the culture of the operation organization (an intangible concept anyway). In reality it is governed mainly by engineering matters: the factual, identifiable technical features of the equipment in question.

Think about how this works in practice. Items of equipment which are simple, which lack engineering complexity, perhaps which are made of common, well understood materials (take as an example a compressed air pipe) have the following characteristics:

- They don't fail or break down *that* often. Granted, there are a handful of failures of compressed air pipes every year but the other 150 million in the world continue to work fine.
- In-service inspections do not generally find lots of problems and defects. There will be some but, because potential failure mechanisms are simple and well understood, they have been 'designed-out', and so rarely occur. This means that the incidence of in-service inspections actually *preventing* failures by the early identification of failure-causing problems is very low, possibly negligible.
- The inspections themselves are straightforward, relying mainly on visual and simple non-destructive examination rather than sophisticated techniques.
- The items have a simple *function*.

These four points fit together with a certain engineering logic – the items themselves are simple, with *simple* design features, so their inspection is *straightforward* and dramatic inspection results are rare. Numerous plant items such as pipework, valves, basic rotating equipment, and batch-produced items such as simple vessels, fire extinguishers, mechanical fittings, and similar items have these characteristics. Owing to the coexistence of the four factors, the result is that the in-service inspection of these types of equipment becomes routine and unexciting. In time, on operating plants that have greater technical issues and problems to worry about, inspection of these items becomes seen as a chore, an imposition. The inevitable result is that the whole ethos of the in-service inspection activity becomes a *defensive* one; it is only done because statute or regulations explicitly say that it *has to be* and, by inference, anybody that does not do it could be adjudged to be not 'duly diligent', if some type of accident were to occur.

Consider now the alternative situation where the equipment has complex high-performance functions, high-technology design features, and is made of many different types of materials, each with their own advanced features of strength, resistance to corrosion, creep, fatigue, high or low temperatures. Such advanced features are accompanied by technical difficulties and uncertainty – damage and degradation mechanisms tend to be unpredictable (experts will inevitably disagree over what they are and how they affect the function of the item). Even the inspection of the item will be subject to uncertainty; it will be difficult to agree the correct degree of stripdown for example (because the item is of complex construction), and the correct inspection techniques to be used, and what type of defects should be looked for. This is a complex picture.

Equipment that has these features is generally sensitive to operate, needs a lot of routine (and expensive) planned maintenance and time-dependent replacement of parts to minimize the chances of breakdown. This will be accompanied by a regular programme of inspection activities recommended by the equipment manufacturer. Some of these inspections will require shutdown and disassembly. Gas turbines, aircraft, complex pumps, and process equipment are typical items that would fall into this category. With this type of equipment, the driving forces behind in-service inspection are related to more positive engineering aspects rather then the blind need for compliance with imposed statutes and regulations. Inspection is seen as a *valid part* of the good engineering care of the item; it is often encouraged, and its scope voluntarily expanded, rather then being treated as a necessary evil.

Another characteristic of the inspection of complex equipment is that it attracts more interest from higher levels of the user's organization. Whereas the inspection of simple pipework etc. is generally delegated to plant inspector and technician level, more complex equipment will attract the attention of site metallurgists, corrosion engineers, performance engineers, and similar. 'Higher management' may even become involved. The result of this is to increase the complexity of inspection decisions – even simple technical situations can evolve into a debate, as everyone tries to decide what could, or could not, happen to the equipment in the future. This is where the character of in-service inspection becomes *predictive* – call it pro-active if you like – rather than being carried out for defensive and self-protective reasons alone.

Figure 1.2 is an attempt to summarize the situation. It is not perfect, and the situation varies from industry to industry, but it fits well with many real businesses.

THINKING OUTSIDE THE BOX – IN-SERVICE INSPECTION VERSUS ASSET MANAGEMENT

In the brave new word of name obscuration, inspection companies reincarnate, phoenix-like, as *asset management* companies. This is because (presumably) inspection can be seen as an organizational activity (a sort of personnel management for the metal), or because it sounds good, or both.

So does good 'asset management' mean more inspection or less inspection? Well, that's like asking if good personnel management means employing more people or less people. The amount of inspection cannot possibly stay the same, (or else there would be no point in changing the name), so it has to be one or the other doesn't it?

Maybe not. More inspectors (asset managers)? Or fewer inspectors (asset managers)? It's not the amount of inspection that's important (say the asset managers' managers), it's the quality of it – how hard you look for asset-worrying defects – that matters and that's where risk-based asset management comes in. Make it more effective, more efficient, and you can have all the reward with absolutely no increase in risk. Before long you won't need any inspection at all – the plant will inspect itself, neatly self-destructing (after giving a suitable audible safety warning of course) at the precise point it reaches the end of its economical life. That sounds like good asset management to me . . .

Fig. 1.2   **Thinking outside the box – in-service inspection versus asset management**

Chapter 2

The inspection business – who does what?

The driving forces

The nature of the in-service inspection business and the character of its main players are not the same in all countries of the world. Although the driving forces for in-service inspection are basically the same, i.e. the need for safety, integrity, and continued fitness for purpose (FFP) of equipment, the way in which these are achieved varies between countries. There are several main influences:

- *The amount of external regulation.* Some countries have firm prescriptive legislation on in-service inspection scope, periodicity, and the regulation of inspection organizations. Others (including the UK) have a more liberal, passive approach verging on self-regulation.
- *The commercial system in use.* Most (but not all) countries whose economy is based on a pure capitalist model tend to have minimum regulation on the qualifications and roles of inspection organizations. Those with more central command-based economies generally have more restrictions on who can inspect what – in extreme cases, inspection duties are restricted to a single government-organized department or technical institution.
- *The level of industrial maturity of the country.* Broadly, the more mature a country's industrial system, the more comprehensive are its in-service inspection practices. This is due mainly to experience – countries that have developed their industrial capability rapidly place emphasis on investment and development rather than the inspection regulations and practice. In such countries, inspection is viewed as a retrospective (even introspective) exercise and rarely attracts serious inward investment or government funding. In contrast, well-developed countries have long experience of plant failures and so have

developed systems of regulation and control to keep them to a minimum.

These driving forces act together to define the character of the in-service inspection industry that exists in a country. Perhaps surprisingly, the final structure that results does not vary *that* much between major industrialized countries. Figure 2.1 shows the main players and the way in which they relate to the inspection task, and each other.

The regulators

The highest level of national regulation of in-service inspection activities is provided by a country's government. It is normally the part of the department of government that deals with safety and/or industry matters. It may masquerade as an 'agency' or some kind of quasi-privatized body but the end result is the same – it has the power to formulate laws that appear in the form of statutory instruments (SIs) and impose them on those that sell, own, or operate industrial plant. In the UK this function is provided by the Department of Trade and Industry (DTI) Standards Regulations Directorate (STRD). Compliance is policed by the Health and Safety Executive (HSE).

Fig. 2.1   The structure of the in-service inspection industry

EU Directives

For countries that are members of the EU, there is a higher level of regulation requirements set by European Directives passed by the European Parliament. These directives are not law, as such, in member countries but have to be *implemented* in each country by national *regulations* – so the effect is the same.

Enforcement

In the UK, compliance with the statutory instruments relevant to in-service inspection is policed by HSE. The situation is similar in most other developed countries, the enforcing body being either an arm of government, or related closely to it. In the UK, enforcement is carried out locally, i.e. the local HSE inspectors 'attached to' a plant, supported by specialist technical laboratories and failure investigators.

The enforcers enforce using a subtle combination of pro-active and reactive methods. Inspectors *do* pro-actively visit plants to do random checks on in-service inspection-related issues but this is rarely the sole reason for their visit. General health and safety issues [predominantly compliance with the Health and Safety at Work Act (HASAWA)] often form the main thrust of HSE inspection visits and time and resource constraints usually mean that the inspection cannot address everything. A lot of HSE enforcement, therefore, ends up being *re*active. This means that the inspectors will be more likely to direct their interest on specific matters *after* an incident has occurred. This targeting is easy, as many types of incident are reportable, i.e. the plant owners or management are obliged to report occurrences possibly involving danger (even if no injuries have occurred) to the HSE. Such reportable incidents may, or may not, result in a visit from an HSE inspector, depending on their severity.

As with all rules and regulations, the quality of enforcement varies from case to case. It is fair to say that HSE inspectors probably have better things to do than review endless in-service inspection reports which report that no defects were found during an inspection and that the plant was reassembled and decommissioned without incident. They are much more interested in questioning what they see as being *wrong* rather than confirming what is *right*.

A similar 'reactive' feel surrounds the situation regarding the competency of organizations and people involved in in-service inspections. Notwithstanding various (mainly voluntary) schemes that exist for accreditation of inspection companies, these schemes do not go all the way in confirming *competence*, as such. In the event of a failure or accident, an organization and individual inspectors can still be called upon to demonstrate their competence, irrespective of any accreditation that they may hold. Whereas accreditation (to, for example, EN 45004: *General criteria for the operation of various types of bodies performing inspection* or any other voluntary scheme) can *help* show competence, it is by no means the final test. This is an important (and frequently misunderstood) point; an organization does not *have* to have accreditation to be competent, and conversely, just because it has accreditation doesn't mean that it *is* competent.

The way in which enforcers police the issue of competence is not always clear. Although there may be some assessment made of the competence of companies and individuals during HSE visits, for example, major questions are more likely to be raised after a breach of regulations has been identified. Some people see this as an 'innocent until proven guilty' principle in action – others take different views. In summary however, the general *principle* in the UK is that the inspection industry should be allowed to be self-regulating, not hindered by tight external regulation on who is competent to do what. Competence only becomes a question, really, when something goes wrong.

Accreditation bodies

In the UK, the main national accreditation body is UKAS (United Kingdom Accreditation Service). UKAS is itself 'accredited' by the Department of Trade and Industry (DTI), and hence given licence to exist, offering accreditation to industry at large. Accreditation is the activity of being checked for compliance with a definitive standard, or set of rules. The overall objective of accreditation is, nominally, to maintain a level of technical, organizational, and administrative capability within an industry and hence provide customers with a degree of confidence in the quality of service provided.

The three key issues of accreditation

The promoters (and maybe also the critics) of accreditation probably agree that there are three key issues at play in accreditation (see Fig. 2.2).

- *The accreditation 'standard'*. By definition, accreditation is a test for compliance against a definitive, tangible (normally published) standard. This means that a standard must exist, and the 'value' bestowed by the accreditation will only be as good as the quality of the standard used.

- *The accreditation process.* Accreditation activity can default two ways: as a search for compliance or as a search for non-compliance. Between these two extremes lie multiple shades of grey, each of which results in a different quality and effectiveness of the accreditation 'product'. The end result is that some accreditation processes are excellent, some just satisfactory, and others poor and of little use to anyone. It is for this reason that the control and licensing of accreditation bodies themselves is so important.
- *Industry norms.* Voluntary accreditation (which most accreditation is) only retains its value if industry gives it credibility. This is particularly relevant to the in-service inspection in the UK where such acceptance has never been total. It is currently not essential to be accredited to EN 45004 in order to perform in-service inspection on pressure plant, under statutory instruments such as the Pressure Systems Safety Regulations (PSSRs). Accreditation may (or may not) help inspection companies get work, but it is not mandatory.

The above three key issues decide the character of the accreditation system that exists in a particular country. Despite the levelling effect of EU rules and regulations, the systems still vary significantly in their

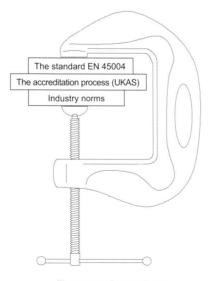

The standard EN 45004

The accreditation process (UKAS)

Industry norms

The pressure for compliance

Fig. 2.2   The three accreditation issues

operation across European countries. It is fair to say, however, that there is a general trend towards convergence of the way that accreditation works (and each country's approach to the three key issues) but the situation is still far from perfect.

Pressure Equipment Directive (PED) 'Notified Body' status

The situation regarding accreditation is different for the qualification of 'Notified Bodies' under the requirements of the European Pressure Equipment Directive (PED) and similar directives. The PED covers new construction (and some types of refurbishment) only and therefore, strictly, has no direct influence on in-service inspection. Organizations that are allowed to allocate the 'CE mark' under the PED are known as 'Notified Bodies' (a formal term mentioned in the PED) and have to be certified as such by the relevant accreditation body in the EU member country (UKAS in the UK). This certification is mandatory, and has the effect of restricting the status to organizations with both a track record of design appraisal and inspection work, and a well-structured and well-documented administration system.

Standards organizations

Standards organizations are divided into those that are purely *national*, and those that are *international*, e.g. with some kind of pan-European structure. Some have the status of government-funded nationalized institutions, with various constraints (supposedly) on what they can and cannot do, while others act almost like private companies, making their own decisions on which services they offer and being free to buy or sell other businesses as they like. Not surprisingly, the status that a standards organization adopts reflects strongly its character. It also affects the technological position of the organization, i.e. some maintain a strong position of up-to-date technical knowledge, demonstrated by the quality of the technical standards and other documents they publish, while others live mainly in the past.

Whatever their business structure, a major function of standards organizations is the coordination of the writing of technical standards. This is done using a committee structure consisting of (largely unpaid) contributors from various interest groups within the industry. For an inspection-related standard the committee would typically involve contributors from equipment manufacturers, contractors, purchasers, and users, as well as inspection companies themselves. Some standards bodies have several hundred new or 'under revision' published standards (and therefore committees) commissioned at any one time, so the process is *slow*.

Standards organizations as inspectors

Some standards organizations offer in-service inspection, mainly on pressure equipment. Such services are unrelated to these organizations' role as coordinators of technical standards, and are sold mainly on the basis of the organization being an independent body. The formalized administration and training structure of standards organizations means that they have little trouble in achieving accreditation, and so easily achieve status in the industry as legitimate players. From a purely technical standpoint, there seems to be little reason why standards organizations should be seen as being necessarily more technically competent than other commercial inspection companies. As mentioned above, the committee structure that produces the technical standards is effectively decoupled from the commercial in-service inspection activity – the people are different, and their way of working is different, so there is no real structural justification for any kind of guaranteed technical superiority.

One topic in which standards organizations may excel, is *design appraisal*. This is an area of in-service inspection that has a strong theoretical nature, particularly in relation to pressure equipment. Design appraisal is necessary to predict the effect of corroded wall thickness, vessel shell distortions, weld defects, and similar. These are mechanisms that can have an effect on the integrity and safety of pressure equipment and need to be properly assessed. In their advanced state, such activities become known as 'fitness-for-purpose assessments' (see Chapter 9). The highly theoretical nature of these assessments means that they are normally carried out by specialists, familiar with the interpretation of relevant standards such as BS 7910 and API 579 that govern this field. Many in-service inspection companies will not employ such specialists (it is not a mainstream business service and the demand is small), so standards organizations are well positioned to take this type of work.

Classification societies

Classification societies are unusual, hybrid organizations that occupy an uneasy role in the world of industrial in-service inspection. Without exception, they have their origin in the world of shipping and marine engineering where their original role was (as their name suggests) to classify ships and other marine vessels. 'Classification' involves checking that ships are designed to a set of 'society rules', written by the society themselves over many years, and then monitoring the construction, materials and safety tests, culminating in the award of a 'certificate of class' (Lloyds Register '100A1' being the best-known example). Classification as an activity has relevance to the shipping world only; there is no 'classification' of industrial plant, as such. Contrary to popular belief, classification societies are *not* insurance companies; in fact they have nothing whatsoever to do with insurance. Insurance of ships is a separate business, a market where marine risks are bought and sold in a market-type situation.

Outside the shipping world, classification societies have the same status as any other organization. They are free to offer industrial design appraisal and in-service inspection services, guided by their own internal constitution (which is a matter purely for them, having no statutory or regulatory significance in most countries). The level of industrial inspection services business varies a lot between classification societies; some offer services across a range of industries: power, petrochemical, manufacturing, etc., while others restrict themselves to plant similar to that found on board ships, perhaps because they understand it better and see it as less risky. In common with other inspection companies, some classification societies have become accredited to EN 45004, and some do not feel it is necessary.

Classification society technical specialisms

Classification societies vary a lot in their competence in specialist technical subjects outside the world of shipping. While most do industrial inspection work, a few take this to the extent that they have, in their structure, a separate *industrial services division*. Others offer it as an 'add-on' to their ship classification and inspection activities. It is fair to say that most classification societies are weaker in land-based industrial technical knowledge than marine-based, and probably secure a lot of their inspection work on the basis of their independence rather than any cutting-edge industry-specific technical knowledge.

Are classification societies truly independent?

Truly, yes they are. It is rare indeed for any of the major classification societies to be deeply involved in the contractual structure of the manufacture, operation, or maintenance of industrial plant. They work hard to maintain their independent 'honest broker' role with no allegiances to designers, contractors, or manufacturers. This invariably means, of course, that they sometimes remain aloof to the real technical or contractual issues in a project – they like to avoid controversy.

The notion of independence is helped along by the fact that some of the major classification societies actually have the status of registered charities. This has a mainly historical basis with its origins, again, in the shipping industry. In practice, it does not seem to affect the business practices of classification societies too much. Of those societies that do profess charitable status, a few have subsidiary limited companies that operate in specialist areas such as non-destructive testing (NDT), information services, design analysis, and similar.

Insurance companies

The position of insurance companies in the world of in-service inspection is frequently misunderstood. The role of the insurance inspector is common to many industries but, again, is a role that is often misconstrued. Perhaps the best way to conceive of the situation is to think of the role of insurance companies as one of *facilitation* – part (but only part) of the overall package of driving forces behind the need for in-service inspection.

What do they do?

Surprisingly the amount of inspection of industrial plant that is actually *necessary* for the purpose of maintaining insurance coverage is fairly small. Most insurance policies operate on the basis of an annual (or similar) inspection, of limited scope, with perhaps an overall 'risk assessment' before the policy is written. The contract of insurance itself, which is the formal document proving the existence of the insurance cover, consists predominantly of legal-type clauses and may say nothing whatsoever about inspection requirements. At first glance therefore, the in-service inspection requirements of plant insurance policies, i.e. the actual inspection requirements specified *directly* by the insurers, appear fairly sparse.

The real power of the insurers lies in requirements that live deeper inside contracts of insurance. They are partly written (probably hidden in the small print that no-one reads) and partly *inferred* – given validity by the long and unexciting tales of precedent that exist in the world of insurance claims. The key points of these requirements are:

- The insured party (known as 'The Insured') *is required to act, at all times, like a 'responsible uninsured'*, i.e. maintain (and inspect) the plant as if they themselves had to bear all the risk if something went wrong. This means that plant in-service inspections that need to be done for technical or engineering reasons have to be carried out, even though they may *not* be expressly mentioned in the contract of insurance.
- *Compliance with legislation, rules, and regulations.* In order not to invalidate the insurance cover, The Insured is expected to comply with whatever statutory legislation, rules, and regulations that are in force in the country where the plant is in use. This will include legislation covering pressure equipment and systems, lifting equipment for people and goods, and some aspects of rotating equipment, all surrounded by an envelope of all-consuming health and safety impositions.

Taken together, these requirements form the 'backdrop' of the structure of inspection requirements as seen by insurers through the eyes of their contracts of insurance. This is why in the UK, for example, in-service inspections under the Pressure Systems Safety Regulations (PSSRs) are often mistakenly understood to be 'insurance inspections' whereas strictly they are not.

The situation is not made easier by the fact that some insurance companies have a *composite* role, in which they act as paid inspectors (under, for example, the PSSRs) even if they are not involved in the insurance of the plant. The role of insurers is not homogeneous; they take different roles, depending on the boundaries drawn (usually by themselves, and a long time ago) for the role of their business. Figure 2.3 shows the situation. The main categories are outlined below.

'Pure' insurers

These are insurers whose only role is to write policies and buy and sell insurance risks. They carry no technical staff and subcontract risk analysis, in-service inspection and post-claim inspection work to other companies. They may be discrete, identifiable companies, or loose, ephemeral groupings or 'syndicates' of insurance companies and individuals. They act in a market situation, buying and selling risks and feeding off (and from) each other. The Lloyds of London insurance market works broadly like this.

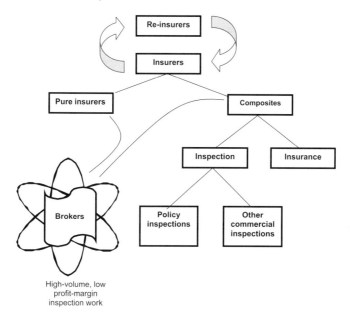

Fig. 2.3   The insurance inspection industry

Re-insurers

Re-insurers are, simply, insurers who take the role of risk *purchasers*; they buy insurance risks (or bits of them) from other insurers in the market, maintaining a percentage holding in a portfolio of different risks, rather than being the lead player. In the true nature of the interrelatedness that is the world of the insurance market, just about all major insurance companies act as re-insurers for all the others. So, most *insurers* are also *re-insurers*, in one way or another.

Composite insurance companies

Composites are a schizophrenic mix of insurance company and inspection organization. They write insurance business (i.e. enter into contracts of insurance) *and* act as inspectors, either in relation to their own insurance contracts and/or other unrelated jobs.

Composites can be difficult organizations to understand – the influence of their parentage as insurers has a visible effect on what they do and say, both at management, and in-service inspection 'surveyor' level. Their inspection tends to be conservative and risk-averse, as perhaps it should be, with emphasis being put on integrity and safety rather than operational fitness-for-purpose (FFP) issues. The assessment of the FFP of complex engineering plant can be a source of controversy when dealing with composites who are active in a purely independent inspection role. There are many situations in which a piece of pressure or lifting equipment can be perfectly fit for its operational purpose but can be surmised to be in a condition (frequently unproven) which represents an increased *risk*. This causes frustration.

From a competency perspective, composites retain their own permanent or contract inspection staff ('surveyors') with a certain degree of technical specialism, mainly in pressure equipment (the more straightforward the better), lifting equipment (cranes, vehicle lifts, fork-lift trucks, passenger elevators, etc.), and a few other minority plant disciplines such as power presses and steam railway locomotives. Composites also like to think they have well-developed surveyor training programmes and good metallurgical laboratories. All of this is bound up in a matrix of risk-averse procedural constraints that makes these hybrid organizations what they are.

Brokers

Brokers are an essential part of the way that the insurance market works. Aside from their main role of selling insurance policies on behalf of insurers, the larger ones also have a role in the organization of in-service inspection activities. The broker will often let an independent inspection contract on behalf of the insurer that underwrites the risk, taking a percentage commission in the process. Broker-commissioned in-service inspection work has the following characteristics:

- It is *high volume and repetitive*, involving multiple inspections at different locations around the country (garages, bakeries, retail outlets, etc.)
- The plant under inspection has *low technical complexity*, i.e. it is mainly small and simple. Typical examples are low-pressure air receivers, hot drinks boilers, fork-lift trucks, vehicle hoists, and small lifting gear.
- It is *low-profit-margin inspection work*, compared with heavier industrial inspection in the power, petrochemical, and process industries. This means that the qualifications of the inspectors used, and incisiveness of the inspections that they are able to do, have to be chosen to suit.
- *It is absolutely defensive*. Most recipients of these inspections, i.e. the small-plant owners/operators themselves, see inspection as an

imposition rather than for whatever engineering or risk-reduction benefits it may have. Many struggle with the hierarchy of reasons behind the need for in-service inspection and, at best, comply only because they have to.

You can see that this type of in-service inspection is not an easy business. It is high-volume, low-fee-rate, risk-bearing inspection work at its best (or worst). So, it attracts that sector of the in-service inspection business that feels best able to cope with this environment. In recent years, many large, highly structured, high-overhead organizations (including composites) have been forced out of this sector of the market, to be replaced by agency-type inspection companies, using freelance technician-level inspectors on an *ad hoc* basis, as the workload demands.

Risk assessors

Risk assessors work on behalf of insurance companies. They assess the 'risk' of a plant either before, or shortly after, an insurance contract is written. Although their role has an inspection function, it mainly involves a broad overview of those aspects of the plant that pose well-known insurance 'perils'. Insurance risk assessors do not, therefore, do detailed engineering in-service inspection as such – they take a much broader, high-level view. Their work rarely conflicts with, or has any influence on, the scope and extent of statutory or purely technical-based in-service inspections.

Loss adjusters

In contrast to risk assessors, loss adjusters only appear on the scene *after* there has been an incident resulting in an insurance claim. They are commissioned by the insurers, or re-insurers (or occasionally both), to investigate the claim and help decide the valid monetary amount, known as 'quantum', of the claim. Although appointed by the insurer, the loss adjuster actually acts in an independent role, with a duty to both parties: the insurer and the insured, to sort out the extent and quantum of claim. For claims including engineering plant and equipment, some type of inspection is often required. This can take the form of inspecting the plant to determine the extent of damage to assess whether it can be refurbished or has to be replaced, and to try to find the cause of the failure or incident. The objective is to identify the 'proximate cause of failure' (a definition with well-defined legal precedent) and to confirm whether or not this is the result of an 'insured peril' such as fire, sudden and unforeseen breakdown, negligence, etc. What is, and is not, an insured peril depends on the wording of the insurance contract or policy document.

Do loss adjusters do technical inspections?

Some do. A few of the major insurance companies, mainly those that act as composites, have their own network of technical and semi-technical loss adjusters who feel comfortable with the technical aspects of engineering claims. These tend to be limited to fairly straightforward, well-precedented, technical failures in which the insurance company has plenty of previous experience. This system seems to work well, with these technical adjusters becoming very specialized in their own industrial area. Equipment items such as transport containers, lifts, and simple pressure vessels are common examples.

For more complex and specialized engineering plant, loss adjusters like to commission engineering specialists to assist with the technical aspects of the inspections relating to the scope and quantum of the claim. This has the dual advantage of using the best technical expertise available for the job, and being seen to invite a third-party view that is independent from both insurer and insured. This is a specialized section of inspection – it has to be more precise and incisive than normal statutory in-service inspection work, and it can have big financial, legal, and liability implications. Typically, it is the more technologically orientated inspection companies and consultants that feel comfortable with this type of inspection work.

Inspection organizations

First, look at Fig. 2.4 and see if you recognize its content. In-service inspection organizations come in many shapes and sizes – they vary from a few large international organizations with offices in 100 or more countries, through to small local inspection companies that have only a handful of permanent staff supplemented by freelance self-employed inspectors who work on a part-time, as-required basis. The quality of the service provided varies significantly – from companies that provide highly competent inspectors with a real knowledge base to those that offer little more than the physical presence of 'an inspector' to meet minimum statutory requirements.

Figure 2.5 summarizes the situation. The figure is an imperfect attempt to show what the world of inspection organizations really looks

---

WHAT ALL INSPECTION ORGANIZATIONS WILL TELL YOU

- We are independent (and even if we're actually not, we promise to act like we are).
- We are quality driven.
- We chose our inspectors carefully and monitor their technical competence and reports.
- We are accredited to some standard or other (and have a small wall full of certificates to prove it).
- We maintain comprehensive internal documentation procedures that are audited regularly by ... someone or other.
- We do not consider ourselves inspectors anymore, we prefer to be called 'asset integrity managers'.
- You will be better off with us – we will be part of your asset integrity management risk-reduction portfolio.

Fig. 2.4   What all inspection organizations will tell you

like. In practice, because they all claim the attributes given in Fig. 2.4 they can be difficult to tell apart, until you have gained some experience in dealing with them. We will look at the three main groups in turn.

Technology-based inspection organizations (Group 1 in Fig. 2.5)

These operate at the upper end of the inspection market and are linked by the fact that they possess a sound technological base. This means that they possess in-house *technical* knowledge that they use in the context of in-service inspection work. Note that this is *technical* knowledge, perhaps on failure mechanisms, advanced materials, special non-intrusive inspection techniques, etc., rather than just knowing how to write an inspection report. This technical knowledge is, in most cases, paired with an understanding of how statutory regulations governing in-service inspection work.

Referring to Fig. 2.5 you can see the types of inspection organization that make up this grouping. They are led by the in-house inspection departments operated by the large petrochemical and process industries. These often operate as a separate organization within a large plant site and many hold their own EN 45004 accreditation. They employ high-level staff, qualified either through the traditional chartered engineer

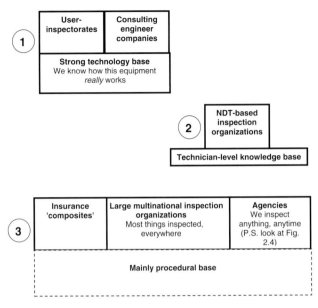

Fig. 2.5   Inspection organizations – in all their glory

(CEng) academic route or sound practical experience reinforced by distance learning, degree-level qualification, and in-house training programmes. These departments rarely dilute their services by becoming involved outside their own organization (although they may provide input to several of their own sites) and hence develop a high level of plant-specific technical experience and familiarity with their plant.

The next residents of the technology-based group are the consultants. There are two types: those that act as consultants for construction of new plant and those that are pure technology consultants. The construction consultants have a lot of experience of plant construction projects worldwide but their sharp-edged technical knowledge is often not *that* well developed. This is because although they may specify the technical requirements of a plant, they rely on engineering contractors to actually design and build it, so they rarely get involved in great technical detail themselves. Equally, they do not as a rule operate plant, so do not have day-to-day experience of the specific process upset conditions, transients, etc. that can cause in-service degradation. On the positive side, their staff academic and training level is high, and well able to adapt to the inspection role. In practice, however, most consultant engineering companies do not treat in-service inspection as a major business area – they concentrate more on 'shop inspection' of new equipment during manufacture.

The pure technology consultants are different. They possess heavily centralized, more specialized technical capabilities centred on their head office. They consider themselves inspection specialists; not in the *procedures* of inspection, but rather the *technical* aspects of the subject. They treat inspection as a discipline rather than an add-on. In contrast to the construction consultants they can be less worldly – finding it more difficult to adapt to different cultures and practices. Their strong academic base sometimes works against them, making them wary of inspection situations involving differences of opinion. Like their cousins the classification societies, they prefer easy technical agreement to contractual conflict.

One characteristic that construction and technology consultants do have in common is their contract and fee structure. Their business relies on large contracts from affluent, well-informed clients who themselves have technological ability and so can appreciate, and pay for, the service that these consultants offer. Ultimately, their work has *low price sensitivity*. Clients in this section of the market will pay well for high-quality technical inspection services with which they feel comfortable.

NDT-based inspection companies *(Group 2 in Fig. 2.5)*

These are the technical journeymen of the in-service inspection business. They are 90 percent NDT contractor and 10 percent inspection company, perfectly happy with this identity, and rarely pretend that they are something that they are not.

They provide their services from a strong technological resource, based almost entirely on high-volume, straightforward NDT activities. These include the standard techniques of dye penetrant (DP), magnetic particle inspection (MPI), ultrasonic testing (UT), and radiographic testing (RT). Occasionally they diversify into more advanced NDT techniques such as eddy current testing, corrosion mapping, and time of flight diffraction (TOFD) but, in revenue terms, this usually makes up only a minority of their business.

The technical power of NDT-based inspection companies lies in their familiarity with NDT techniques, the identification and classification of defects, and the competence of their staff. Their inspectors are virtually all technician-level NDT operators with recognized qualifications such as Personnel Certification in Non-Destructive Testing (PCN), American Society for Non-Destructive Testing (ASNT), or Certification Scheme for Welding and Inspection Personnel (CSWIP), which cover all the major NDT techniques. Within the technical boundaries of published codes and standards for structural items, pressure equipment, etc., these companies provide a competent technical service.

At the higher technical level, NDT-based companies have less capability. They are happier working within a predetermined inspection plan that tells them the scope of NDT and defect acceptance criteria to be used. The task of *deciding* the scope of inspection, or the choice of defect acceptance criteria to be used, is more difficult for them. The situation is even more awkward when it involves deciding what to do once defects have been found. It is uncommon for NDT companies to possess sufficient skills in fitness-for-purpose (FFP) assessment to decide whether a pressure equipment item containing a significant defect can safely remain in service or whether it has to be repaired. At this point, they like to regress to their role as NDT contractor, and call for specialist technical advice from elsewhere. This can mean other organizations becoming involved, so the chance of differences of opinion, and the resulting 'indecisive consensus answers' increases.

The work of NDT-based organizations, because of their technician-level technological base, is highly competitive. Universally recognized NDT qualifications mean that technician staff can move frequently between employers. This leads to almost 'perfect market' conditions in which multiple companies offer basically identical services, each with a similar overhead structure, and with full knowledge of what services the others provide. The result of this is to cause this market to be very competitive; cost cutting is commonplace, with many contracts being awarded on the basis of price alone.

One way in which some NDT-based inspection companies *try* to differentiate themselves is in their systems of administration procedures and reporting. The more sophisticated ones have obtained EN 45004 and ISO 9000 accreditation and use this as justification for working, under subcontract, for inspection organizations in the technology-led group. This is perhaps how they fit in best, providing the role of the skilled NDT contractor (which is what they really are), dealing with the black and white world of NDT techniques and code-specified defect acceptance criteria, while leaving the 'shades of grey' decisions to someone else. Surprisingly, despite their quantitative skills, NDT companies shun inspection-based arguments, which limits their effectiveness in the true in-service inspection role.

Procedurally based inspection organizations (Group 3 in Fig. 2.5)

This grouping contains all the other organizations, large or small, organized or anarchic, that do in-service *statutory* inspection. Their overriding characteristic that sets them easily apart from the first two groups is that they have no real technical base. They are an absolute technical eggshell – crack the outer shell of the business (you can see it in Fig. 2.5) and there is nothing substantial *inside* that does not commonly exist *outside*.

So what are they selling? The answer is that they are selling *procedural compliance*. Their business, which is more than 95 percent administration, is based around a set of procedures that organize the paperwork and housekeeping activities that surround in-service statutory inspection activities. Many of these systems are accredited to some standard or other. These systems can plan periodic inspections, allocate an inspector to visit, audit the production of the resulting reports for administrative content (report number, date of visit, type of inspection, etc.), and provide all the necessary paperwork to match. The benefit to the passive plant-owning client is a level of comfort that statutory requirements have been complied with, and can be demonstrated, by waving all the necessary correct bits of paper, in unison, at the right time.

What about the technical quality of the inspections performed by organizations in this grouping? The situation seems to be that it is *highly variable*, being almost entirely dependent on the characteristics of the individual inspectors, rather than any system of organizational values. There are good, bad, decisive, and indecisive inspectors who work in this sector of the inspection business. Some like the technical freedom, while putting up with the procedural constraints, which they see as anodyne – others just like an easy life.

The character of the inspection business at this level is shaped by cut-throat price competition. At this level, however, the scenery is a bleak one – of competition that has long since lost its benefit to the client in keeping prices at a sensible level. It has gone past that, into the chill regions where the provision of a good quality, competent technical service to the client has been relegated to second or third place; inspection work is sold at rock bottom rates and excuses are made afterwards.

Opinions vary about the effectiveness of in-service inspection organizations that trade at this procedurally based level. They frequently meet all the necessary requirements for accreditation and competency to perform statutory inspections on pressure plant and lifting equipment. Although many are, in reality, little more than agencies, they maintain competence records for a network of freelance self-employed inspectors who have no direct link to anyone in the organization, other than that of a paid provider of inspection services. Virtually the only certainty is that the technical quality of the inspections is going to be variable.

Conclusion – the in-service inspection business

There we have it – an imperfect view of the UK in-service inspection business. The situation is not too different in other European countries where there is a similar degree of self-regulation allowed for statutory compliance. Be warned that the situation is different in the USA, where the same degree of self-regulation is generally not allowed and specific in-service inspection codes by ASME, API, and similar bodies are accepted as a statutory requirement. It also varies from state to state.

Is the character of in-service inspection changing? In most cases, yes – there is an increasing trend towards risk-based inspection (RBI), which effectively *increases* the degree of self-regulation that plant owners and operators are allowed. In theory, this should mean that the pressure is on an inspection organization to increase their technology level so they can deal with the greater technological requirements of advanced non-intrusive inspection techniques. Maybe this is happening, but there also seems to be a discernible shift toward more inspection organizations operating the purely procedural-based way. Multi-skilling of inspectors (such as is happening in the offshore industry) is a visible indication of this.

The only question left about the inspection business is: how do you control it all and what are the ways to manage an inspection organization to best effect? That is discussed in Chapter 4.

Chapter 3

Roles and duties of the in-service inspector

The most important aspect of in-service inspection is that of the role of the individual, i.e. the inspector. Irrespective of the organization for which they work, inspectors have a role which involves a wide and varied scope of technical and engineering issues and the problems and situations that accompany them.

The role of the in-service inspector of mechanical plant is far from straightforward. It is a squarely technical discipline, rooted firmly in the logical rules and practices of mechanical engineering. It is also an unashamedly practical subject, which takes place in operating *sites* rather than in the back room or design office. The role is complicated by the context in which it is applied – inspection is a type of *policing* function, it involves checking (and influencing) the actions of others. It is this resultant mixture of conflict and cooperation that gives inspection its full character. So what is the role of the inspector – what do they do, and to whom do they owe responsibility?

The inspector's duty of care

Figure 3.1 is an attempt to show the situation. The role of the in-service inspector is not as straightforward as in some other engineering disciplines. The main difference is the way in which an inspector's duties extend *outside* the normal confines of an inspector/client relationship. Figure 3.1 shows how this extension exists in two separate, complementary parts:

- Extension in *scope* of responsibility. An inspector has responsibility to more than just the immediate client that is paying their inspection and reporting fees.
- Elongation of the *timescale* that the inspector's responsibility lasts. Inspection brings with it responsibility that lasts well after the

inspection itself (and any immediate actions that result from it) have been completed. It involves a degree of forethought, perhaps even *foresight* about what possibly could happen to a piece of plant or equipment in the future, during its working life.

Look also in Fig. 3.1 at the external influences and aspects that influence the role of inspectors – it is part of their job to try to anticipate happenings and build safeguards into their inspection practices, and make recommendations to reduce unpredictable effects on plant safety and integrity. The amalgamation of responsibility of the inspector to both the client and wider aspects of public safety, are rolled together under the generic term of 'duty of care'. There is probably a well-precedented legal definition for this but, seen simply in layman's terms, it encapsulates the duty of the inspector to perform the inspection task 'thoroughly and to the best standard possible', in order to protect everyone involved from danger.

Some beneficiaries of the inspection are nearer the inspection activity than others. Direct customers of the protection afforded by regular inspection are the plant user and their staff. Other more remote parties such as plant insurers, owners, investors, and suchlike, benefit from the feeling of comfort that inspections bring. Insurers in particular (see Chapter 2), may use inspection reports prepared for statutory reasons (e.g. PSSRs) for their own comfort instead of commissioning their own risk assessment inspections under a contract of insurance.

Hot on the heels of the inspector's duty of care comes the spectre of *liability*. This is the bit that no-one likes. The extent to which inspectors are liable when an accident or failure incident occurs in plant they have inspected can only be decided, by a court, on a case-by-case basis, but there are a few generic 'ground-rules' that apply (see Fig. 3.2). Some inspectors and inspection organizations allow the existence of their duty of care, be it real or implied, to constrain their entire approach to the inspection business. They become highly risk averse, designing their technical activities and procedures primarily to avoid liability, rather than provide a high-quality inspection service to their client. Others treat their liability in a more balanced way, seeing it as a legitimate part of the business that they have chosen to be in. This approach to inspection risk results in a freer, more open approach, less restricted by the defensiveness that so constrains the provision of a good, responsive technical service.

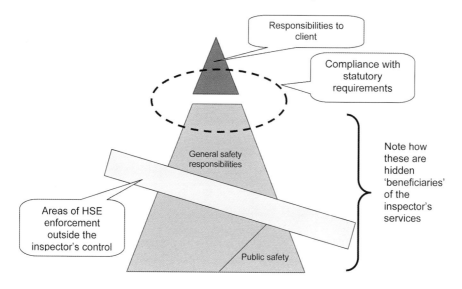

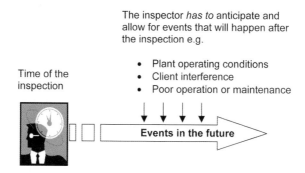

Fig. 3.1   Roles and responsibilities of the in-service inspector

Liability – inspector versus employer

The way in which liability is apportioned between an inspection employee and their employer can vary, depending on the type of inspection that is involved. Figures 3.2 and 3.3 show the situation in most cases; the corporate liability for the quality of the inspection task lies firmly with the inspection employer. The organization, with its separate legal identity, is liable for the work that it does. In time, the

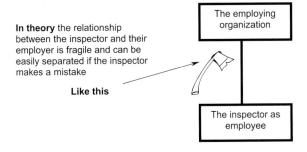

In theory the relationship between the inspector and their employer is fragile and can be easily separated if the inspector makes a mistake

Like this

The employing organization

The inspector as employee

In practice the employer is tied firmly into the situation, and obliged to take responsibility for the actions of their inspectors. They are more firmly locked together than many of them realize.

Like this

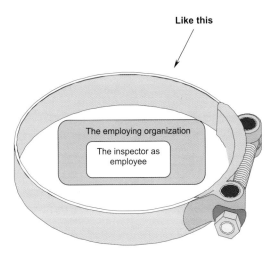

The employing organization

The inspector as employee

Fig. 3.2   The liability of the inspector and their employer

organization delegates certain levels of responsibility to its individual inspectors. The exact form and extent of this delegation depends on several factors including:

- The form of employment contract between an inspector and their employer.
- The pressure of various legal requirements and precedents on the employer, which means that some responsibility remains with the employer, however much they try to 'delegate' it away.

When an accident or failure incident occurs on plant that you, as the inspector previously inspected, you can expect:

- To be asked **some serious questions** by various parties including:
  - your client;
  - your client's insurers;
  - your own managers and their managers;
  - possibly the Health and Safety Executive (HSE); and
  - maybe a few others as well.

- The questions will be about **what you did do** (fairly easy to answer) **and what you did not do** and why you did not do it (difficult to answer).
- **Criticism** of what you did do, and what you did not do. Most of this criticism will seem grossly unfair to you, because you will feel that you genuinely tried to do a good job.

Fig. 3.3   Incidents and questions

Each case is different, but two things are certain:

- If there is an investigation following an inspection-related incident, relationships between employers and their inspectors become *strained*.
- Each party will attempt to avoid as much liability as they possibly can, perhaps even stretching credibility to the limit while doing so.

Inspection organisations are little different to most other employers – some are paternalistic and protective towards their employees, and make a policy of indemnifying their staff against any claims of negligence. Others are mainly defensive and self-serving and are more likely to withdraw rapidly from supporting an inspector if a serious negligence incident occurs.

Inspector competence

The issue of inspector competence is one of the big current issues in the in-service inspection industry. Competence should obviously be a prerequisite for any job but in the world of in-service inspection, with its implications for the integrity and safety of potentially dangerous plant, it is of critical importance.

Why is inspector competence a difficult issue?

There are two main reasons. First, the technical scope of the subject is so wide – an inspector may be required to inspect anything from a simple low-pressure garage air receiver to a highly dangerous process reactor operating under highly corrosive condition at pressures of 200 bar or more. Second, most of the statutory legislation that governs in-service inspection relies on the inspection industry practising a high degree of self-regulation, i.e. it is charged with deciding the competence of its inspectors *itself* rather than requirements being externally imposed. The resulting freedom means that each inspection organization takes a different view of what *competence* means to them, and will defend this view vigorously against their peers who may see things differently.

The net result is that there is no single, agreed system for the verification of the competence of in-service inspectors. There are several different approaches which are used as an attempt by various interested parties to bring some uniformity, but the reality is that they vary widely in scope, technical level, and effectiveness. Figure 3.4 shows the current situation.

Inspector competencies and career pathways

Notwithstanding the (various) formal requirements for 'competence' shown in Fig. 3.4 the competency skills of in-service inspectors fall into a few well-defined patterns. These patterns reflect the career routes commonly taken by inspectors. Their common feature is that they are strongly *experiential* i.e. reliant on a foundation of technical experience rather then based solely around formal technical qualifications. Figure 3.5 shows the situation; note the main career paths 1, 2, and 3.

The NDT technician route

A lot of inspectors graduate into inspection from an NDT technician background. They have formal NDT qualifications (CSWIP, PCN, ASNT, etc.) and have gained practical experience by involvement with welding, fabricated structures, and pressure equipment. NDT technicians also benefit from the experience of dealing with plant manufacturers, contractors, and operators and have an appreciation of the ways that they all interact with each other. Areas of weakness may include:

• Lack of experience of the operational aspects of engineering plant.

Location/industry	Competence assessment method
USA industrial plant (in most states).	This is a tightly controlled system. In-service inspectors are formally certified to the construction/in-service codes. Pressure vessels: ASME code/API 510. Steam/process pipework systems: ANSI B31/API 570. Atmospheric storage tanks: API 650/API 653.
UK pressure systems under the Pressure System Safety Regulations (PSSRs).	There is no formal examination or certification to qualify an inspector to inspect on behalf of a 'Competent Person' organization. Organizations choose their inspectors using their own interpretation of what 'competence' means to them. If they are certified to EN 45004 they have to demonstrate to UKAS that their system of competence assessment is working in compliance with the company's own procedures, and meets the requirements of EN 45004.\
\
The PSSRs give guidance (clauses 35–43 of the PSSRs ACoP (see Chapter 5) on what constitutes a Competent Person *organization* but do not go into detail about the assessment of the competence of individuals within the organization.\
\
The final (indeed only) real test of inspection competency under the PSSRs would be in a court of law, following a claimed breach of the regulation, or similar investigation into a serious incident or accident.
UK inspection organization undergoing certification by UKAS to EN 45004.	EN 45004 certification is not a mandatory legal requirement for an organization. It is used by some (mainly large) organizations as a visible demonstration of their compliance with EN 45004.\
\
EN 45004 does, however, have a link with UKAS document RG2 (see later in this chapter) that gives a set of 'qualifications categories' for in-service inspection personnel.

Fig. 3.4 In-service inspection competence – the current situation

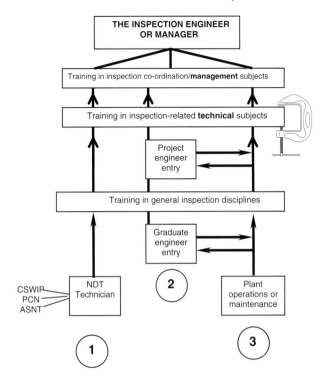

Fig. 3.5   Inspector competence – the main career paths

- Uncertainty of technical knowledge in some areas of plant design, degradation/failure mechanisms, and academic topics such as fitness-for-purpose (FFP) assessment.
- Difficulty in adapting from the world of NDT, which is based around 'hard-edged' and well-defined techniques, procedures, and defect acceptance criteria to the more judgemental, 'multiple shades of grey', world of in-service inspection.

The plant operator route

Traditionally, many plant inspectors started their careers in plant operations in power, process, or marine engineering. The high levels of plant experience and academic achievement required form a sound technical background for an eventual move into plant inspection. Although it has not disappeared completely, this route has depleted steadily over the past 20 years. Weaknesses in inspectors taking this route may include:

- Poor knowledge of important design codes, regulations, and statutory aspects.

- Limited experience of industry-specific technical issues (materials, designs, etc.), i.e. they are often generalists with a wide, but shallow, knowledge base.

In recent years career benefits (mainly salaries) for competent operations technicians and engineers in power/process/petrochemical industries have become quite attractive, thereby discouraging salary-related career moves into plant inspection.

The graduate/project engineer route

This is the most modern career route into plant inspection. It will probably grow, and perhaps become the dominant route in the future. It has its root in the graduate engineer, employed by construction, contracting, or process/petrochemical utility organizations. Graduates will typically start their engineering career in a specific project engineer role, or as part of an in-company graduate training scheme involving experience of several different jobs over a period of a few years. At some point, they take on the role of project engineer for a specific project, involving the construction, refurbishment, or operational aspects of plant.

It is the wide responsibilities of the project engineer that often kick-starts a graduate engineer's interest in plant inspection. Inspection plays a part in most plant projects and the technical complexity (and difficulties) of it soon become apparent. In common with the other inspection career routes, the graduate entry route has weaknesses. Typically, these are:

- Lack of hands-on engineering experience.
- Poor initial appreciation of the ways that technical and management disciplines interleave together rather than act alone.

In practice, these weaknesses have a short timescale – shorter than almost anyone thinks. Graduates that have survived the selection procedures of major companies are well motivated and have the cognitive ability to absorb large amounts of technical information. This means that they *learn very quickly*. They also have the advantage that they are not encumbered by the restricted mindset of having worked in a single role. These clear *advantages* have to be taken in context. In many cases, the latent technical power of graduates is never fully received into the world of in-service inspection; instead it is redirected towards other disciplines such as project management or similar. This is the inevitable downside of graduate/project engineers' participation in inspection: it is frequently temporary – used as a technical stepping-stone towards progression in other disciplines that are felt to have a higher management profile.

Taken together, these three routes make up the majority of paths taken by engineers who become inspectors. Although the routes themselves have different backgrounds and involve different types of people, the technical skills that have to be acquired to do the job of inspection do not vary much.

The technical skills of the inspector

The skills required by an in-service inspector are characterized by their breadth. In order to feel confident in inspection situations it is necessary to have a technical understanding of a wide variety of engineering plant. Knowledge that is superficial is not sufficient – the complexity of most inspection situations can soon outstrip the knowledge of any inspector who can recite the correct words, but does not know what they actually mean. Inspection requires knowledge that is *applied* and robust. It must be able to withstand scrutiny, disagreement, and criticism.

Figure 3.6 is an attempt to summarize the skill-set of the in-service plant inspector. Look how it is constructed – the topics nearest the centre of the circle are those core activities of greatest importance to the in-service inspector, while those further from the centre are used in less depth, or less often. Now look at the radial sections of the diagram – it is divided into three *technical* sections and two *non-technical* sections. The technical sections cover the topics of how plant is manufactured, operated, and then the mechanisms by which it fails. This shows the breadth of the discipline of in-service inspection – it involves an understanding of all three of these aspects, rather than a single one in isolation. Note also how those aspects (in all three sectors) with an influence on plant *integrity* reside near the centre of the skill-set diagram. Integrity and lack of integrity (otherwise known as *failure*) are closely related.

The remaining two sectors involve the two non-technical aspects: the tactics of inspection and understanding commercial realities. The tactics of inspection are a set of actions that make the task of in-service inspection more effective and help it to run more smoothly. In contrast, the skills relating to the commercial aspects are mainly passive – it is the understanding that is important, rather than any actions which necessarily result *from* this understanding. Both, however, share the ability to make the task of in-service inspection a bit easier, at least some of the time.

The topics nearer the centre of the circle are those
core skills of greatest benefit to the in-service
inspector

Fig. 3.6   Skills of the in-service inspector

The relationship between Figures 3.1 and 3.6 shows the skill-set subjects that should fit into the three levels of inspection training inferred in Fig. 3.5 Together they form a structured approach to inspection training. Some inspector training schemes address this breakdown of inspection training subjects while others steer well clear, relying instead on reference to academic levels and institution membership grades, surrounded by statements of generality.

The inspector as technical specialist?

Most in-service plant inspectors are not technical specialists. They are more likely to be generalists with a wide general knowledge of plant engineering subjects (those in Fig. 3.6) and an appreciation of how to apply it to best effect in an inspection situation. Although some engineers that have trained in specialized, technical subjects (i.e. metallurgy, NDT, design, or similar) do move across into plant inspection work, their numbers are not large. They are also fairly unlikely to have been operating as true, high-level specialists before their move; more probably they have been involved on the fringe of specialist technical activity, or in some sort of support role.

Specialists in subjects which are predominantly non-technical, e.g. quality assurance (QA), contract management and procurement, are also found in inspection. Although it is not uncommon to find in-service inspectors with this type of background, they frequently struggle to meet the level of technical appreciation necessary to deal with inspection issues that arise in complex plant, and find the job difficult and stressful. They tend to adapt better to the pre-purchase 'shop' inspection of new equipment, which is more reliant on documentation and procedure-related issues.

Inspection as its own specialism?

The model of the inspector as *absolute generalist* is, fortunately, also imperfect. The scope of inspection is wide and general but there are aspects to it which are definitely inspection-specific, and do not relate well to other aspects of engineering. In-service inspection in particular is best thought of as a discipline of its own, with its own set of unwritten (except in this book) rules and norms. It also has its own *values*, summarized earlier in this chapter and in Chapter 2. Inspection certainly seems to work better, if you think of it like this.

UKAS Document RG2

Document RG2: *Accreditation for in-service inspection of pressure systems and equipment* is one of several documents published by UKAS as guidance for their accredited inspection bodies. There are two parts to its main content: a set of qualification categories for in-service inspection personnel and a set of guidelines for how inspectors should be supervised. The content of RG2 is heavily cross-referenced to EN 45004: *General criteria for the operation of various types of bodies performing inspection* and UKAS document E2: *UKAS regulations to be met by inspection bodies*, hence linking RG2 firmly to the process of accreditation to EN 45004. One further link is that to the Pressure Systems Safety Regulations (PSSRs) SI 128 (see Chapter 5). Although the PSSRs are a statutory instrument rather then a UKAS document, RG2 uses the PSSRs definitions of categories of pressure systems (major, intermediate, and minor) as a basis for its inspector qualification categories and supervision requirements.

The RG2 qualification categories

The philosophy of the inspector qualification categories is based on the general principle (used in EN 45004, the PSSRs, and elsewhere) that the choice of inspectors to carry out particular types of inspections should be based on them having suitable *qualifications, training, experience, and knowledge* of the requirements of the inspection to be carried out. This rather general form of words also appears in the PSSRs; however, RG2 goes slightly further by splitting this into six categories based on academic qualification and experience, and linking them to their suitability for performing inspections on major, intermediate, and minor pressure systems. This is the limit of RG2's incisiveness on the suitability of inspectors for various tasks – it does not differentiate between the specific type of inspection or plant items, nor draw any distinctions between different industries such as power, petrochemical, general engineering plant, etc.

The fact that RG2 is neither equipment nor industry-specific can weaken the specific applicability of its inspector qualification categories in some industries. There are many situations where the major/intermediate/minor pressure system categorization does not truly reflect either the technical complexity and difficulty of the inspection task, or the damage consequences of a failure. In practice most good inspection organizations categorize their inspection competence levels more stringently, using a competence 'matrix' showing competence/experience levels relevant to specific items of engineering plant. This does not prevent inspectors still being allocated an RG2 qualification category rating, however, to comply with EN 45004.

The message is, then, that RG2 qualification categories are linked mainly to the self-consistent world of EN 45004 and its audit requirements, and are often further refined and interpreted during implementation in different industries. Under the current system, inspection organizations 'self-certify' their own inspectors as possessing a level of competence. The ways in which they do it differ – some organizations insist on equipment-specific training courses, while others rely more on experience. Either way, RG2 is normally used as a basis only, not a full statement of 'competence' requirements.

Figure 3.7 shows the six inspector qualification categories (Category 1, highest, to Category 5, lowest) as defined in UKAS document RG2. Table 3.1 shows the RG2 requirements for qualification and supervision

Category 1. Chartered Engineer as defined by the Engineering Council or equivalent (e.g. appropriate degree with relevant experience, NVQ Level V Engineering Surveying) including at least 3 years' experience within an engineering discipline associated with in-service inspection of pressure systems.

Category 2. Incorporated Engineer as defined by the Engineering Council or equivalent (e.g. appropriate HNC with relevant experience, NVQ Level IV Surveying Engineering) including at least 5 years experience within a relevant engineering discipline of which at least 1 year* shall have been spent working within an engineering discipline associated with in-service inspection of pressure systems.

Category 3. Person employed prior to the date of application for accreditation in the in-service inspection of pressure systems with less than Incorporated Engineer qualification but meeting the criteria of Category 4 below.

Category 4(a). Engineering Technician as defined by the Engineering Council or equivalent (e.g. appropriate ONC with relevant experience) having a minimum of 5 years' experience within a relevant discipline of which at least 1 year shall have been spent working within an engineering discipline associated with the in service inspection of pressure systems.

Category 4(b). Person trained[†] in a relevant engineering discipline with a recognised and documented engineering apprenticeship with a minimum of 5 years' experience within a relevant discipline of which at least 1 year shall have been spent working within an engineering discipline associated with the in-service inspection of pressure systems.

Category 5. Person employed prior to the date of application for accreditation in the inspection of pressure systems with less than tradesman's apprenticeship but with a minimum of 5 years spent working with or within the industry associated with pressure systems and has general knowledge of pressure systems and its operating environment. Personnel shall be placed on recognized training courses with appropriate documented tests in in-service inspection of pressure systems. The minimum age for this category is 21 years.

---

* Where a person meets the minimum requirement for a specific discipline and is to be trained in a second discipline, it may not be necessary to have experience of at least 1 year in the second discipline provided that the required competence can be demonstrated.

† Persons in Categories 4(b) and 5 shall pass a qualifying test, established by the Inspection Body, associated with the particular inspection activities relating to pressure systems/equipment and this should cover relevant knowledge of the law, codes of practice, and inspection techniques. UKAS should be given the opportunity to review the content of such tests prior to implementing it.

Fig. 3.7   UKAS document RG2 inspection qualification categories

Table 3.1   RG2 requirements for qualification and supervision of inspectors performing inspection of pressure systems

Pressure system	RG2 grade	Supervision category	Constraints
Major systems (including steam)	1	Occasional	Inspection or associated activities in technology outside the field of competence is prohibited except by formally documented consultation
	2	Occasional	The above constraint plus prohibition on any non-routine repairs, modifications, changes to operating parameters, changes to inspection methods, calculations not defined in recognized standards except with specific approval by an appropriately qualified person (e.g. Metallurgist, Designer, Process engineer)
	3	Occasional	Permitted only for testing and examination to identify defects, within the limits specified by Category 1 or 2 person. Any decisions involving limits of acceptability, repairs or modifications shall be approved by authorized persons qualified to Category 1 or 2
Intermediate systems (excluding steam)	1, 2, 3	Occasional	Same constraints as for major systems stated above for respective categories
	4, 5	Frequent	Permitted only for carrying out routine, repetitive, and well-defined examinations on a specific range of installations
Intermediate systems (steam only)	1, 2, 3	Occasional	Same constraints as for major systems stated above for respective categories
Minor systems (excluding steam and pipelines)	1, 2	Occasional	Same constraints as for major systems stated above for respective categories
	3	Occasional	Same constraint as for Category 2 person stated above under major pressure systems
	4	Frequent	Same constraint as for Category 3 person stated above under major pressure systems
	5	Frequent	Permitted only for carrying out routine, repetitive and well-defined examinations on a specific range of storage installations
Minor systems (steam only)	1, 2	Occasional	Same constraints as for major systems stated above for respective categories
	3	Occasional	Same constraint as for Category 2 person stated above under major pressure systems
	4	Frequent	Same constraint as for Category 3 person stated above under major systems

Major systems

Major systems are those, which because of size, complexity or hazardous contents require the highest level of expertise in determining their condition. They include steam-generating systems where the individual capacities of the steam generators are more than 10 MW, and any pressure storage system where the pressure–volume product for the largest pressure vessel is more than 10^5 bar litres (10 MPa. m³). Pipelines are included if they constitute a major hazard.

Intermediate systems

Intermediate systems include the majority of storage systems and process systems which do not fall into either of the other categories. Pipelines (as defined in the PSSRs: 2000) are included unless they come within the major system category.

Minor systems

● Systems containing non-corrosive and non-flammable liquids, steam, pressurized hot water, compressed air, inert gases, or fluorocarbon refrigerants which are small and present few engineering problems. The pressure shall be less than 20 bar (2.0 MPa) above atmospheric pressure except in systems with a direct-fired heat source, when it shall be less than 2 bar (200 KPa).

● The pressure–volume product for the largest vessel shall be less than 2×10^5 bar litres (20 MPa m³). The temperatures in the system shall be between -20 and $250\,°C$ except in the case of smaller refrigeration systems operating at lower temperatures which will also fall into this category. No pipelines are included in this category.

Pipelines and pipework

The definition should be in line with the PSSRs: 2000.

Fig. 3.8   RG2 definition of pressure systems

of inspectors performing in-service inspection of pressure systems under the UK PSSRs. Remember that Table 3.1 is not an overt requirement of the PSSRs (RG2 is not a statutory instrument), it merely represents UKAS's view of the situation and probably has no special legal status. The definitions of major, intermediate, and minor pressure systems are shown in Fig. 3.8 – these are exactly the same as those given in the PSSRs and the associated Approved Code of Practice (ACoP) that explains the regulations in greater depth.

**THINKING OUTSIDE THE BOX – 1. INSPECTING THINGS –
COOPERATION OR CONFLICT?**

Step 1. Put aside any preconception you may have that:

   (a) *Cooperation* is either desirable, undesirable, inherently good, or inherently bad.

   (b) *Conflict* is either desirable, undesirable, inherently good, or inherently bad.

Step 2. Make sure that the preconceptions mentioned in Step 1 don't keep creeping back, because they have a huge tendency to do so. Try again to kill them off ... absolutely dead.

Step 3. Think carefully about the paragraph below (be as critical as you want), and decide carefully if you agree with it (while remembering Steps 1 and 2 of course).

Inspecting things – cooperation or conflict?
An inspection situation, where you, as the inspector, are inspecting and passing comment, and adjudicating on things that affect other people (and other people's work) needs *both* cooperation and conflict to make it work properly. If the situation is *all conflict* you will not be very effective, because you will not find out properly what is really happening. If you make the situation *all cooperation* you will get precisely nowhere, because the real technical issues and problems are sly; they hide, as if in a box, *only conflict* will open the box. Then you need cooperation to enable you to understand and sort out the problems that are in the box. So you need both.

Fig. 3.9   Thinking outside the box –1. Inspecting things – cooperation or conflict?

**THINKING OUTSIDE THE BOX – 2. INSPECTING THINGS –
DIPLOMAT OR ANTAGONIST?**

'Well, I suppose it has to be a bit of both, doesn't it? ... if that last figure is to be believed ... I'm not sure I agree with it totally though, I favour the diplomatic approach to inspection.'

'Do you?'

'I said I did, didn't I, aren't you listening to me?'

'And of course the politics needs to be looked after, it's like looking after a cat, careworn feline coats left smoothed and preened so it feels good ... you need to be a people person to be an inspector – use a bit of tact...'

'Absolutely.'

'But ... hold on ... what happens when something wakes a tiger up? Does it trot over, offer its paw, welcome you generously to its part of the tundra and extend a demure invitation to share its dinner?'

'You're being silly, there clearly aren't any tigers in the world. All the plant owners I've met have happily told me about all the difficult technical issues in their plant, explaining the risks to fitness for purpose, integrity, safety to personnel – all that stuff ... and have confirmed that they've all been rectified, even before I arrived to do the inspection – that shows the power of diplomacy.'

'No technical problems that you had to find yourself?'

'Not one.'

'Not even in old chemical plant pipework under all the old loose, wet lagging?'

'No need to ask for lagging to be taken off, you won't find rust under there, and plant owners don't like it'

'I'm not sure that's what API 570 *In-service inspection of pipework* says ...'

'Can't say I've even read it ... but even if it does say you need to inspect pipework under damaged lagging, it probably doesn't really mean that. It probably means remove lagging, only if you don't annoy anyone by asking.'

'Hold on, I'll just get the document out and check ...'

Fig. 3.10   Thinking outside the box – 2. Inspecting things – diplomat or antagonist?

Chapter 4

The management of in-service inspection

Like any business, there are many aspects to the management of an in-service inspection (ISI) organization. The characteristics of the management role, variable though they inevitably are, are influenced by the basic format of the business – as an almost pure *service industry*. Even those ISI companies that combine their independent inspection role with the provision of non-destructive testing (NDT) services are still firmly anchored in the service industry sector. NDT may involve the use of capital equipment and consumables, but it is still a service.

There is little new about the concept of a service industry that involves technical activities. All manner of technical service companies operate worldwide, competing with each other on a contract-by-contract basis. Rarely a day goes by without every single one of them expressing annoyance that they have to compete with similar companies who have the affront to bid for the same work that they do, while 'obviously not being capable of doing such a good job'.

So what about the *management* of a technical service company where the service is in-service inspection? As with any other service business, the nature, features, and shape of the service give the business its character – it has to because these are the only places that true business-specific influences can come from. Once the character of the business is defined, then the form of the objectives of the business can follow. Forget the notion that there is anything particularly complex or difficult about the formulation of management *objectives* for ISI businesses – it is actually easier than for many other types of service business. This is because the ISI product is not particularly complex – it is objective and absolutely sharp-edged with well-defined features that are not difficult to see, if you look carefully through the normal day-to-day business complexity that surrounds them. There are two basic models that apply: the ISI organization form based on *technical competence* (TC), and its competitor, the company model based on *procedures, paperwork, and audits* (PPA). These models lie behind everything that happens in an ISI company. We can look at these in turn.

The technical competence (TC) model

The TC model is a model of an ISI organization that bases its business strategy on its *technical strength*. It prides itself on the technical abilities of its staff and, as an organization, owns a solid technical resource that others do not have. This resource may be in the form of client-specific knowledge, metallurgical skills and experience, or highly applied knowledge of individual items of engineering plant. The common feature is that this resource is real and *scarce*. It means that the organization has a genuine technical centre rather than just thinking that it has – if you split the organization open, you would find a metal core, not a paper one, or hot air.

So what characteristics go with a business model like this and what effect do they have on the management objective of the business? Figure 4.1 shows the situation. Note the characteristic at the bottom of the figure – that of the ISI organization implementing its technical expertise through its *inspectors*. This is the key to the technical power of this model of organization; the implementation of its technical power is demonstrated at client-facing inspector level. This means that the inspectors operate with a high, almost absolute, degree of real technical autonomy on site. They make the most of their own on-site inspection decisions and are not hindered from doing so by restrictive practices or the numbing requirement to 'consult head office' on every technical decision that they make.

One significant, well-hidden, characteristic of the technical competency model of an ISI company is its ability to see through the apparent technical complexity of the inspection task as it relates to many different types of engineering equipment. Its managers view the sorting of technical complexity as a valid task of management. The good ones are adept at sorting the technical issues into manageable chunks, then selecting and training their inspectors to deal easily with each one.

The model in Fig. 4.1 is not a theoretical one – those ISI companies that do manage to achieve this high technical competence level reap the rewards of a large share of the high-value ISI market. Part of the reward is that in-service inspection at this level, (centred mainly on the power, petrochemical, and process industry sector) has *low* price sensitivity. This is because the *barriers to entry* are high, with organizations needing high-quality, usable technical resources before they can join the game.

THE TECHNICAL COMPETENCE (TC) ORGANIZATION

- Is technically perceptive. It knows what technical problems are going on and has a view on them.
- Does not rely, for its *success*, on procedural control by ISO 9000, EN 45004, or similar. It may posses such accreditations, but:
  - sees them purely as a necessary burden that they have to comply with; and
  - never tells anyone this.
- Chooses to remain in the ISI business, in the knowledge that it *could* move up the technology business chain if it wanted to.
- Has a technical core; and above all
- Implements its technical expertise *through its inspectors*, rather than backroom technical staff that no-one ever sees.

Fig. 4.1   The 'technical competence' ISI business model

ISI organizations with a good presence at this level can therefore return good profit margins on their inspection activities.

Success is not easy, so players who have managed to enter, and remain successful in, this section of ISI go about their business quietly and assuredly, sometimes tinged with a carefully considered touch of risk taking where necessary. They tend to take a back seat in the arena of trade associations, technical committees, and similar, under the assumption that their peers have little to offer them in terms of advice, guidance, or technical knowledge.

The procedures, paperwork and audits (PPA) model

The PPA model is a picture of the opposing type of ISI organization. In contrast to the TC model, the PPA inspection company relies, for its strength, on its ability to write and comply with *procedures*, generate the *paperwork* to facilitate their implementation, and submit to *audits* to show that the generation, the complying, and the facilitation has worked. There are three strategic characteristics to this type of ISI:

- *Compliance*. PPA-type inspection organizations feel driven to comply with every possible external certification and accreditation scheme that could possibly be construed as being something that they *should* be complying with. They will comply with (and vocally defend) such schemes, even if they, from the outside, seem to have little effect on their commercial business objectives.

- *They are defensive.* They have a strategy that is centred on *defence.* They are defensive perhaps because they are frightened of the market, other organizations and parts of themselves. Paradoxically, although such organizations seek multiple appraisals through accreditation and audits – their level of self-confidence remains low – they feel under continual pressure from the market, their critics, and, ultimately, the technology that shapes the nature of the inspection work that they do.
- *They love technical neutrality.* This is the seed that shapes their role as a PPA organization, the traditional (and essential) role of the inspection organization as an independent viewpoint – a neutral. True PPA inspection organizations, as well as demonstrating *organizational* neutrality, allow the influence to extend to their *technical* capability. Over time they shave off their own technical cutting edge, adopting the stance of the lowest technical common denominator.

The end result is that ISI organizations like this consciously avoid making controversial technical decisions. This conscious avoidance leads to the avoidance of controversy becoming entwined as one of the organization's core values (unwritten, of course). Technical decision-making elements that *do* exist find themselves a victim of the organization's pressure to rebalance itself internally, so they slowly drift away, to be replaced by PPA-orientated managers and departments that the rest of the organization finds less threatening. Soon, the whole enterprise becomes a proponent of PPA values and norms – and almost totally technically neutral.

The effects of neutrality

Comfortable though the idea of technical neutrality might be, it brings with it hidden consequences to the business and its management. The biggest one is an environment of *price competition.* Unlike the alternative model, which is based on technical competence, the PPA organization uses as the basis of its presence the existence of its procedures, paperwork, and audits, enclosed in their wrapper of technical neutrality. The resultant problems are threefold:

- *PPA is not a scarce resource.* PPA capability is like buying a long-term capital asset – once you've bought it, it is yours to use – it doesn't deteriorate quickly, or immediately drift away if you neglect it. Contrast this with technical competence, which does.

- *Barriers to entry are low.* It is easy to develop a small ISI company using a transient resource of freelance inspectors into one which can demonstrate PPA compliance, including various forms of accreditation. Once gained, the status is rarely taken away. The reality is, therefore, that the barriers to entry are low.
- *Competition.* Low barriers to entry *always* result in an environment of strong competition. PPA-based inspection companies are therefore always in strong price competition with other PPA-based inspection companies. That is the only way they can compete, because they have, unconsciously perhaps, relinquished their chance to differentiate themselves on the basis of their technical competence, decisiveness, or willingness to make authoritative technical decisions. Their PPA resource is unfortunately, undifferentiable, so all that is left is competition on price. This leads to endemic cost cutting, undercutting and underbidding in order to obtain ISI projects to stay in existence. This is competition in its worst and most destructive form.

In business terms then, the in-service inspection organization that structures itself on the PPA model lives squarely in the world of *low added value*. The amount of value it can offer is limited to the cohesion of the procedures, paperwork, and audits that form the 'product' that it sells. The technical added value that it can offer is severely limited because its technical power is low. Many ISI organizations structure themselves like this, and even manage to remain in business, competing daily with their peer organizations that view the world in the same way.

TC versus PPA – decision or balancing act?

In terms of an organizational ideal to work to, TC and PPA models provide two feasible, if opposing, possibilities (see Fig. 4.2). Perhaps the best management solution (and which managers are not searching for that?) would be to choose the best attributes from both models, combining them together (using a bit of delicate top-management craftsmanship) to form an in-service inspection company that has all the strengths, but without the weaknesses, of both types. This seems like a good idea. Unfortunately, organizational reality gets in the way because:

- The two models *cannot mix* – they are incompatible. So
- If you are defining the management identity (and then subsequently the strategy) of an in-service inspection organization, *you have to adopt one model or the other*. You cannot have both.

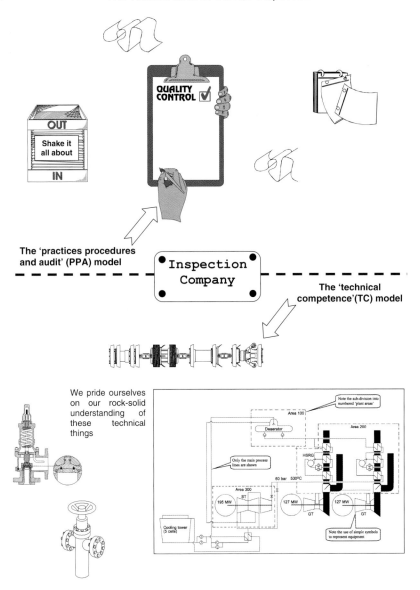

Fig. 4.2   Types of inspection company ... decision time

The choice of the TC versus PPA model is therefore a *decision* rather than a balancing act. Fine and persuasive balancing technicians some managers may be, but this one will defeat most of them. ISI companies that attempt to balance TC and PPA approaches end up as an organization that struggles badly or, at best, suffers a long drawn-out fade into mediocrity, struggling for a clear identity, hunting from fashion to fashion as it tries to decide what it wants to be. Take a look into the world of in-service inspection organizations and see if you can see them.

And now – the task of ISI management

The rest is easy, once you have set the framework for the ISI organization, by deciding which of the two models you want, all the other tasks of management will follow, almost by themselves. The bidding for contracts, organizations of inspections, coordination and checking of reports, employing of inspectors, writing of procedures, ordering of paperclips, and all the other 'tasks' of management will flow easily. Between them, though, they are unlikely to govern whether the organization becomes a success or a has-been. They are peripheral – easy and uncontroversial to act on, but in reality, sufficiently remote from the central core of business identity not to matter that much. The real task of ISI management is to decide which identity of organization you want to be.

THINKING OUTSIDE THE BOX – INSPECTOR COMPETENCE

'It's all very simple you see, if you look at Chapters 2, 3 and 4 together. It depends who they work for, the well-organized technical inspectors work for the well-organized technology-based inspection organizations in Group 1, and the rest get involved with the others, the organizations in Groups 2 and 3.'

'Groups 2 and 3, that's those that suffer the price competition that constrains everything they do, right?'

'That's them, the price wars squeeze the last drop of technical enthusiasm out of them, ... they have to use *cheap* inspectors if they want to get the contracts.'

'So how do they stay in business then? Surely if their technical service is poor, clients will soon see this and go elsewhere?'

'Well, they can't *all* be poor, technically, some must have specialist technical knowledge, these Group 2 and 3 inspection organizations are not only procedural you know, there's a strong technical bias to all they do look, it says over there, in Fig. 2.4.'

'But the price sensitivity is less, almost non-existent perhaps, in Group 1 organizations, so the fees *must* be higher in that group?'

'Yes, two, or three or even four times in some cases.'

'So ... what?'

'So why don't all these inspectors that have hot technical knowledge quickly leave Groups 2 and 3 and all end up in Group 1 technology-based companies, for double or more their salary?'

'Natural selection ... Maslow's hierarchy of inspectors' personal needs ... and all that stuff ... things will never balance out totally you see.'

'Any altruism thrown in? ... that's a-l-t-r-u-i-s-m by the way, a sort of selfless serving for the general good.'

'Of course, just think of all those exploding boilers if they weren't there ...'

Fig. 4.3   Thinking outside the box – inspector competence

Chapter 5

The Pressure Systems Safety Regulations (PSSRs): 2000

The Pressure Systems Safety Regulations (PSSRs) are *the* main driving force behind the in-service inspection of most pressure equipment in the UK. Without the PSSRs, in-service inspection would probably not be as widespread, or comprehensive, as it is. The PSSRs also set a framework for the in-service inspection industry via their designation of the role of the 'Competent Person'.

The PSSRs – what are they?

The PSSRs are not a technical or engineering document or standard – they are a statutory instrument known as SI 128: 2000 or, in other words, they are the law. They are applicable to certain types of pressure plant used in the UK. It is important to note what they are *not*. The PSSRs are:

- Nothing whatsoever to do with any European Directive, such as the Pressure Equipment Directive (PED).
- Not applicable to equipment which is used outside the UK, even though it may be manufactured in the UK.
- Not an all-encompassing safety document. There are many safety aspects of plant that are not addressed by the PSSRs.
- Not directly related to fitness for purpose. It is not difficult for equipment to be fully compliant with the PSSRs but totally useless for the technical purpose for which it was intended.
- Unrelated to ISO 9000: 2000 (Quality Management Systems). ISO 9000 is not even mentioned in the PSSRs.

The PSSRs take their place as part of the raft of legislation covering engineering plant but they are limited to their own particular scope and application – they do not cover everything.

PSSRs – their legislative background

Much of what is written about the PSSRs in government documentation, leaflets, etc. relates to their history. Most of this is of purely academic interest and is weighed down by multiple acronyms relating to previous legislation. For the record, the outline of the situation is as follows:

- The PSSRs: 2000 came into force on 21 February 2000.
- They revoke and re-enact with minor changes, the Pressure Systems and Transportable Gas Containers legislation (PSTGC): 1989.
- Since the PSTGC regulations came into effect they had been subject to various amendments including part of them (Regulations 16–23) having been revoked by The Carriage of Dangerous Goods – Classification Packaging and Labelling (CDGCPL2) regulations. These deal with the supply and use of transportable gas containers [now called transportable pressure receptacles (TPRs), see Chapter 17] which were effectively removed from the PSTGC scope.

The end result is that the PSSRs: 2000 now stand alone and do not cover transportable pressure receptacles.

The PSSRs and ACoP

The PSSRs themselves are a legal document and so are written in a quasi-legal format, which does not make the text particularly easy to understand. They are supplemented, therefore, by a separately published document 'The Approved Code of Practice (ACoP L122)'. This contains:

- The full text of the regulations and guide notes.
- More text which constitutes the ACoP (which has special legal status).
- Additional 'guidance notes' on the ACoP subjects.

Note the difference between these separate parts of the ACoP publication (they are printed in different colour codes in the publication itself). The full regulation text is the law, while the ACoP text itself has special legal status in that it gives practical advice on how to comply with the law. The guidance notes are not compulsory *but* are generally taken as illustrating 'good practice'. The inference is that, by following the guidance, you will normally be doing enough to comply with the law.

Legal aspects apart, the ACoP for the PSSRs does provide some useful information which helps explain the PSSRs. In line with ACoPs published to support other pieces of legislation, it is helpful but not perfect.

Implementation precedents – what are they?

Implementation precedents are site-specific documents prepared by plant owners and users. They are prepared to help avoid misunder-standings about which systems or parts of systems on the site fall within the jurisdiction of the PSSRs and which do not. They are not mandatory documents (and have no special legal status), their main purpose being to state the plant user's interpretation of the PSSRs and ACoP at their site-specific application.

Implementation precedents are found on large, well-managed operating sites and commonly include the following equipment and systems:

- Protective devices, e.g. what is defined as a protective device and what is not.
- Instrumentation (pressure gauges, transmitters, etc.).
- Anticipated events of 'unintentional pressurization' (a specific requirement of the ACoP) and how these are interpreted in terms of possible releases of stored energy.
- Properties of fluids such as hydrogen products, including their vapour pressure.
- Piping systems connected to transportable pressure receptacles (gas cylinders).

These are common areas of controversy when applying the PSSRs. Most responsible plant users allow the PSSRs the benefit of the doubt when writing their implementation precedent documents. This is a more defensible position, in the event of a legal investigation following an accident or incident involving the unintentional release of stored energy.

Smaller or less competent plant operators, who may view the PSSRs as more of an imposition, rarely have implementation precedent documents in place. The main reason is normally lack of knowledge rather than any deeper disrespect for the law. A surprisingly large number of plant operators misunderstand the PSSRs – believing them to be something to do with insurance inspections, and have no knowledge of the ACoP. In such situations, it is often left to the in-service inspector to advise informally on implementation precedents.

Figures 5.1 and 5.2 provide 'quick reference' information on the PSSRs. The following section looks at the content of individual regulations. Cross-references are provided to the numbered clauses of the PSSRs ACoP.

Content of the PSSRs

The PSSRs are subdivided into a number of regulations (1 to 19) supplemented by various schedules. The ACoP is structured in a similar way as follows.

- Notice of Approval.
- Preface.
- Introduction.
- *Part I: Introduction*
  Regulation 1.   Citation and commencement.
  Regulation 2.   Interpretation.
  Regulation 3.   Application and duties.
- *Part II: General*
  Regulation 4.   Design and construction.
  Regulation 5.   Provision of information and marking.
  Regulation 6.   Installation.
  Regulation 7.   Safe operating limits.
  Regulation 8.   Written scheme of examination.
  Regulation 9.   Examination in accordance with the written scheme.
  Regulation 10.   Action in case of imminent danger.
  Regulation 11.   Operation.
  Regulation 12.   Maintenance.
  Regulation 13.   Modification and repair.
  Regulation 14.   Keeping of records etc.
  Regulation 15.   Precautions to prevent pressurization of certain vessels.
- *Part III: Miscellaneous*
  Regulation 16.   Defence.
  Regulation 17.   Power to grant exemptions.
  Regulation 18.   Repeals and revocations.
  Regulation 19.   Transitional provisions.
  Schedule 1.   Exceptions to the regulations.
  Schedule 2.   Modification of duties in cases where pressure systems are supplied by way of lease, hire, or other arrangements.
  Schedule 3.   Marking of pressure vessels.

Included in the PSSRs	Excluded from the PSSRs
The following pressurized systems are likely to be *included*:	The following pressurized systems are likely to be *excluded*:
• A compressed air receiver and the associated pipework where the product of the pressure times the internal capacity of the receiver is greater than 250 bar litres.	• An office hot water urn (for making tea).
• A steam sterilizing autoclave and associated pipework and protective devices.	• A machine tool hydraulic system.
• A steam boiler and associated pipework and protective devices.	• A pneumatic cylinder in a compressed air system.
• A pressure cooker.	• A hand-held tool.
• A gas-loaded hydraulic accumulator.	• A combustion engine cooling system.
• A portable hot water/steam cleaning unit.	• A compressed air receiver and the associated pipework where the product of the pressure times the internal capacity of the receiver is less than 250 bar litres.
• A vapour compression refrigeration system where the installed power exceeds 25 kW.	• Any pipeline and its protective devices in which the pressure does not exceed 2 bar above atmospheric pressure.
• A narrow-gauge steam locomotive.	• A portable fire extinguisher with a working pressure below 25 bar at 60 °C and having a total mass not exceeding 23 kg.
• The components of self-contained breathing apparatus sets (excluding the gas container).	• A portable LPG cylinder.
• A fixed LPG storage system supplying fuel for heating in a workplace.	• A tyre used on a vehicle.

Fig. 5.1   Quick reference – inclusions/exclusions from the PSSRs

What is a WSE?

It is a document containing information about selected items of plant or equipment which form a pressure system, operate under pressure, and contain a 'relevant fluid'.

What is a 'relevant fluid'?

The term 'relevant fluid' is defined in the PSSRs and covers:

- Compressed or liquefied **gas** including air above 0.5 bar pressure.
- Pressurized **hot water above 110 °C**
- *Steam* at any pressure.

What does the WSE contain?

The typical contents of a written scheme of examination should include:

- Identification number of the item of plant or equipment.
- Those parts of the item which are to be examined.
- The nature of the examination required, including the inspection and testing to be carried out on any protective devices.
- The preparatory work necessary to enable the item to be examined.
- The date by which the initial examination is to be completed (for newly installed systems).
- The maximum interval between one examination and the next.
- The critical parts of the system which, if modified or repaired, should be examined by a competent person before the system is used again.
- The name of the Competent Person certifying the written scheme of examination.
- The date of certification.

The plant items included are those which, if they fail, 'could unintentionally release pressure from the system and the resulting release of stored energy could cause injury'.

Who decides which items of plant are included in the WSE?

The user or the owner of the equipment.

The written scheme of examination must be 'suitable' throughout the lifetime of the plant or equipment so it needs to be reviewed periodically and, when necessary, revised.

What is a 'Competent Person'?

The definition in the PSSRs is: *'competent person' means a competent individual person (other than an employee) or a competent body of persons corporate or unincorporated.*

This is probably the best definition you are going to get.

What does the 'Competent Person' do ?

- Advise on the nature and frequency of examination and any special safety measures necessary to prepare the system for examination; and/or

- Draw up and certify as suitable the written scheme of examination; or

- Simply certify as suitable a written scheme of examination prepared by the user or owner.

Users (or owners) of pressure systems are free to select any Competent Person they wish, but they are obliged take all reasonable steps to ensure that the Competent Person selected can actually demonstrate competence.

For further information look at:
The HSE booklet: *A Guide to the Pressure Systems and Transportable Gas Containers Regulations 1989.*
Guidance on the selection of Competent Persons is given in the HSE leaflet: *Introducing Competent Persons* IND(S)29(L): 1992(ISBN 0-7176-0820-4)

Fig. 5.2   Quick reference – written schemes of examination (WSEs)

Appendix 1.   User/owner decision tree.
Appendix 2.   Major health and safety legislation.

Table 5.1 shows the regulations arranged in diagrammatic form. Note how there are some provisions for exemption.

Part 1: Introduction

This gives general introductory material explaining some areas of technical background to the PSSRs. ACoP paragraphs 6 and 7 as follows explain the concept of the PSSRs in terms of the approach to the risk of release of stored energy.

6. The aim of the PSSRs is to prevent serious injury from the hazard of stored energy as a result of the failure of a pressure system or one of its component parts. The regulations are concerned with steam at any pressure, gases which exert a pressure in excess of 0.5 bar above atmospheric pressure, and fluids which may be mixtures of liquids, gases, and vapours where the gas or vapour phase may exert a pressure in excess of 0.5 bar above atmospheric pressure.

7. The PSSRs are concerned with the risks created by a release of stored energy through system failure. With the exception of the scalding effects of steam, the regulations do not consider the hazardous properties of the contents released following system failure. The stored content's properties are of concern only to the extent that they may be liable to accelerate wear and cause a more rapid deterioration in the condition of the system, so leading to an increased risk of failure. The risk from steam includes not only any possible deterioration in the condition of the system, which could increase the risk of failure, but also its scalding effect in the event of release. The PSSRs do not deal with all the hazards arising from the operation of such a system. The contents may be highly toxic or the plant may form part of a major hazard site. These aspects are all subject to separate legislative requirements and duty holders will need to consider these other aspects when deciding on the level of precautions required.

Note how these ACoP clauses set the limitation on the scope of the PSSRs. They do not attempt to cover all situations of danger; only those relating to stored energy and the scalding effect of steam.

Table 5.1 The PSSRs – exemptions

Regulation no. →	1 Citation and commencement	2 Interpretation	3 Application and duties	4 Design and construction	5 Provision of information and marking	6 Installation	7 Safe operating limits (SOLs)	8 Written scheme of examination (WSE)	9 Examination in accordance with WSE	10 Action in case of imminent danger	11 Operation	12 Maintenance	13 Modification and repair	14 Keeping of records	15 Overpressurization precautions	16 Defence	17 Power to grant exemptions	18 Repeals and revocations	19 Transitional provisions	Sch. 1 Exceptions to the PSSRs	Sch. 2 Lease/hire arrangements	Sch. 3 Marking of pressure vessels
Equipment compliant with PERs			E*	E																		
PV[†] <250 bar litres with relevant fluid (but not steam)					E			E	E													
PV <250 bar litres containing steam at any pressure										No exemptions: all regulations apply												

The following equipment is exempted from all PSSRs: systems on ships, aircraft, weapons, pipelines <2 barg, research equipment, engine cooling, prime movers (turbines etc.) refrigeration units <25 kW and systems covered by various other regulations (see ACoP Schedule 1 for full list).

* E, Exempt

[†] PV, Pressure × Volume

ACoP paragraph 10 introduces the idea of pressure and volume 'product'. This is the way that the PSSRs designate the stored energy capacity of vessels.

10. The amount of stored energy in a vessel is generally considered to be directly related to the volume of the vessel and the pressure of the contents. The measure of the stored energy is expressed by multiplying the pressure by the internal volume ($P \times V$), i.e. the pressure–volume product. If the units used are bar for pressure and litres for volume, the measure (or product) is given in bar litres.

Interface between the Pressure Equipment Directive and the implementing regulations in the UK: the Pressure Equipment Regulations 1999 (PERs)

The PERs are those UK regulations which implement the European Pressure Equipment Directive. ACoP paragraph 12 tries to explain the links between them.

12. The PERs cover the supply and putting into service of equipment that would also form whole or part of a pressure system and fall within the scope of the PSSRs. It must, however, be appreciated that the scope of the PERs and the PSSRs are not exactly the same. For example, equipment containing steam at or below 0.5 bar gauge does not fall under the scope of the PERs but does fall within the scope of the PSSRs.

Figure 5.3 outlines the differences between these regulations – it is easy to get confused.

Regulation 1.   Citation and commencement

This is a legal clause confirming that the regulations may be cited as the Pressure Systems Safety Regulations 2000 and came into force on 21 February 2000.

Regulation 2.   Interpretation

This regulation contains a list of definitions of terms used in the PSSRs. Some of the definitions are complex, so the ACoP 'guidelines' provide further guidance. ACoP paragraphs 17, 18, 19, 24, 25, and 26 are the more important ones. Their contents are shown below:

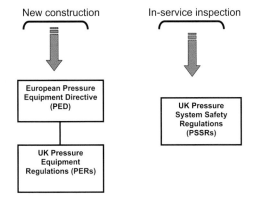

Fig. 5.3   The PED, PERs, and PSSRs

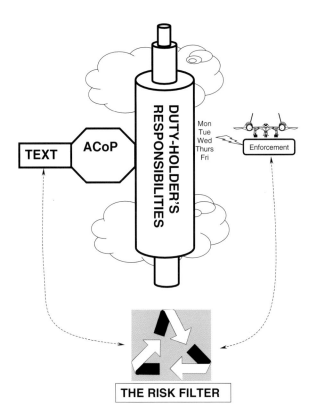

Fig. 5.4   Thinking outside the box – risk filter

Competent Person

17. The term 'competent person' refers not to the individual employee who carries out duties under the PSSRs but to the body which employs the person charged with those duties. Thus, the definition of competent person makes it clear that the legal duty to comply rests with a competent person's employer, and not with an individual, unless that person is self-employed.

Danger

18. The PSSRs are concerned with reasonably foreseeable danger to people from the unintentional release of stored energy. In addition, they deal with the scalding effects of steam which is classed as a relevant fluid irrespective of pressure. Leaks of gas, for instance, which do not have the potential to cause injury from stored energy, are not covered. Risk of injury from escapes of toxic or flammable materials are covered under other statutory provisions.

Examination

19. In the context of the PSSRs, the term *examination* relates solely to examinations carried out under the written scheme of examination, i.e. ones conducted to assess the condition of those parts of the system which may give rise to danger (as defined) in the event of an uncontrolled release of stored energy.

Pressure system

24. The PSSRs define three types of system:

(a) a system comprising a pressure vessel, its associated pipework, and protective devices. It is necessary for there to be a pressure vessel in the system for the PSSRs to apply under this definition. Where there is more than one system on the premises, whether interconnected or not, the user/owner is responsible for deciding where the boundaries for each system occur;

(b) pipework with its protective devices to which a transportable pressure receptacle is, or is intended to be, connected. A transportable pressure receptacle on its own is not a pressure system as defined. Pipework containing a relevant fluid (other than steam) at pressure of 0.5 bar or less is outside the scope of the PSSRs;

(c) a pipeline with its protective devices.

Protective devices

25. 'Protective device' includes any protective control or measuring equipment which is essential to prevent a dangerous situation from arising. Instrumentation and control equipment would be classed as a protective device in the following situations:

(a) where it has to function correctly in order to be able to protect the system; and

(b) where it prevents the safe operating limits being exceeded in situations where no other protective device is provided (for example, where the relevant fluid is so toxic that it cannot be released to atmosphere). In these cases the control equipment is itself the protective device.

Protective devices which protect a system which contains or is liable to contain a relevant fluid are covered by the PSSRs even if they are located on a part of the system which does not contain a relevant fluid.

Relevant fluid

26. The following conditions have to be fulfilled for a fluid to be a relevant fluid within the scope of the PSSRs:

(a) the pressure has to be greater than 0.5 bar above atmospheric (except for steam). Where the pressure varies with time, then the maximum pressure that is normally reached should be the determining factor;

(b) either the fluid should be gas or mixture of gases under the actual conditions in that part of the system or a liquid which would turn into a gas if system failure occurred. Therefore the PSSRs will cover compressed air (a mixture of gases) as well as other compressed gases such as nitrogen, acetylene, and oxygen. The definition will include also hot water contained above its boiling point at atmospheric pressure (pressurized hot water) or aqueous solutions where a vapour pressure above 0.5 bar (gauge) is generated. Classification of gases is given in standards in

BS 5045 which lists gases under separate headings as permanent gases and liquefiable gases.

AcoP paragraphs 35 to 43 cover the role of the 'Competent Person,' explaining the different possible 'grades' of Competent Person. The explanations are necessarily long-winded because of the fact that the Competent Person is not a 'person' as such, but an organization or body.

Competent Persons

35. The term 'competent person' is used in connection with two distinct functions:

    (a) drawing up or certifying schemes of examination (Regulation 8); and
    (b) carrying out examinations under the scheme (Regulation 9).

    The general guidance at paragraphs 35 to 41 relates to both functions. ACoP paragraphs 103 to 119 deal with schemes of examination and paragraphs 130 to 137 deal with examinations themselves.

36. Although separate guidance is given on these functions, this does not mean that they have to be carried out by different competent persons. In addition, the user/owner may seek advice from a competent person on other matters relating to these regulations. For example, advice could be sought on the scope of the written scheme (see ACoP paragraphs 110 and 111). In such circumstances, a competent person would be acting solely as an adviser, rather than as a competent person as defined.

37. It is the responsibility of the user/owner to select a Competent Person capable of carrying out the duties in a proper manner with sufficient expertise in the particular type of system. In some cases, the necessary expertise will lie within the user's/owner's own organization (but see ACoP paragraph 40 for guidance on independence). In such cases, the user/owner is acting as Competent Person and is responsible for compliance with the PSSRs. However, small or medium-sized businesses may not have sufficient in-house expertise. If this is the case, they should use a suitably qualified and experienced independent competent person. Whether the competent person is

drawn from within the user's/owner's organization or from outside, they should have sufficient understanding of the systems in question to enable them to draw up schemes of examination or certify them as suitable.

38. A competent person capable of drawing up schemes of examination or examining a simple system may not have the expertise, knowledge, and experience to act as competent person for more complex systems. For a number of systems, including the larger or more complex ones, it is unlikely that one individual will have sufficient knowledge and expertise to act on their own. A competent person should be chosen who has available a team of employees with the necessary breadth of knowledge and experience.

39. In general terms, the competent person should have:

    (a) staff with practical and theoretical knowledge and actual experience of the relevant systems;
    (b) access to specialist services;
    (c) effective support and professional expertise within their organization; and
    (d) proper standards of professional probity.

40. Where the competent person is a direct employee of the user's/ owner's organization, there should be a suitable degree of independence from the operating functions of the company. In particular, where the staff are provided from an in-house inspection department and carry out functions in addition to their competent person duties, they should be separately accountable under their job descriptions for their activities as competent persons. They should act in an objective and professional manner with no conflict of interests and should give an impartial assessment of the nature and condition of the system.

41. The competent person is responsible for all examinations. For example where ancillary examination methods (e.g. non-destructive testing) are undertaken by another person or body, the competent person should accept responsibility for these tests and their interpretation.

42. It is for users to select a competent person to carry out the duties required by the PSSRs. The following bodies may provide competent person services:

    (a) a user company;
    (b) a third-party organization/external company; or
    (c) a self-employed person.

43. Accreditation to BS EN 45004 1995 *General criteria for operation of the various types of bodies performing inspection* is an indication of the competence of an inspection department, organization, or self-employed person. Accreditation is carried out on behalf of Government by the United Kingdom Accreditation Service (UKAS). Accreditation to BS EN 45004 is recommended for bodies acting as competent persons engaged to draw up or certify a written scheme of examination or conduct examinations for major systems [as detailed in ACoP paragraph 105(c)]. Users/owners may also wish to consider using accredited bodies for other categories of system. Accreditation to BS EN 45004 is voluntary.

Note how ACoP paragraph 43 clears up one of the main misconceptions about the PSSRs – the Competent Person *does not have to be accredited* to BS EN 45004 – it is optional.

Safe operating limits (SOLs)

52. These are the limits beyond which the system should not be taken. They are not the ultimate limits beyond which system failure will occur. In establishing the limits within which a system should be operated, there may be a need to take account of matters other than pressure energy and the likelihood of system failure. Small steam generators, for example, present a risk from scalding as opposed to stored energy.

This paragraph is not *entirely* prescriptive – it is left to the Competent Person to interpret it. In some cases the SOLs will coincide with the pressure safety valve (PSV) lift pressure but not always.

Regulation 3.   Application and duties

Regulation 3 is best summarized by ACoP guidance paragraph 54 which pinpoints the duty of people to comply with the PSSRs.

54. The primary purpose of the PSSRs is to secure the safety of people at work. The PSSRs therefore apply to pressure systems used, or intended to be used, at work (but see Schedule 1 for exceptions to all or part of the PSSRs). The duties imposed relate to activities at work. They also cover the self-employed, for example a self-employed installer of a pressure system, or a self-employed Competent Person.

Regulation 4. Design and construction

This regulation places duties on designers, manufacturers, and any person who supplies equipment or a component intended to be part of a pressure system to ensure that it is fit for purpose, so as to prevent danger. Equipment (including assemblies) supplied in accordance with the Pressure Equipment Regulations 1999 is considered to meet these requirements and is, therefore, excepted from this regulation. ACoP paragraphs 57 and 58 outline the general situation for equipment which is covered.

57. Designers and manufacturers should consider at the manufacturing stage both the purpose of the plant and the means of ensuring compliance with the PSSRs.

58. The designer, manufacturer, importer, or supplier should consider and take due account of the following, where applicable:

   (a) the expected working life (the design life) of the system;
   (b) the properties of the contained fluid;
   (c) all extreme operating conditions including start-up, shutdown, and reasonably foreseeable fault or emergency conditions;
   (d) the need for system examination to ensure continued integrity throughout its design life;
   (e) any foreseeable changes to the design conditions;
   (f) conditions for standby operation;
   (g) protection against system failure, using suitable measuring, control, and protective devices as appropriate;
   (h) suitable material for each component part;
   (i) the external forces expected to be exerted on the system including thermal loads and wind loading; and

(j) safe access for operation, maintenance and examination, including the fitting of access (e.g. door) safety devices or suitable guards, as appropriate.

Protection against failure

ACoP paragraph 64 addresses the question of safety devices, e.g. PSVs. Note how it states general principles only, rather than give detailed information on acceptable or prohibited designs.

64. Every plant item in which the pressure can exceed the safe operating limit (i.e. those which have not been designed to withstand the maximum pressure which can be generated within the system) should be protected, whenever operated, by at least one pressure-relieving or pressure-limiting device. The device should be suitable for its intended duty and should be fitted as close as practicable to the plant item it is designed to protect. Sufficient devices should be fitted at other points to ensure that the pressures inside the system do not exceed the safe operating limits. In the event of a pressure relief device operating, the design should enable the contents to be released in as safe a manner as is practicable.

Construction materials

The PSSRs do not give detailed requirements for construction materials. ACoP paragraphs 71 and 72 address a few specific issues.

71. Materials used in construction should be suitable for the intended use. For example, steam boiler stop (crown) valves made from flake graphite (grey) cast iron are not recommended. Account should be taken of the intended duty of the valve, including pressure, temperature, size, frequency of use, nature of contents, and any particular foreseeable fault conditions, when selecting valves. The direction of opening and closing should preferably be indicated on valves.

72. Plastic pipes are often used on compressed air systems. However, not all plastics are suitable for use where there is the possibility of their becoming brittle or otherwise damaged due to exposure to heat or other adverse conditions.

External forces

External forces are a common cause of failure in pressure equipment. ACoP paragraph 73 gives general guidance.

73. Account should be taken of any external forces which could affect the integrity of the equipment. These may include the forces exerted on pipework from thermal expansion and contraction, externally applied loads or any reasonably foreseeable vibration or shock loading, for example from water hammer. Suitable expansion bends and/or joints and drains should be incorporated in the pipework as necessary.

Regulation 5.   Provision of information and marking

An extract from Regulation 5 says:
- Any person who:

  (a) designs for another any pressure system or any article which is intended to be a component part thereof; or
  (b) supplies (whether as manufacturer, importer or in any other capacity) any pressure system or any such article,

shall provide sufficient written information concerning its design, construction, examination, operation, and maintenance as may reasonably foreseeably be needed to enable the provisions of the PSSRs to be complied with.

ACoP paragraphs 77 and 81 give useful information about the type of information required.

77. The aim of this regulation is to ensure that adequate information about any pressure system subject to the Pressure Systems Safety Regulations 2000 is made available to users/owners by designers, suppliers, or those who modify or repair equipment. Basic information about pressure vessels should be permanently marked on the vessel. This information is listed in Schedule 3 of these regulations.

81. Additional information about pressure vessels and information relevant to the whole system (apart from that already marked on the vessel under Regulation 5(4) should be provided in writing. The purpose is to provide users/owners with sufficient information on the design, construction, examination, operation, and maintenance of the equipment to enable them to comply with the requirements of the PSSRs. The designer or supplier should use their judgement, knowledge, and experience to decide what information is required. Although it is not possible to give a complete list of all the information which

might be needed the following items should be considered where relevant:

(a) design standards used and evidence of compliance with national/European international standards or documentation showing conformity;

(b) design pressures (maximum and minimum);

(c) fatigue life;

(d) design temperature (maximum and minimum);

(e) creep life;

(f) intended contents, especially where the design has been carried out for a specific process.

(g) flow rates and discharge capacities;

(h) corrosion allowances;

(i) wall thickness.

Regulation 6.   Installation

An extract from Regulation 6 says:

'The employer of a person who installs a pressure system at work shall ensure that nothing about the way in which it is installed gives rise to danger or otherwise impairs the operation of any protective device or inspection facility.'

This is expanded in ACoP paragraph 86. Note the high level of detail in this paragraph.

86. When planning an installation, the employer of the installer should ensure that all of the following items which are relevant to the system are actioned (this list is not exhaustive and additional actions may be needed depending on the type of system, its location, and planned operating conditions):

(a) ensure that those doing the installation have the required training, skills and experience.

(b) provide adequate supervision, taking into account the complexity of the system being installed;

(c) prepare suitable foundations to support the system, taking into account the nature of the ground and design loads such as the weight of the system and many likely external forces;

(d) decide on the most suitable method of lifting and handling the vessel(s), protective devices, and pipework so as to avoid accidental damage;

(e) check for signs of damage in transit;

(f) protect the system from adverse weather conditions before and during installation;

(g) remove any protective packaging carefully before commissioning;

(h) ensure that any hot work such as welding or cutting will not affect the integrity of the system;

(i) ensure that protective devices are clear of obstruction and operate correctly without hindrance or blockage and that the discharge is routed to a safe place;

(j) ensure that any access doors/hatches are clear of obstruction and operate correctly;

(k) ensure that any labels or markings attached to the system are clearly visible;

(l) provide adequate access for maintenance and examination purposes;

(m) provide suitable physical protection against mechanical damage, e.g. accidental impact by vehicles;

(n) allow sufficient space for access around and beneath valves, in particular drain valves.

(o) clear away any debris such as metal shavings or dust arising from the installation process; and

(p) have the installation work checked and approved on completion by a suitably qualified person.

Regulation 7. Safe operating limits

The designation of safe operating limits (SOLs) is one of the most important engineering parts of the PSSRs. This Regulation complements Regulation 5, which makes the designer, manufacturer, and supplier responsible for providing adequate information about the system or its component parts. It prohibits the user/owner from operating the system, or allowing it to be operated before the safe operating limits have been established. ACoP paragraphs 90 to 92 give further details.

Establishing the limits

90. Where the system consists of a standard production item, the designer/manufacturer should assess the safe operating limits and pass the relevant information to the user/owner. In these circumstances, the user/owner will not always need to carry out

the detailed work required to establish the safe operating limits of the system. In cases where the user/owner has specified the design, the responsibility for establishing the safe operating limits rests with user/owner.

91. If the user/owner does not have sufficient technical expertise to establish the safe operating limits, an organization which is competent to carry out the task should be used.

92. The exact nature and type of safe operating limits which need to be specified will depend on the complexity and operating conditions of the particular system. Small, simple systems may need little more than the establishment of the maximum pressure for safe operation. Complex, larger systems are likely to need a wide range of conditions specified, e.g. maximum and minimum temperatures and pressures, nature, volumes, and flowrates of contents, operating times, heat input or coolant flow. In all cases the safe operating limits should incorporate a suitable margin of safety.

Regulation 8.   *Written scheme of examination (WSE)*

An extract from Regulation 8 says:

(1) The user of an installed system and owner of a mobile system shall not operate the system or allow it to be operated unless he has a written scheme for the periodic examination, by a competent person, of the following parts of the system, that is to say:

(a) all protective devices;
(b) every pressure vessel and every pipeline in which (in either case) a defect may give rise to danger; and
(c) those parts of the pipework in which a defect may give rise to danger.'

ACoP paragraphs 103 to 108 explain the requirements in some detail. Paragraph 105 gives the three categories of pressure system that form the basis of the subdivision of the WSE.

103. Before a pressure system is operated, the user/owner must ensure that a written scheme of examination has been prepared. The written scheme of examination should be drawn up by someone other than a competent person,

certified as suitable by a competent person. See ACoP paragraphs 35 to 41 for general guidance on the role of the competent person and paragraphs 107 to 119 for guidance on the function of the competent person in relation to the written scheme of examination.

Attributes and role of the Competent Person

104. The level of expertise needed by the competent person depends on the size and complexity of the system in question. To illustrate the level of expertise, knowledge and experience needed in different circumstances, pressure systems are divided into three categories as described in ACoP paragraph 105. However, in practice there are no clear dividing lines. The three categories should be taken as an indication of the range of systems covered rather than providing clear cut divisions. Each system should be individually assessed and an informed decision made on which of the categories is the most appropriate.

105. The three categories are as follows:

(a) *Minor systems* include those containing steam, pressurized hot water, compressed air, inert gases, or fluorocarbon refrigerants which are small and present few engineering problems. The pressure (above atmospheric pressure) should be less than 20 bar (2.0 MPa) [except for systems with a direct-fire heat source when it should be less than 2 bar (200 kPa)]. The pressure–volume product for the largest vessel should be less than 2×10^5 bar litres (20 MPa m^3). The temperature in the system should be between -20 and $250\,^\circ$C except in the case of smaller refrigeration systems operating at lower temperatures which will also fall into this category. Pipelines are not included.

(b) *Intermediate systems* include the majority of storage systems and process systems which do not fall into either of the other two categories. Pipelines are included unless they fall into the major system category.

(c) *Major systems* are those which because of their size, complexity or hazardous contents require the highest level of expertise in determining their condition. They include steam-generating systems where the individual capacities of the steam generators are more than 10 MW, any pressure storage system where the pressure–volume product for the largest pressure vessel is more than 10^6 bar litres (100 MPa m^3) and any manufacturing or chemical reaction system where the pressure–volume product for the largest pressure vessel is more than 10^5 bar litres (10 MPa m^3). Pipelines are included if the pressure–volume product is greater than 10^5 bar litres.

106. The attributes needed for competent persons who draw up or certify schemes of examination relating to minor, intermediate, and major systems are as shown below.

(a) *Minor systems*:
   – *Staff*
   At least one member of staff qualified to incorporated engineer level with adequate relevant experience and knowledge of the law, codes of practice, examination and inspection techniques, and understanding of the effects of operation for the system concerned.
   – *Specialist services*
   Established access to basic design and plant operation advice, materials engineering and non-destructive testing (NDT) facilities.
   – *Organization*
   Sufficient organization to ensure a reasonable document storage and retrieval system with ready access to relevant law, technical standards and codes.

(b) *Intermediate systems*:
   – *Staff*
   Depending on the complexity of the system, at least one senior member of staff of chartered engineer or equivalent status in each relevant discipline and supported by technically qualified and experienced staff with knowledge of the law, codes of practice, examination and inspection techniques, and understanding of the effects of operation for the system concerned.
   – *Specialist services*
   In-house or clearly established access to materials engineering, NDT, design and plant operating advice.

- *Organization*

Clear supervisory arrangements with an adequate degree of formal organization. Appropriate document storage and retrieval system with ready access to relevant law, technical codes and standards.

(c) *Major systems*:

- *Staff*

Depending on the complexity of the system, at least one senior member of staff of chartered engineer or equivalent status in each relevant discipline and supported by technically qualified and experienced staff with knowledge of the law, codes of practice, examination and inspection techniques, and understanding of the effects of operation of the system concerned.

- *Specialist services*

In-house or clearly established access to the full range of relevant specialist services in the fields of materials engineering, NDT, design, and plant operation.

- *Organization*

Formal structure and clear lines of authority and responsibility set out in a written statement. Formal recruitment and training policies for staff. Effective document storage and retrieval system with ready access to relevant law, technical codes and standards.

Drawing up the written scheme of examination

107. The written scheme of examination can be written and certified as suitable either by an independent competent person or by the in-house competent person. For either function, the criteria for competent persons in ACoP paragraph 106 should be met, depending on the category of the system.

108. Where the appropriate technical expertise exists in-house, the written scheme may be drawn up by the user of the system and certified as suitable by a competent individual within their own organization provided they fulfil the requirements for a competent person. Alternatively, there may be sufficient in-house expertise to draw up the scheme but not certify it, in which case the user should employ an independent competent person to carry out the certification.

The content of WSEs is, in practice, very variable. For this reason, the PSSRs provide broad guidelines only (see ACoP paragraphs 112 to 117).

Content

112. At least the following information should be included in the written scheme of examination:

    (a) those parts of the system which are to be examined;
    (b) identification of the item of plant or equipment;
    (c) the nature of the examination required, including the inspection and testing to be carried out on any protective devices;
    (d) the preparatory work necessary to enable the item to be examined safely;
    (e) specify what examination is necessary before the system is first used, where appropriate;
    (f) the maximum interval between examinations;
    (g) the critical parts of the system which, if modified or repaired, should be examined by a competent person before they are used again;
    (h) the name of the competent person certifying the written scheme; and
    (i) the date of the certification.

113. The nature of the examination should be specified in the written scheme. This may vary from out-of-service examination with the system stripped down, to in-service examination with the system running under normal operating conditions. Some systems (for example fired equipment) may need to undergo both out-of-service and in-service examinations. The competent person may need to seek advice from the manufacturer/supplier on appropriate methods of testing, particularly where internal examination is difficult.

First examination

114. Where appropriate, the requirement for an examination before the system is first taken into use should be specified in the written scheme of examination. For equipment supplied in accordance with the Pressure Equipment Regulations 1999, the person who draws up or certifies the written scheme should consider whether an initial examination is appropriate

and the form that any such examination should take. This consideration should take account of the results of the conformity assessment to which the equipment was subject before it was placed on the market. In general, further assessment of the equipment under the written scheme should be judged on the merits of each individual case.

Periodicity

115. When deciding on the periodicity between examinations, the aim should be to ensure that sufficient examinations are carried out to identify at an early stage any deterioration or malfunction which is likely to affect the safe operation of the system. Different parts of the system may be examined at different intervals, depending on the degree of risk associated with each part.

116. Protective devices should be examined, at least, at the same time and frequency as the plant to which they are fitted. Some protective devices may need to be examined at more frequent intervals. The examination should include checks that the devices function correctly and are properly calibrated or, alternatively, that they have been replaced by recently tested units.

117. All relevant factors should be taken into account when deciding on the appropriate interval between examinations, including:

   (a) the safety record and previous history of the system;
   (b) any generic information available about the particular type of system;
   (c) its current condition, e.g. due to corrosion/erosion, etc. (internal and external);
   (d) the expected operating conditions (especially any particularly arduous conditions);
   (e) the quality of fluids used in the system;
   (f) the standard of technical supervision, operation, maintenance, and inspection in the user's/owner's organization; and
   (g) the applicability of any on-stream monitoring.

Regulation 9. Examination in accordance with the written scheme

An extract from Regulation 9 says:

... the user of an installed system and the owner of a mobile system shall:

(a) ensure that those parts of the pressure system included in the scheme of examination are examined by a competent person within the intervals specified in the scheme and, where the scheme so provides, before the system is used for the first time; and

(b) before each examination take all appropriate safety measures to prepare the system for examination, including any such measures as are specified in the scheme of examination pursuant to Regulation 8(3)(b).

ACoP paragraphs 126 and 127 attempt to explain the examination characteristics.

126. Although this Regulation (a) places duties on the competent person in relation to carrying out the examination, there is a clear duty on users/owners to ensure that the equipment is not operated beyond the date specified in the current examination report.

127. The words 'as soon as is practicable' in Regulation 9(3) are intended to ensure that where repairs have to be carried out within a short timescale, there is no delay in producing examination reports. However, where examinations are concentrated into a short period of time for reasons of efficiency, for instance on large integrated chemical plant, it may be unreasonable to expect all the reports to be completed at the same time. In these circumstances, the competent person should complete the reports and forward them within 28 days of the completion of the final examination in that series.

WSE reports

WSE inspection reports content and format are, essentially, left to the discretion of the competent person. Various ACoP paragraphs, 132 to 137 and 144, give an outline of what is required but the guidance is very broad. In practice, good in-service inspection reports are more comprehensive than the minimum requested in the ACoP.

132. The competent person should examine and report on all parts of the system covered by the written scheme of examination. The competent person should be satisfied that, as a result of the examination, the condition of the parts included in the written scheme and their fitness for continued use has been properly assessed. The following points (although not exhaustive) should aid the competent person's decisions:

   (a) the age and known history of the part or system;
   (b) the nature of the relevant fluid;
   (c) the conditions of use;
   (d) the length of time since the last examination; and
   (e) the expected operating conditions and maintenance regime until the date of the next examination.

134. The report should be based on the actual condition of the system as found during the examination. If repairs are carried out as a result of the examination, the report should include details of the fault and the remedial action taken even if the repair works are finished before the examination has been completed.

135. The competent person should consider whether any changes are needed in the safe operating limits, or in the scheme of examination. The report may state that continued use of the system is dependent on altering certain specified operating conditions or undertaking specific maintenance tasks.

136. The competent person may decide that the risk of danger may be significantly increased if the next examination is delayed until a date set in accordance with the current written scheme. In these circumstances, the written scheme should be reviewed and an earlier date set beyond which the system should not be operated without a further examination.

137. At the end of the examination, the competent person should be satisfied that the protective devices, especially any safety valves, have been tested and set correctly. Where protective devices which have been removed during an examination are found to be defective, the cause of the problem should be investigated further by the user/owner and the

necessary corrective measures taken (see also ACoP paragraphs 145 to 150 for guidance on 'action in case of imminent danger').

Format of reports

144. No particular format is laid down for the report as systems vary in size and complexity. The format can be chosen to fit in with the record keeping systems of the user/owner and competent person. Computer-generated reports with appropriate validation are acceptable and this is likely to speed up the despatch by inspectors. Any secure system that can be validated as the sole record-keeping medium is acceptable. Suggested items for inclusion in the report include the following:

(a) name and address of owner;
(b) address, location of system, and name of the user (if different from the owner);
(c) whether subject to a written agreement under Schedule 2;
(d) identification of system or parts examined;
(e) condition of system or parts examined;
(f) parts not examined;
(g) result of the examination;
(h) any repairs needed and the time scale for completion;
(i) any changes in the safe operating limits and the date by which they should be made;
(j) any change in the written scheme of examination;
(k) date by which the next examination must be completed;
(l) other observations;
(m) where the most recent examination due was postponed in accordance with Regulation 9(7), the names of appropriate members of the competent person's and the user's/owner's organization, the date of giving the relaxation, and the new date by which the examination was to be completed;
(n) date examination took place;
(o) name and address of competent person;
(p) signature; and
(q) date of report.

Regulation 10.   Action in case of imminent danger

This regulation applies only to 'serious defects requiring immediate attention'. That is, where there is a risk of imminent failure of the system if immediate repairs are not undertaken or other suitable modifications are not made to the operating conditions. The essence of the regulation is given in ACoP paragraph 146.

> 146. The user/owner should be notified immediately of those defects which the competent person considers could cause imminent failure of the system. Therefore, the written report to the user/owner should be made immediately. It is separate from and does not replace the report of the examination under the written scheme of examination required by Regulation 9.

There is a requirement (often a controversial one) to notify the HSE – see ACoP paragraphs 148 and 150.

> 148. Notification to the enforcing authority (i.e. the Health and Safety Executive or the Environment Health Department of the Local Authority) is also a separate action required of the competent person. The period of 14 days to make this report to the enforcing authority is to allow sufficient time for a formal written report to be sent.

> 150. The sequence of events for reporting imminent danger is given below.
>
> (a) The competent person immediately produces a written report identifying the system and specifying the repairs, modifications, or changes required and gives it to the user/owner.
>
> (b) The user/owner ensures that the system (or, if the report only affects a discrete part of the system, that part) is not operated until the necessary repairs, modifications, or changes have been carried out.
>
> (c) The competent person sends a written report to the relevant enforcing authority within 14 days.
>
> (d) The competent person produces a report of the examination under the written scheme (Regulation 9) and sends it to the user/owner within 28 days.

Regulation 11. Operation

An extract from Regulation 11 says:

'The user of an installed system and the owner of a mobile system shall provide for any person operating the system adequate and suitable instructions for:

(a) the safe operation of the system; and
(b) the action to be taken in the event of any emergency.'

ACoP paragraphs 152, 153, and 156 give useful details:

152. The instructions provided to operators by the user/owner should cover:

(a) all procedures and information needed so that the system can be operated safely; and
(b) any special procedures to be followed in the event of an emergency.

153. Information provided by manufacturers or suppliers such as instruction sheets and operating manuals may form part or all of the instructions developed to meet the PSSRs. To fulfil this role they should be sufficiently comprehensive, cover the particular installation and its safe operation, and be consistent with the site operating conditions.

Content

156. The instructions should contain all the information needed for safe operation, of the system including:

(a) start-up and shutdown procedures;
(b) precautions for standby operation;
(c) function and effect of controls and protective devices;
(d) probable fluctuations expected in normal operation;
(e) the requirement to ensure that the system is adequately protected against overpressure at all times; and
(f) procedures in the event of an emergency.

Regulation 12. Maintenance

An extract from Regulation 12 says:

'The user of an installed system and the owner of a mobile system shall ensure that the system is properly maintained in good repair, so as to prevent danger'.

ACoP paragraph 165, 168, and 169 give guidance as follows:

165. This regulation builds on the more general duties in the HSW Act and Regulation 5 of the Provision and Use of Work Equipment Regulations (PUWER) 1998 which require that work equipment is maintained so that it does not give rise to risks to health and safety.

168. The type and frequency of maintenance for the system should be assessed and a suitable maintenance programme planned.

169. A suitable maintenance programme should take account of:

   (a) the age of the system;
   (b) the operating/process conditions;
   (c) the working environment;
   (d) the manufacturer's/supplier's instructions;
   (e) any previous maintenance history;
   (f) reports of examinations carried out under the written scheme of examination by the competent person;
   (g) the results of other relevant inspections (e.g. for main-tenance or operational purposes);
   (h) repairs or modifications to the system; and
   (i) the risks to health and safety from failure or deterioration.

Regulation 13.   Modification and repair

An extract from Regulation 13 says:

'The employer of a person who modifies or repairs a pressure system at work shall ensure that nothing about the way in which it is modified or repaired gives rise to danger or otherwise impairs the operation of any protective device or inspection facility.'

The essence of this is that repair or modification of non-pressure containing parts of the system should be carried out so that the integrity of the pressure system is not adversely affected. This should ensure that any repairs or modifications (including extensions or additions) do not affect the operation of any protective devices.

Regulation 14.   Keeping of records

The requirement is summarized in ACoP paragraph 183:

> 183. The user/owner should keep the following documents readily available:
>
>   (a) any designer's/manufacturer's/supplier's documents relating to parts of the system included in the written scheme;
>   (b) any documents required to be kept by the Pressure Equipment Regulations 1999;
>   (c) the most recent examination report produced by the competent person under the written scheme of examination;
>   (d) any agreement or notification relating to postponement of the most recent examination under the written scheme; and
>   (e) all other reports which contain information relevant to the assessment of matters of safety.

Regulation 15.   Precautions to prevent pressurization of certain vessels

An extract from Regulation 15 says:

> '(1) Paragraph (2) shall apply to a vessel
>
>   (a) which is constructed with a permanent outlet to the atmosphere or to a space where the pressure does not exceed atmospheric pressure; and
>   (b) which could become a pressure vessel if that outlet were obstructed.
>
> (2) The user of a vessel to which this paragraph applies shall ensure that the outlet referred to in subparagraph (a) of paragraph (1) is at all times kept open and free from obstruction when the vessel is in use.'

The purpose of this regulation is to prevent an unintentional build-up of pressure in a vessel which is provided with a permanent outlet to atmosphere, or to a space where the pressure does not exceed atmospheric pressure.

Regulation 16.   Defence

An extract from Regulation 16 contains wording to the effect that:

(1) In any proceedings for an offence for a contravention of any of the provisions of these regulations it shall be a defence for the person charged to prove:

    (a) that the commission of the offence was due to the act or default of another person not being one of his employees (hereinafter called 'the other person'); and

    (b) that he took all reasonable precautions and exercised all due diligence to avoid the commission of the offence.

This is a legal-orientated clause. It is frequently used by parties who are being prosecuted for alleged non-compliance with the PSSRs.

Schedule 1. Pressure systems excepted from all regulations

This gives a comprehensive list of pressure systems that are exempt from the PSSRs (see Fig. 5.1 for a summary).

Part II. Pressure systems excepted from certain regulations
This covers pressure systems which are partially exempt from the PSSRs (see Fig. 5.1 and Table 5.1 for a summary).

Schedule 2. Modifications of duties in cases where pressure systems are supplied by way of lease, hire, or other arrangements

Schedule 3. Marking of pressure vessels

The minimum information required for marking of a pressure vessels is:

1. The manufacturer's name.
2. A serial number to identify the vessel.
3. The date of manufacture of the vessel.
4. The standard to which the vessel was built.
5. The maximum allowable pressure of the vessel.
6. The minimum allowable pressure of the vessel where it is other than atmospheric.
7. The design temperature.

Bibliography

A Guide to the Health and Safety at Work Act 1974 L1:1992 (HSE Books, ISBN 0-7176-0441-1).

A Guide to the Pipelines Safety Regulations 1996, Guidance on Regulations L82: 1996 (HSE Books, ISBN 0-7176-1182-5).

Management of Health and Safety at Work. Management of Health and Safety at Work Regulations 1999. Approved Code of Practice and Guidance L21: 2000, Second edition (HSE Books, ISBN 0-7176-2488-9).

Safe Use of Work Equipment. Provision and Use of Work Equipment Regulations 1998. Approved Code of Practice and Guidance L22: 1998, Second edition (HSE Books, ISBN 0-7176-1626-6).

Safety in Pressure Testing GS4: 1998, Third edition (HSE Books, ISBN 0-7176-1629-6).

Workplace Health, Safety and Welfare. Workplace (Health Safety and Welfare) Regulations 1992. Approved Code of Practice and Guidance L24: 1992 (HSE Books, ISBN 0-7176-1413-6).

Safe work in confined spaces. Confined Spaces Regulations 1997. Approved Code of Practice, Regulations and Guidance L101: 1997 (HSE Books, ISBN 0-7176-1405-0).

Further reading

Automatically Controlled Steam and Hot Water Boilers PM5: 1997 (HSE Books, ISBN 0-7176-1028-4).

Essentials of Health and Safety at Work: 1994 (HSE Books, ISBN 0-7176-0716-X).

Pressure Systems Safety and You INDG261: 1997 (HSE Books, ISBN 0-7176-1452-2). Single copies free, multiple copies in priced packs.

Safe Management of Ammonia Refrigeration Systems PM81: 1995 (HSE Books, ISBN 0-7176-1066-7).

Safety at Autoclaves PM73: 1998 (HSE Books, ISBN 0-7176-1534-0).

Safe Work in Confined Spaces INDG258: 1997 (HSE Books, ISBN 0-7176-1442-5). Single copies free, multiple copies in priced packs.

Written Scheme of Examination: Pressure Systems and Transportable Gas Containers Regulations 1989 INDG178: 1994 (HSE Books). Free publication.

Chapter 6

Non-destructive testing (NDT)

Non-destructive testing (NDT) is a subject which has major influences on the role of the in-service inspector. It is a wide technical topic incorporating everything from straightforward visual inspection to advanced technologies using complex, bespoke equipment. NDT is responsible for a significant proportion of the overall inspection cost of plant during shutdown – costs are incurred not only for the NDT activities themselves but also for the associated scaffold, lagging removal, and cleaning/preparation activities that go with them

Figure 6.1 shows, conceptually, the extent of NDT techniques used in in-service inspection. Note how the majority of NDT activity 'volume' (i.e. the bottom of the triangle) is straightforward visual examination and surface crack detection, dye penetrant (DP) and magnetic particle inspection (MPI) techniques. The basic radiographic testing (RT) and ultrasonic testing (UT) techniques lie in the middle, with advanced NDT technologies such as time of flight diffraction (TOFD) and similar at the top end. Hence the reality is that advanced techniques represent only *a very small proportion* of NDT work carried out, despite their predominance in technical papers and NDT company presentation literature.

How reliable is site NDT?

Probably not as reliable as you think. Despite various schemes for certification of NDT technicians and a wide range of published technical standards, it seems that site NDT is not as reliable as perhaps it should be. A major study was carried out in 1999, funded by the UK Health and Safety Executive [Reference: The Programme for the Assessment of NDT in Industry (PANI) study]. This was a wide-ranging and comprehensive study, of direct relevance to the role of the in-service inspector of pressure equipment and other mechanical plant. Figures 6.2 to 6.4 summarize the details and the key findings and conclusions.

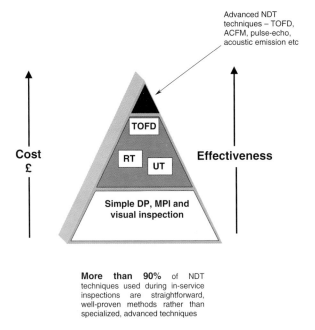

Fig. 6.1   In-service inspection and NDT – the situation

The results of the PANI study have had implications in several large industries. Particularly affected have been those large process and petrochemical plant operators that act under EN 45004 as their own 'user inspectorate' to meet their obligations under the PSSRs. Actions taken by such organizations have included:

- *Increasing* their own monitoring of their NDT subcontractors.
- *Specifying* more closely the NDT techniques and defect acceptance criteria to be used, rather than relying totally on their NDT subcontractors' judgement.
- *Increasing the scope of NDT* on in-service pressure equipment, to increase the overall probability of detection (POD) of defects that are there.

One thing that the PANI study has not done, however, is to increase fundamentally the application of advanced NDT techniques such as TOFD and similar. The effectiveness of these techniques was not questioned by PANI to the same extent as the more standard ones, so the conclusions have been less dramatic. General consensus seems to be that the effectiveness of such advanced techniques is not fundamentally in doubt – it is their inconvenience and cost that places the practical limitation on their site use.

Outline

The PANI study was an HSE-funded exercise in which qualified NDT technicians were tested on their ability to detect, classify, and size defects under simulated site conditions.
   Test pieces used were all of low-carbon steel and included:

- a set-on partial penetration nozzle;
- a set-through full penetration nozzle;
- a partial penetration tee-weld;
- a vessel longitudinal seam weld;
- pipe-to-pipe and pipe-to-elbow welds;
- an ex-service package steam boiler.

All these contained known defects of various types.
   Techniques used were: manual ultrasonic testing (UT) (BS 3923–1: $0°$, $45°$, $60°$, $70°$ probes with 14 dB sensitivity) and magnetic particle (MPI), accompanied by supporting time of flight diffraction (TOFD). Operators were qualified to at least PCN/ASNT Level 2 in the relevant techniques.
   Practical difficulties simulated during the tests were:

- Time pressure to complete the tests in a predetermined time.
- Awkward geometries making it difficult to take measurements.
- Limited access.
- Cold conditions ($\cong 14\,°C$).

Fig. 6.2   The PANI (Programme for the Assessment of NDT in Industry) study 1999

On balance, the conclusions of the PANI study have not had overly dramatic consequences on the general role of the in-service inspector. The intuitive feelings of most inspectors is that NDT is a process that requires a thorough, careful approach if it is to be effective – a view reinforced absolutely by the PANI conclusions.

The inspector as NDT technician?

This is a fertile area for confusion because the role of the in-service inspector is heavily intertwined with the work of NDT technicians, and the monitoring of their results. Many in-service inspectors evolve from NDT career backgrounds so they like to be involved in the mechanics of

Following a rigorous and scientific assessment of the NDT operators' written reports and verbal debriefings it was concluded that:

- Despite having good NDT qualifications, the performance of the NDT operators deteriorated once the geometry of the test pieces became more complex.
- Detection of individual defects was **lower than 20%** in some cases. About 50–70% was the norm. Defects associated with weld lands proved the hardest to detect.
- The sizing of the defects was **poor**. Small defects tended to be over-sized while large defects (6 mm+) tended to be under-sized.
- Even experienced, well-qualified NDT operators **did not always give reliable results** when faced with new situations.
- The effectiveness of TOFD is severely limited by complex weld joint geometries.

These conclusions have important implications for the in-service inspector. The message is that you cannot rely 100% on individual NDT results, without considering seriously the limitations of the techniques and operators. Multiple or repeat tests may be needed to ascertain the full situation.

Fig. 6.3   The PANI study – results

the NDT activity. Strictly however, the in-service inspector *is not* a practitioner of NDT. Figure 6.5 shows the situation.

Priority? What priority?

This can be a controversial concept – the idea that in-service inspectors can purposefully prioritize their involvement in NDT matters, in the belief that technical knowledge in some area can consciously be ignored, while concentrating on what is seen (by the inspectors themselves) as knowledge which is somehow more important. The saviour of this argument is, however, the (hopefully) indisputable truth that an inspector cannot be an expert in everything. This is particularly relevant to NDT where the field is so wide. So two simple conclusions can be made:

- An inspector does not have to be an expert in everything to do with NDT.
- An inspector does not have to *witness all* NDT activities to be effective. In practice, it will only be possible to actually witness, in

Based on the PANI study findings – and a bit of experience and intuition – some general guidance points for the in-service inspector are:

- **Monitoring NDT operators.** It is the inspector's duty to monitor NDT operators to reduce the risk of poor results as far as is possible.
- **Access and working conditions.** These have a big effect on the reliability of the NDT results that you can expect from an individual job. Aspects such as access scaffolding, sufficient lagging removal, etc. are important.
- **Weld types and geometries.** UT is less reliable on thin plate (< 10 mm) and pipe than on thicker sections. Pay particular attention to monitoring this type of UT activity.
- **Expected defect types.** NDT operators are much more effective at finding defects when they know what types of defects to *expect* in a particular testing programme. Find out, and then tell them.
- **Unwitnessed NDT reports.** Not all NDT reports give a full representation of all the defects that are present, or are accurate in terms of sizing the defects either. You need to use your experience, and intuition, to decide what level of credibility to give reports covering NDT activities that you did not witness yourself.

And if anyone suggests that site NDT is a 100% reliable technique – suggest that they read the PANI study.

Fig. 6.4   The PANI study conclusions – some implications for in-service inspectors

person, a small sample of all relevant activities performed – it is a question of prioritizing the possible witnessed activities to best effect.

Prioritizing types of NDT

This is one of the key skills of the in-service inspector – deciding which types of NDT to witness and which types to restrict their input to reviewing of results. There are two basic approaches: allocation of priority depending on whether the NDT technique is a *surface or volumetric technique,* or an allocation based on *application criticality.* Figure 6.6 summarizes the situation. Note how the two basic approaches in Fig. 6.6 are not mutually exclusive, so the most beneficial way to decide what NDT to witness is to use them together as far as possible, rather than separately. Remember that the idea is to allocate witnessing priority in a way that will give you, the inspector, the *highest*

Some guidelines

- An in-service inspector *is not* an NDT technician.
- An inspector needs to understand the mechanics of NDT techniques
  – but not necessarily as well as the technician who is doing them.

However

- *An inspector must be* more knowledgeable *than the NDT technician on:*

  – The *objectives* of the NDT technique, i.e. what type of defects are
    being looked for.
  – Which *defect acceptance criteria* to use, to fulfil the requirements
    of the particular type of fitness-for-purpose (FFP) assessment that
    is being done.
  – Interpreting the criticality of the NDT results.
  – Deciding in which areas the NDT should be done in the first place.

Fig. 6.5   NDT technician or inspector?

---

1. **Surface or volumetric techniques**
   On balance, volumetric techniques is (RT/UT) are subject to greater
   all-round uncertainty than surface techniques (DP/MPI). So inspectors should witness more volumetric NDT than surface NDT
   activities.
2. **Application criticality**
   Inspection witness time is best spent on those components that
   have the highest risk of mechanical failure if defects are missed, e.g.
   boiler headers are higher risk than boiler drums (which rarely
   actually fail).

Fig. 6.6   Two ways to decide what NDT to witness

chance of finding any defects that are present, independent of the
competence (or otherwise) of the NDT technician who is doing the
actual testing.

The above guidelines parallel, broadly, the overall findings of the
PANI study. The PANI study provides excellent coverage of some of
the technical rationale behind the results. It also reinforces the general

intuitive feel of most experienced inspectors about which NDT activities are inherently less robust than others, and so represent a higher risk to integrity and fitness for purpose.

NDT interpretation

Although in-service inspectors are not NDT technicians, they certainly act as *interpreters*. It is a common requirement for inspectors in all industries to be able to:

- Decide what NDT needs to be done.
- Make sure it is done properly.
- Interpret the results.

The interpretation stage consists of the four substages shown in Fig. 6.7; note how it is actually a programme of closely spaced, chained events. The problem with this programme (apart from its technical difficulty) is that it only makes full sense in its entirety. An inspector who reports a defect then abrogates responsibility, is introducing significant weaknesses into the chain. Even if the decision on defect classification and sentencing are left to a more highly qualified engineer, the chain is still broken because it is the inspector who has found and seen the defect. and has first-hand experience of the context in which it exists. By far the most effective method of in-service inspection is for the inspector to play a leading role in the *full chain* of events shown in Fig. 6.7. By all means invite the views of design engineers, corrosion engineers, managers, and suchlike but remember that the overriding objective must be to maintain the integrity of the chain.

The final stage in the NDT interpretation chain is the *sentencing* of the component containing a defect. Sentencing means nothing more than deciding what to do. While an in-service inspector will rarely be the only party involved in making sentencing decisions, it is important that the inspector's input is maintained at this stage. In many cases the solution will not be as straightforward as a simple decision on whether to continue to use the component or scrap it – there will be options to repair, re-rate, or do further tests on the component to establish its condition. All of these solution options are likely to involve a requirement for inspectors at some time in the future, so there is a clear need for inspectors to maintain their involvement.

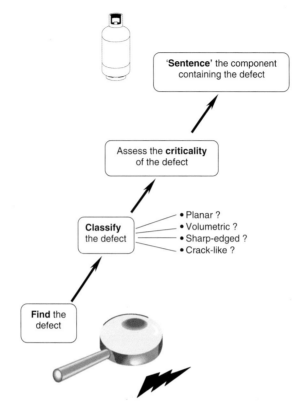

Fig. 6.7   The chain of NDT interpretation

Visual inspection

Visual inspection is the simplest form of NDT. It is also the most common; more than 90 percent of inspections involve nothing more sophisticated than simply looking carefully at an item and checking for corrosion, cracks, or other defects. The most common type of activity is the visual examination of welds. For in-service inspection (as opposed to 'shop' inspection of new equipment under manufacture), the main emphasis is on inspecting welds which are already complete, rather than at the weld preparation or tack-up stage.

Visual weld inspection standards

There are two main 'sets' of published technical standards relating to the visual inspection of welds: those included in application standards (e.g. pressure vessel, boiler structural steelwork codes, etc.) and those that are stand-alone, i.e. not related to any specific type of equipment. Figure 6.8 shows the main standards. Note that none of these standards address the inspection of any type of defects that a weld has developed in-service, they are all limited to defects resulting from the initial welding process itself.

'Stand-alone' generic standards

BS 5289: 1976 Visual inspection of fusion welded joints
This short standard (ten pages) is now formally withdrawn, although it is still used as a reference in some industries. It covers the basic terminology and definitions of visual defects such as root defects, contour defects (overconvexity etc.) undercut, overlap and cracking, but does not give any acceptance criteria for the inspection to check against. It was replaced by the European standard EN 970.

EN 970: 1997 Visual examination of welds
This is a similar style and length of standard to BS 5289 but actually contains less information about weld defects and their terminology. Instead, it explains about different types of weld gauges used for measuring weld profile, root gap, undercut, etc. It contains nothing at all itself about acceptance criteria but cross-references the main EN standard that does, i.e. EN 25817: *Arc welded joints in steel – Guidance on quality levels for inspections.* It similarly references the equivalent standard for aluminium welding which is EN 3004 *Arc-welded joints in aluminium and its weldable alloys – Guidance on quality levels for inspections.*

EN 25817: Arc welded joints in steel – Guidance on quality levels for inspection
This developed from a previous standard ISO 5817 (the text is almost identical) and is becoming the most widely used technical standard for the visual acceptance of welds in pressure equipment. It is now referenced in pressure equipment regulations in European countries and is similarly used for other high-integrity applications such as cranes, structural steelwork, and some rotating equipment. It provides a detailed breakdown of approximately 30 types of visual weld defects and specifies three levels of acceptance criteria. These are:

- Level B: 'High integrity'.
- Level C: 'Medium integrity'.
- Level D: 'Low integrity'.

Note that there is no level 'A'. One problem with EN 25817 is that there are large differences in acceptance criteria between the three levels

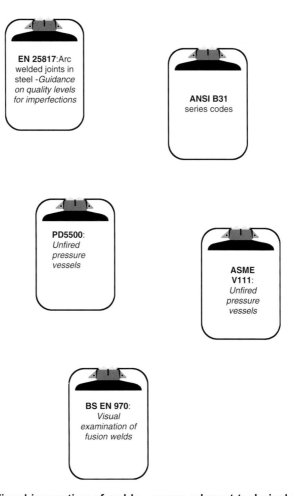

Fig. 6.8   Visual inspection of welds – some relevant technical standards

– Level B is suitable for true high-integrity applications but Levels C and D are much looser, suitable for lower-integrity or non-structural applications. For this reason, many pressure equipment purchasers specify Level B or C as a minimum requirement.

'Application standards' (codes)
Nearly all application standards (codes) for pressure vessels and other types of pressure equipment provide their own acceptance criteria for visual defects in welds. They are more definitive than the generic technical standards and relate closely to fitness for purpose of items of pressure equipment. Figure 6.9 shows some key points.

- **Code scope.** Surprisingly, some parts of pressure equipment can be exempt from visual acceptance criteria and still not 'contradict' the code (see ASME I for examples of this). This can cause awkward situations when weld defects are found during pre-commissioning or in-service inspections.
- **Code compliance.** Some codes allow visual acceptance criteria to be overruled 'by agreement between manufacture and purchaser'. This is not common, but it does happen.
- **Defect definitions.** All codes do not use the same visual defect terminology and definitions. This can lead to confusion.

Fig. 6.9   Visual weld acceptance criteria in pressure equipment application standards (codes) – some key points

In-service inspection of welds

'Code acceptance' levels intended for new construction assessment can also be used as a crude set of acceptance criteria for the visual inspection of welds in-service. Their use, however, is limited because most of the types of visual weld defects mentioned in codes are pure manufacturing defects, rather than those likely to be caused by in-service use. The main exceptions to this are cracks, which are outlawed by all high-integrity pressure equipment codes anyway.

What acceptance criteria exist for in-service weld defects?

There are very few published technical standards that even try to give acceptance criteria for in-service visual defects such as general electrochemical corrosion, pitting, galvanic weld attack, erosion, cap porosity, etc. This is because these mechanisms are so complex and variable as to make meaningful definition of accept/reject criteria very difficult. Some standards that are of (limited) use are:

- API 510. *'In-service inspection of vessels'*. This contains broad acceptance criteria for the acceptance of pitting and general corrosion. A similar approach is used in API 653 *In-service inspection of atmospheric storage tanks* (see Chapter 14 for more details).
- BS 7910. This covers fitness-for-purpose (FFP) assessments of defective pressure equipment (components and so contains comprehensive methods of assessing cracks and other in-service defects (see Chapter 9)).

- API 579. This is the American approach to FFP assessment. It is more detailed in parts than BS 7910 and uses slightly different assessment methodologies. (See Chapter 9 for comparisons between API 579 and BS 7910).

The difficulty with the above standards is *applying them* in a meaningful way. The criteria they provide have their roots in specific types of components and industry applications; for example the API standards are biased heavily towards onshore petroleum plant. Hence they have limited application for other types of plant such as boilers, and power and nuclear plant where the degradation mechanisms and defects may be very different. A further difficulty is that these standards based on FFP analysis are too complicated for day-to-day site use. They use advanced methodologies and techniques, heavily qualified by a raft of technical assumptions and boundary conditions. The end result is that the in-service inspection of welds cannot always be carried out against an easily agreed set of acceptance criteria. Inspectors therefore have to apply a lot of *judgement* – based on their experience and general technical understanding of the issues surrounding the integrity of welds and their degradation mechanisms.

Remote visual inspection (RVI) of welds

For many in-service inspections, direct visual examination of welds is impossible owing to access limitations. Pressure plant and boilers are particularly problematic with many of the critical pressure envelope welds being inaccessible from the outside without cutting of major components, which would itself cause further technical problems. The situation is even more difficult in complex rotating machinery such as large gas turbines, where full access to critical components, such as blades, can only be obtained by disassembling the machine. In such situations, inspections are often carried out using remote visual inspection (RVI) equipment. Such services require specialized, expensive equipment, with specialist contractors to operate it. The in-service inspector's role is normally limited to monitoring the RVI procedure and interpreting the results in the context of fitness for purpose, rather than operating the equipment.

Two basic methods of RVI are used: flexible video endoscopes and miniature video cameras. Figure 6.10 shows a comparison.

Flexible endoscopes
Flexible video endoscopes or *videoprobes* are available in lengths up to about 35 m and vary in diameter from 5 to about 35 mm. For most

Video endoscopes	Miniature cameras
• Manoeuvrability and flexibility • Small diameters, down to 6 mm • Built-in illumination • High price and maintenance costs • Moderate image quality • Limited illumination ability	• Limited manoeuvrability • Reach up to 30 m • External light source needed • Relatively low price • High image quality

Fig. 6.10   RVI equipment comparisons

inspection jobs on pressure systems, boilers, etc., a probe diameter of about 10 mm is the most common.

Illumination

Illumination is provided via a fibre optic light built into the probe. This transmits the light to the probe tip from a source housed in a box situated at the control end of the probe. A variety of light sources are available, depending on the type of observation that is required. When inspecting welds on stainless steel components for discolouration etc., the choice of the correct light bulb source is very important in order to obtain the correct 'white balance'. Despite the fairly high power of the light source (up to 1000 W) the range of illumination from a video endoscope is limited due to the small light outlet opening in the probe tip and the transmission loss in the probe.

The probe tip can be either straight (with a field of view of approximately 100°) or angled up to 150°, in two or four orthogonal directions. This allows the direction of view and angle of incidence of the light to be changed to suit the situation. It also increases manoeuvrability, particularly when entering a pipe system with bends.

Range

All videoscopes have a limited range. A standard probe is normally capable of travelling round up to five bends, depending on the pipe diameter and bend radius and whether the bends are in one plane or change planar direction. In practice, however, it is rare to attempt more than three bends, to avoid the risk of the probe getting stuck. The effectiveness of videoscopes is restricted by the internal diameter of the pipe, header, etc. being inspected; for diameters greater than about

70 mm it is normal to either use some kind of centring device to hold the probe centred in the pipe.

Remote video cameras

Remote cameras are available in diameters down to as small as 5 mm. They provide pictures with a resolution in excess of 0.5 megapixels and offer improved quality of image compared with video endoscopes. The cable length between camera and control unit can be a limiting factor; in some applications the cable length is limited to less than 5 m. The camera is connected to a control unit that processes the signals and provides output in either digital video (DV) or VCR-compatible format. Light is provided by a ring of low-voltage miniature halogen bulbs mounted around the front lens of the camera.

To provide an assembly with sufficient stiffness to allow insertion into a pipe by pushing, the cables supplying the camera and the bulbs are passed through a corrugated plastic tube. This allows the camera to be pushed up to 20–30 m into a straight horizontal pipe, or a lesser distance if bends are present.

Main plant applications of RVI

Figure 6.11 shows the main plant applications of RVI. Table 6.1 shows its scope on the specialist application of gas turbines

Surface crack detection

Surface crack detection is the simplest NDT technique used during in-service inspection. It is quick and easy compared with other NDT techniques and gives fairly reliable, although not perfect, results. The disadvantage is that it can only be used on accessible surfaces and will only detect cracks which are surface breaking, i.e. have either started from the surface or have started from somewhere else and have reached the surface. There are two main generic types of surface crack detection in common use: dye penetrant (DP) and magnetic particle inspection (MPI). Opinions vary on the relative effectiveness of each technique but there is general consensus that both are predominantly enhanced visual techniques, i.e. they will not detect much that cannot be detected by eye during a close visual examination, aided by a magnifying glass. Some of the volumetric NDT techniques (see later) have some capability to detect surface-breaking defects but DP/MPI techniques account for more than 90 percent of the surface crack detection activities performed in practical in-service inspections (see Fig. 6.1).

Application	Inspection
Air blast circuit breakers	Checking for internal corrosion and the condition of contacts.
Air receivers	Weld condition, corrosion, water ingress.
Alternators	Winding insulation defects, for laminate cracking in pole pieces.
Desuperheaters	Integrity of spray nozzles.
Boiler drums	For scaling, corrosion, debris.
Boiler feed pumps	Internal condition, impact damage, seal condition, clearance checks.
Boiler headers	Ligament cracking, weld condition, attachment condition, retrieval of debris after modification.
Boiler tubes	For blockage, corrosion, erosion and leaks.
Diesel generators	To check bore condition and carbon build-up.
Furnaces	On-line inspection, fans, duct dampers, refractory linings for general condition and corrosion, internal support integrity.
Gas turbines	Compressor blades, combustion system – fuel nozzles, combustion chambers, instrumentation, turbine nozzles, blades, power turbine, bearings, gearboxes.
Generator stator	Access to windings and other complex areas during failure investigation. Cooling slots – looking for rotor lamination pieces following failure.
Heat exchangers and condensers	Internal tubes for blockage, erosion, corrosion; external tubes and baffle plates for fretting damage.
Motor windings	Inspection through cooling holes to find faults and monitor condition.
Oil burners	Examine burner heads for erosion.
Pipework	Cleanliness, weld configuration, valve seats, erosion, corrosion.
Stacks, flues	Refractory linings.
Steam chests	Cracking, weld condition, corrosion monitoring.
Steam turbines	Blades for erosion, corrosion, mineral deposit build-up, magnetite deposit build-up, blade impact damage.
Tanks	On-line inspection; tank floor, walls, the roof condition; attachment integrity, columns, and central roof supports.
Transformers	Internal condition without draining the oil, support integrity.
Underground piping	Blockages, leaks, corrosion monitoring, debris build-up.
Wind turbine blades	Internal condition.

Fig. 6.11   Remote visual inspection: main plant applications

Table 6.1   Specialist gas turbine applications

REMOTE VISUAL INSPECTION OF AERODERIVATIVE-TYPE GAS TURBINE		
Inspection area	*Access position*	*Subject of inspection*
Compressor (front)	Air intake	Intake guide vanes, early stages of compressor, rotor and stator blades
Compressor (rear)	Burner ports and combustion air casing ports	Outlet guide vanes (using burner ports only), 15th stage rotor blades, air swept surfaces
Flame tube snouts	Burner ports and/or tappings in side of combustion air casing	Snout bosses and dowels
Flame tubes	Burner ports and/or combustion air casing borescope ports	Flare, swirlers, combustion air interconnectors, rear casing suspension, and bridge ports pieces, general internal condition
Discharge nozzles	Burner ports and/or combustion air casing borescope ports	Internal condition and location with flame tubes Condition of securing brackets
Turbine (front)	Burner ports and/or combustion air casing borescope ports	High-pressure (HP) turbine blades (leading edges), HP guide vanes (leading edges)
Turbine (rear)	Removal of rear transition section and/or removal of thermocouple	Intermediate-pressure (IP) turbine blades (trailing edges), Low-pressure (LP) turbine blades (trailing edges)

Typical equipment used: high-intensity light source; 10 mm diameter, 55 cm long, 50°/110° borescopes; 8 mm diameter, 3 m long videoprobe; guide tubing; video recording unit.

DP testing

DP testing uses a set of three aerosols: a cleaner (clear), a penetrant (red), and a developer (white). These are applied to the component in a specific sequence and surface-breaking defects reveal themselves as a thin red line. Figure 6.12 shows the procedure. Figure 6.13 shows some practical tips for witnessing DP tests, and some guidelines on the limitations.

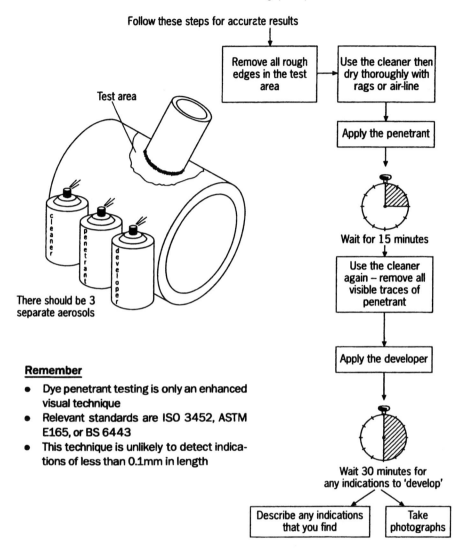

Fig. 6.12   DP testing procedure

DP technical standards

These divide into two main families: those aligned to European standards and those aligned to USA-based industry standards (ASME, API, AWS, and similar). The main ones of relevance to in-service inspectors are:

- EN 1289: 1998 *Non-destructive examination of welds – Penetrant testing – Acceptance levels.*

Crack size. Under site conditions DP will only detect cracks > approximately 0.5 mm.

Check the surface condition. If the surface is too rough (e.g. weld spatter) the results will be poor.

Witness the procedure. Make sure that the sequence (and times) are followed carefully – particularly the dwell time for the red penetrant and white developer.

The DP technician. There are formal qualifications for DP technicians (e.g. CSWIP/PCN/ASNT Levels 1, 2, and 3). Technicians do not have to have these qualifications to do a good job, but most company accreditation procedures will require it.

Published standards. DP techniques are covered by several sets of comprehensive technical standards (see below). Practically, there is rarely a need for inspectors to refer to these during in-service inspections.

Some advanced points on DP testing

'Surface blinding'. Some stainless steel and high-nickel alloys (e.g. C 276 Hastelloy) form an invisible oxide film on their surface that conceals ('blinds') cracks, so they do not show up under DP testing. The solution is to mechanically clean the surface with an abrasive disc or flap-wheel to remove the film, before doing the DP test.

Chloride-free DP materials. Sensitive stainless steels require special chloride-free DP materials so as to reduce the risk of causing stress corrosion cracking (SCC).

The problem of weld toes. DP testing can give spurious indications from weld toes. Slight undercut, overlap, etc. can easily be mistaken for cracks.

Fig. 6.13   Witnessing DP testing – some key points

- EN 571–1: 1997 *Penetrant testing – General principles.*
- ASTM E165: 1982 *Dye penetrant examination.*

DP test acceptance criteria

Test acceptance criteria are complicated by the fact that DP testing can reveal other crack-like defects, i.e. that are not strictly cracks. Published technical standards do not all contain acceptance criteria. Of those that do, the following general 'rules of thumb' apply.

- *Cracks.* These are outlawed by all construction codes for high-integrity pressure equipment. Some cracks may be acceptable in in-service equipment but only after a FFP assessment (e.g. to BS 7910/

AP 579 (see Chapter 9)), rather than by comparison with a set of simple acceptance criteria.

- *Microcracks*. Micro-indications (e.g. stress corrosion cracking, creep cracks, and similar) will only show up under DP testing when they have propagated to a significant length. Again, acceptance is governed by an FFP assessment, rather than any simple published acceptance criteria.
- *Other defects*. Defects such as weld undercut, elongated surface porosity, etc. are much less likely to be caused by an in-service degradation mechanism, so in-service DP acceptance criteria are not an important issue.

Despite the operational limitation of DP testing, it is widely used during in-service inspection of all types of industrial plant. It has the advantage that it can be used on all types of metallic materials (not only magnetic ones) and, if circumstances require, can be used by non-specialist people.

Magnetic particle inspection (MPI) testing

This is referred to in USA codes as magnetic testing (MT). It is, again, an enhanced visual technique that can be used in a site inspection situation. It works on the principle of creating a magnetic flux in the component material then applying a magnetic ink or powder which shows where the lines of flux have been disturbed by a surface-breaking defect. MPI has two main differences from DP testing:

- MPI can only be used on magnetic materials so non-ferrous and most types of austenitic stainless steels cannot be tested, as they are non-magnetic.
- In an ideal situation, MPI can detect defects which are slightly subsurface, as well as surface-breaking ones. This is difficult, however, in practical site inspection conditions where component access, orientation, and surface finish may not be ideal.

Types of MPI testing

There are three main types of MPI technique: black magnetic ink, fluorescent magnetic ink, and dry magnetic powder (red or blue). All work on the same principles but have different limitations on where their use is most effective. Figure 6.14 shows the techniques. Figure 6.15 gives some practical tips on witnessing MPI testing, and some guidelines on limitations. Figure 6.16 shows the technique in progress.

MPI gives an equally accurate representation of defect size and shape as DP testing. During the MPI test the material is locally magnetized. The ferrous magnetic particles in the ink are attracted to any breaks in the magnetic field and so show their size and shape. The best results are achieved when the lines of magnetic flux are fully perpendicular to the defect, the efficiency of detection reducing as the angle reduces, which is why two field directions are required.

MPI technical standards

As with other types of NDT, MPI technical standards divide roughly into those aligned to European standards and those aligned to USA-based industry standards (ASME, API, etc). For MPI, there is little technical difference between the standards that will affect the role of the in-service inspector.

The situation with acceptance criteria is the same as for DP testing, i.e. the published MPI standards rarely contain simple acceptance criteria. Defects have to be analysed using an FFP assessment (e.g. BS 7910/API 579, see Chapter 9). MPI has the practical disadvantage that, unlike DP, the defect disappears once the magnetic field is removed. Defects therefore need to be recorded using photographs to give a permanent record.

Eddy current testing (ECT)

Eddy current testing (ECT) is a newer technique that is capable of detecting surface-breaking defects (as well as a few other types). It is suitable for specialized applications in tube-type heat exchangers and some specialist types of welds in the offshore petroleum industry. It is also used in automatic pipeline scanning (see Chapter 13) and in the on-line automatic NDT of pipe spools during manufacture in the tube mill. It is much less widely used in general in-service weld inspection.

ECT advantages and disadvantages

ECT has several disadvantages that restrict its large-scale use in site in-service inspections. These are balanced by some distinct advantages. Figure 6.17 shows the details.

In recent years, ECT has developed in the offshore industry for the crack detection of ferritic steel welds through paintwork coatings. The technique uses special probes which overcome traditional problems caused by changes in material properties in the heat-affected zone (HAZ) of welds. The main advantage of this system is one of cost, i.e there is no need to remove paintwork coatings before testing. This gives

1. Black magnetic ink and white contrast paint

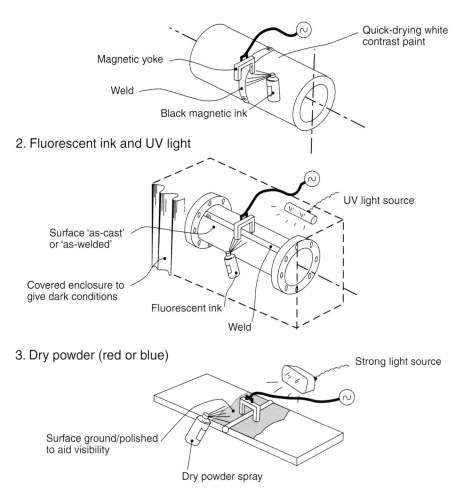

Quick-drying white contrast paint

Magnetic yoke

Weld

Black magnetic ink

2. Fluorescent ink and UV light

UV light source

Surface 'as-cast' or 'as-welded'

Covered enclosure to give dark conditions

Fluorescent ink

Weld

3. Dry powder (red or blue)

Strong light source

Surface ground/polished to aid visibility

Dry powder spray

Fig. 6.14   The three types of MPI technique

a huge cost saving, as this has traditionally been the most expensive part of the procedure. Testing is possible with coatings up to about 2 mm thick. As with other types of ECT, the technique will mainly find surface-breaking defects. Various probes are used and the results displayed on a computer screen. There are strict limitations on the weld joint design and weld contour geometries that can be successfully tested. Figures 6.18 and 6.19 show typical ECT arrangements for heat exchanger tubes and welds. Figure 6.20 outlines some key points to note when witnessing tests on ferritic welds.

Crack size. As for DP, MPI under site conditions will only detect cracks > approximately 0.5 mm in length.

Visibility of defects. Overall, the fluorescent MPI technique gives the best visibility of defects. The dry powder technique tends to be the worst, unless the metal surfaces are ground to a shiny finish so that the red or blue powder shows up under a good light.

The MPI technique. Some points to watch for are:

• The operator must use two field directions orientated at (or near) 90%.
• Field strength should be checked using a test indicator.
• If electric 'prods' are used (because access is not available for a yoke), watch for arcing marks left on the metal. Serious marks need to be polished smooth.
• Some operators use a permanent magnet instead of an electro-magnet. If so, the strength of the permanent magnet should be sufficient to lift an 18 kg test weight (note this is more than the 5 kg test weight used for an AC electromagnet).

Fig. 6.15   Witnessing MPI testing – some key points

ECT has limited application on non-ferritic welds. Testing of other materials is not impossible, but is subject to severe practical restrictions which limit the accuracy available under site in-service inspection conditions. Techniques are available for ECT of various specialized items (e.g. in the aerospace industries) but are rarely seen in routine inspection of industrial plant.

ECT technical standards
The main standard in use in Europe is EN 1711: 1999 *NDE of welds – Eddy current examination of welds by complex plane analysis*. This recent standard summarizes the following areas relating to ECT testing of ferritic welds:

• Certification of ECT operators.
• Equipment calibration.
• Weld inspection procedures.
• ECT report proformas.

One of the strengths of EN 1711 is its coverage of ECT scanning patterns, particularly those for detecting defects in the weld HAZ. It also contains typical defect characteristics, as they appear on the display screen.

Fig. 6.16   MPI technique in progress

General points

Calibration. The effectiveness of an ECT process relies heavily on the way the process is calibrated. Without proper calibration, the testing procedure can be worthless.

Operator competence. There is a qualification scheme for ECT operators (PCN/CSWIP, etc.) but there are fewer qualified operators than for DP/MPI. ECT is *highly sensitive* to operator skill.

ECT of heat exchanger tubes

Limitations. ECT will detect tube wall thinning and surface-breaking defects; however it is limited in its application in the curved part of U-tubes.

Cleanliness. Heat exchanger tubes must be very clean (water jetted/ mechanically cleaned, etc.) before ECT can be effectively used.

Fig. 6.17   Witnessing ECT – some key points

Volumetric NDT

Volumetric NDT techniques are those that have the ability to detect defects in the *body* of a component material rather than only those

ECT – how it works

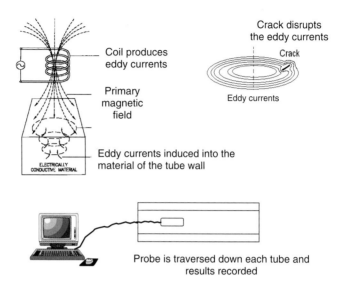

Crack disrupts the eddy currents

Coil produces eddy currents

Primary magnetic field

Eddy currents induced into the material of the tube wall

Crack

Eddy currents

Probe is traversed down each tube and results recorded

Typical ECT 'tubemap' results showing degrees of wall thinning and defect depth

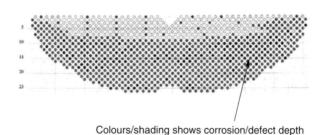

Colours/shading shows corrosion/defect depth

Fig. 6.18   Heat exchanger eddy current testing (ECT)

which are surface breaking. Volumetric NDT techniques are therefore used in conjunction with surface crack detection methods, so as to provide a full picture of the defects present in a component. Volumetric defects that can *develop* during in-service use of an engineering component include:

• *Fatigue cracks* can propagate through the body of a material.
• *Lamellar tearing* and hydrogen-related defects.
• *Creep voids.*
• *General corrosion.*

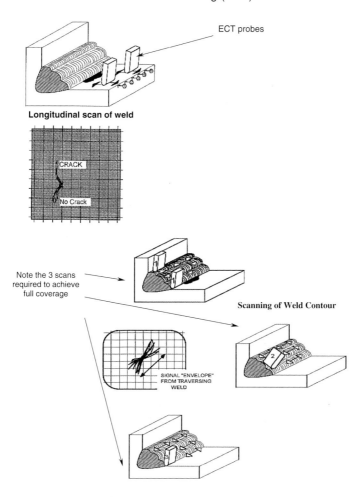

Fig. 6.19   Eddy current testing(ECT) of welds

- *Galvanic corrosion* as a result of dissimilar materials microstructures around weld/HAZ regions.

In order to perform a proper inspection for fitness for purpose, inspection techniques need to be chosen that will enable both surface and volumetric defects to be detected. The two main families of volumetric NDT techniques are ultrasonic testing (UT) and radio-graphic testing (RT). Each is subdivided into several variants, with their own specific on-site applications. Table 6.2 shows how volumetric NDT techniques fit into the overall scope of NDT methods used in in-service

Key points to note when witnessing weld ECT are:

- **Procedure.** Should follow the standard EN 1711 *NDE of welds –
  Eddy current examination of welds by complex plane analysis.*
- **Frequency.** Should be at least 100 kHz for weld inspection.
- **Geometry.** As with UT, the weld joint geometry should be known, to
  enable accurate interpretation of the ECT results.
- **HAZ examination.** Special care is required to use the correct probe
  angle when scanning the HAZ (see EN 1711 for details).

When defects are found using ECT it is normal to confirm the findings
using MPI.

Fig. 6.20   ECT of ferritic welds – key points

inspection. Tables 6.3 to 6.7 outline the strengths and weaknesses of
various common NDT methods in inspection applications.

Ultrasonic testing (UT)

UT is the quickest and least problematic of the volumetric NDT
techniques. It is used as the basis for several advanced NDT techniques
such as time of flight diffraction (TOFD) and pulse-echo methods. The
basic UT technique consists of introducing high-frequency sound beams
into a material or weld on a predictable path. The beam reflects off a
defect (known as 'discontinuity') and is picked up by emitting probe.
The time taken for the beam to return is calculated, enabling the
location of the discontinuity to be identified. Similar methods are used
to 'size' the defect and determine its shape i.e. whether it is planar,
volumetric, blunt, sharp-edged, etc. The two following basic UT
techniques form the basis of most in-service inspections.

0° 'Compression probe' testing

This is the simplest and most limited ultrasonic test. Figure 6.21 shows
the process. It can detect the following in-service defects:

- *Reduced material thickness* (e.g. on vessels or pipework) caused by
  internal or external corrosion. The reduction can be either overall or
  localized.
- *Major laminar defects* (laminations, tearing, or cracks in the plane of
  the main material axis). These are more often manufacturing rather
  than in-service defects.

Table 6.2 **NDT for in-service inspections – main techniques used**

Damage type	NDT method/technique	Capability/limitations
Corrosion/erosion (internal)	Visual inspection (vessels only) – internal	Good detection capability but requires internal access. Limited sizing capability (depth/remaining wall thickness)
	Manual ultrasonic testing/ 0° probe – external	Generally good detection and sizing capability (can be poor if corrosion isolated, particularly the detection of pitting)
	Automated ultrasonic testing/ 0° probe mapping – external	Very good detection and sizing capability (application limited to pipe sections/vessel walls where simple manipulation can be facilitated). Corrosion maps allow accurate comparison of data between repeat inspections. Comparatively slow technique to apply
	Continuous ultrasonic monitoring – external	Good detection and sizing capability (at specific monitoring locations)
Weld root corrosion/ erosion	TOFD – external	Very good detection and sizing capability (depth/remaining wall thickness). Access to both sides of weld cap required
	Manual/automated ultrasonic testing/0° probe – external	Good detection and sizing capability but requires extensive surface preparation, i.e. removal of weld cap
	Manual/automated ultrasonic testing/angle probe – external	Detection and sizing capability but can be unreliable
Hot hydrogen attack(HHA Internal)	Ultrasonic testing – external 0° probe/high sensitivity	Detection capability/base material but can give false indications. Use of mapping system facilitates monitoring. For welds, removal of cap is required
	Angle probe(s) medium sensitivity	Detection capability/welds but cannot detect microscopic stages of HHA. Use of automated system facilitates monitoring of macrocracking
	TOFD	Good detection capability on welds but discrimination between microcracking and other weld defects can be difficult problem. Establishment of a baseline helps monitoring of microcracking
Hydrogen-induced cracking (HIC, stepwise cracking)	Ultrasonic testing – external • 0° probe • 45° angle probe	Good detection at later stages, but there are no proven early warning (susceptibility to cracking) tests for on-site inspection
Creep damage	Surface testing	Surface replication can be used to examine microstructure. This is a well-proven technique

Table 6.2 (Continued)

Damage type	NDT method/technique	Capability/limitations
	Ultrasonic testing	Methods developed for detection of early stages of creep have not been proven in the field. Standard ultrasonic testing techniques are effective only at detecting later stages of creep
Fatigue cracking (internal external)	Magnetic particle testing	Good detection capability but requires access to fatigue crack surface. Good length sizing capability. Some surface preparation usually required
	DP/ECT	As above, for non-magnetic materials
	Ultrasonic testing/angle probe	Good detection and sizing capability (length and height), enhanced by use of automated systems – TOFD gives very accurate flaw height measurement and allows in-service crack growth monitoring
	Alternating Current Field Measurement – ACFM (can be used in lieu of surface techniques stated above)	Good detection capability but requires access to fatigue crack surface. Length and some depth sizing capability. Unlike MPI, ACFM does not usually require surface preparation and can be used through coatings. Better for inspecting welds than ECT
Stress corrosion cracking/SCC (internal/external)	Surface testing	DP/MPI (not austenitic) ECT (not ferritic) techniques – good detection capability but access required to crack surface
	Ultrasonic testing – external	Fair detection capability; can be used on-line. Specialist techniques have some capability to determine crack features [orientation and dimensions (including height)]
	Acoustic emission – external	On-line detection of growing SCC in large-component systems too complex to be inspected by other techniques. Extraneous system noise can produce false indications

Table 6.3 Detection and sizing capability of the main NDT methods

	NDT method					
	Visual testing (VT)	Dye penetrant (DP) testing	Magnetic particle inspection (MPI)	Eddy current testing (ECT)	Radiographic testing (RT)	Ultrasonic testing (UT)
Detection capability						
Cracking (open to surface)	✓	✓	✓	✓	✓	✓
Cracking (internal)					✓	✓
Lack of fusion					✓	✓
Slag/inclusions					✓	✓
Porosity/voids					✓	✓
Corrosion/erosion	✓				✓	✓
Sizing capability						
Flaw location	✓	✓	✓	✓	✓	✓
Flaw length	✓	✓	✓	✓	✓	✓
Flaw height				✓		✓
Component thickness					✓	✓
Coating thickness				✓		✓

Table 6.4   Generally accepted methods for the detection of accessible surface flaws

Material	NDT method			
	VT	DP	MPI	ECT
Ferritic steel		✓	✓	✓*
Austenitic steel		✓		✓*

✓* Indicates that the method is applicable with some limitations.

Table 6.5   Generally accepted methods for the detection of internal flaws in full-penetration welds

	Parent material thickness (t) (mm)		
	t≤8 mm	8<t≤40 mm	t > 40 mm
Ferritic butt weld	RT or (UT)	UT or RT	UT or (RT)
Ferritic T-weld	(UT) or (RT)	UT or (RT)	UT or (RT)
Austenitic butt weld	RT	RT or (UT)	RT or (UT)
Austenitic T-weld	(UT) or (RT)	(UT) or (RT)	(UT) or (RT)

* or ( ) Indicates that the method is applicable with some limitations

Some common applications of $0°$ probe UT are to detect:

- Head or shell internal corrosion on pressure vessels, pipework, valves, and boiler headers.
- Weld root erosion (internal) or pipework (after grinding the external weld cap flat).

Angle probe testing

Angle probes use ultrasonic 'shear waves', projected at an angle into the component under examination. This provides a sensitive technique for the detection of cracks and other discontinuities in the body of the component. Full scanning techniques involve the use of different angle probes, e.g. $30°$, $45°$, $60°$ angles, in order to find defects lying in various planes in the material. Probes are available using different ultrasound frequencies to give optimum results on various types and thicknesses of material.

Table 6.6　Effectiveness of main NDT methods – summary

Defect type Inspection method	Inspection effectiveness (low, medium, high)	
	Detection	Sizing
General wall thickness loss		
UT pulse-echo	H	H
UT mapping 'MapScan'	H	H
UT EMAT	H	H
Pulsed eddy current	H	H
Film radiography	M	M
Real-time radiography	M	M
Magnetic flux leakage	L	—
Saturated low-frequency eddy current	L	—
Local wall thickness loss, pitting		
UT pulse-echo	L	H
UT mapping 'Mapscan'	H	H
UT EMAT		
Film radiography	H	L
Real-time radiography	H	L
Magnetic flux leakage	H	—
UT CHIME	M	—
UT long range	M	—
Saturated low-frequency eddy current	H	—
Blisters and embedded horizontal cracks, delamination		
UT pulse-echo	H	—
UT mapping 'MapScan'	H	—
UT TOFD	H	H
Acoustic emission	L	—
Surface breaking cracks		
Eddy current ACFM	—	L
Liquid penetrant	H	—
Magnetic particle inspection	H	—
UT pulse-echo angled beam	H	—
UT TOFD	H	H
Acoustic emission	M	—
Embedded cracks		
UT pulse-echo angled beam	H	—
UT TOFD	H	—
Acoustic emission	M	—
Embedded volumetric voids		
Film radiography	H	—
Real-time radiography	H	—

Table 6.7 Common NDT methods – limitations and accuracy

Inspection method	Wall thickness (mm)	Material	Temperature range	Surface finish	Accuracy
UT pulse-echo angled beam	6–300 mm	Reduced capability for austenitic and duplex welds	Up to 250 °C using special probes	Uniform coating up to 1.5 mm. Roughness 6.3 μm maximum. Free of scale, slag, rust, oil and grease at probe location.	±3 mm mean error for weld inspection
UT pulse-echo straight beam	2–300 mm	Reduced capability for austenitic and duplex welds	Up to 250 °C using special probes. Higher temperatures possible for spot checks	Uniform coating up to 1.5 mm. Roughness 6.3 μm maximum. Free of scale, slag, rust, oil and grease at probe location.	±0.1 mm digital thickness gauge
UT mapping 'MapScan'	2–300 mm	As pulse-echo UT	Up to 125 °C (longer contact)	As pulse-echo UT	0.1 mm
UT TOFD	8–300 mm	Restricted for fine grain material	As pulse-echo UT	As pulse-echo UT	±1 mm on sizing and position
UT CHIME	Up to 40 mm	As pulse-echo UT	As pulse-echo UT	As pulse-echo UT	
UT long range	6–25 mm	As pulse-echo UT	–10 to 150 °C	Probe contact area as pulse-echo UT, examined area not restricted	30–50% wall thickness. 10 mm on position
UT EMAT	2–150 mm	As pulse-echo UT	–200 to 460 °C	Rough surface acceptable	As pulse-echo UT

Method	Penetration/Depth	Material	Temperature	Coating	Sensitivity/Accuracy
Eddy current ACFM	N/A	All	Up to 150 °C using special probes	Coating allowed with restrictions	Crack length > 10 mm
Pulsed eddy current	6–60 mm (through maximum 150 mm insulation)	Low alloy C steel, restricted by ferro magnetic sheeting	−100 to 500 °C contact temperature maximum 70 °C	Non-contact, through insulation	Accuracy ±5% wall thickness relative measurement
Saturated low frequency eddy current	Up to 35 mm	All	Up to 60 °C	Non-contact, maximum 8 mm coating	Accuracy ±15% wall thickness
Liquid penetrant	N/A	All, non-porous	10–50 °C	Free of scale, slag, rust, oil, grease or paint	Anything visible
Magnetic particle inspection	N/A	Ferro magnetic	Up to 100 °C	Free of scale, slag, rust, oil, grease or paint. Smooth surface increases reliability	Anything visible
Magnetic flux leakage	4–10 mm	Ferro magnetic	Up to 60 °C	Maximum 3 mm coating, clean surface	Minimum 30% WT or 20 mm^3
Thermography	Surface	N/A	Non-contact technique, −20 to 1000 °C	Depends on the surface emissivity	Temperature variations of 0.2 °C at 1 m; 10 °C at 100 m
Film radiography	Up to 150 penetrated thickness	All	Maximum 40 °C	N/A	

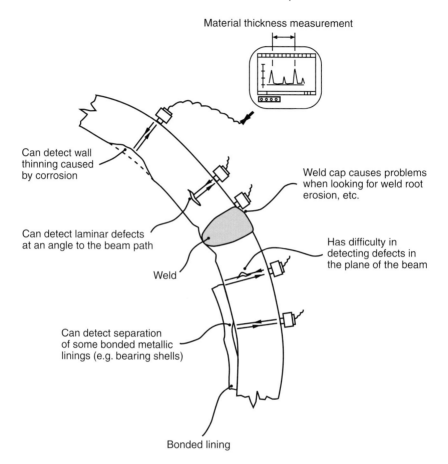

Material thickness measurement

Can detect wall
thinning caused
by corrosion

Weld cap causes problems
when looking for weld root
erosion, etc.

Can detect laminar defects
at an angle to the beam path

Has difficulty in
detecting defects in
the plane of the beam

Weld

Can detect separation
of some bonded metallic
linings (e.g. bearing shells)

Bonded lining

Fig. 6.21   UT techniques using a 0° compression probe

FFP assessment

In order for an in-service UT programme to provide a sufficiently comprehensive search for all defects that can threaten the FFP of a component (particularly welds), the programme must follow specified procedures. Typical procedures specify the number and extent of scans to be carried out and the angle probes to be used. These procedures are specified in published standards such as BS 3923 (now withdrawn but still used in many industries) and EN 1712/1713/1714 *Ultrasonic examination of welds*. These, strictly, are standards covering new equipment under manufacture, but they are also used for in-service defects.

A variety of scanning techniques are required, depending on the application. Figure 6.22 shows typical procedures for typical types of welds found on a pressure vessel. These are equally applicable for similar weld joint geometries found on other types of pressure equipment, e.g. valves, pipework, etc. Note that these standards specify the testing technique *only*, and do not recommend acceptance criteria once defects are found. Figure 6.22 shows techniques from a BS 3923 (recently superseded) Level 2 FFP UT assessment – a level that gives adequate testing coverage to assess the FFP of the types of welds shown. This is an important point when specifying UT on in-service welds – the testing must be to a *minimum level*, if the test is to be properly recognized as a test for FFP. Less extensive testing (i.e. fewer scan angles, shorter scan lengths, etc.) may still reveal defects, but will not constitute a full FFP assessment. Common examples of this are vessels and pipework where it is not possible for time, cost, or access reasons, to grind off weld caps to enable a proper longitudinal scan of the welds. Hence only a partial set of scans is possible and the test does not qualify as a full FFP assessment.

Specialized scanning techniques
Specialized UT techniques may be required when looking for particular defect types or in high-risk/high-integrity plant situations. These techniques are normally developed by plant users themselves, working in conjunction with specialist plant integrity and NDT contractors, rather than standards bodies, who limit themselves to more general coverage. Such techniques therefore have very narrow application.

Pipe spools with longitudinal seams that have proved problematical in service, are a good example of the application of such a technique. Figure 6.23 shows a typical example for SA 335-P22 $2\frac{1}{2}$ Cr-alloy material.

Echo evaluation
The evaluation of the ultrasonic echo is a skilled process – it is strictly the remit of the UT operator rather than the in-service inspector. It is however, the role of the inspector to oversee the process, perhaps making sample checks to ensure that the interpretation is being done correctly. The principles of echo evaluation are as follows:

• A distance amplitude correction (DAC) is defined. This allows an echo coming from a defect to be compared with that from a known defect (termed 'a reflection'). It is done using a pre-prepared calibration block.

Set-on nozzle with both bores accessible

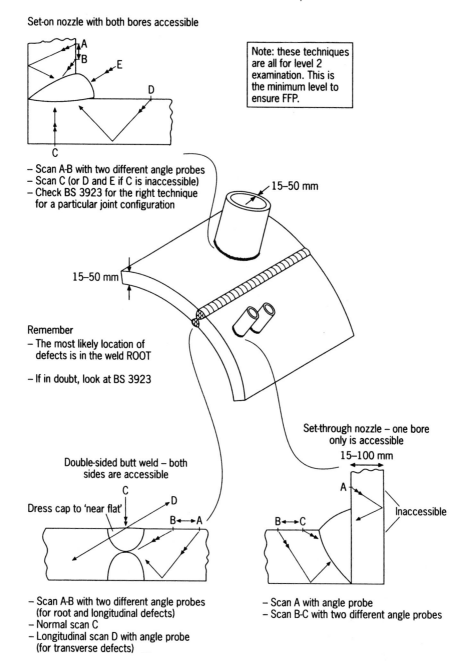

Note: these techniques are all for level 2 examination. This is the minimum level to ensure FFP.

- Scan A-B with two different angle probes
- Scan C (or D and E if C is inaccessible)
- Check BS 3923 for the right technique for a particular joint configuration

15–50 mm

15–50 mm

Remember
- The most likely location of defects is in the weld ROOT

- If in doubt, look at BS 3923

Double-sided butt weld – both sides are accessible

Dress cap to 'near flat'

- Scan A-B with two different angle probes (for root and longitudinal defects)
- Normal scan C
- Longitudinal scan D with angle probe (for transverse defects)

Set-through nozzle – one bore only is accessible

15–100 mm

Inaccessible

- Scan A with angle probe
- Scan B-C with two different angle probes

Fig. 6.22   UT scanning techniques using angle probes

Preparation

- Identify any disguised weld seams using an acid etch.
- Grind off all weld caps and polish area for a distance of 4 × material thickness either side of the weld centre-line.

Technique

- Use 0° (normal) compression probe over all areas first, to find any 'linear-deflection' defects.
- Use 45°, 60°, and 70° angle shear wave probes (1–5 MHz) to give full scanning in all planes (in excess of the requirement of e.g. ASME V).
- Scan using a minimum +14dB sensitivity.
- Use a scanning overlap of at least 10% and a scanning speed < 150 mm/second.

Defect sizing

- Defect sizing performed by time of flight diffraction (see Fig. 6.29). Aspect ratio of defect to be measured.

Fig. 6.23   Specialized UT technique for SA 335-P22 $2\frac{1}{2}$% Cr seamed high-pressure pipework

- The comparison works using the height of the echo on the display screen.
- UT beams react differently in all materials, so calibration needs to be material specific.
- Echo evaluation is made difficult if there is weld spatter or dirt on the surface, or if the surface finish is too rough.
- Any 'defect' that causes an indication greater than a certain percentage of DAC curve has to be 'sized' (possibly using several different UT probes).
- Sizing is done using an 'intensity drop technique'. The normal level is 6 dB drop, although other levels are used if greater sensitivity is required.
- For defects discovered by UT which prove difficult to size, other techniques may be used to assist, e.g. time of flight diffraction (TOFD), 'creeping wave', and 'bimodal methods' using dual-element tandem probes. These techniques require properly trained and experienced operators.

Defect acceptance criteria

For UT-detectable defects, acceptance criteria for in-service inspections are mainly found in the codes or application standard for the equipment under examination, i.e. they are highly equipment specific. Examples are found in ASME VIII, ANSI B31, EN 13445 and similar. The acceptance criteria vary significantly between these documents. The most stringent criteria (for example, for high-speed rotating equipment components) are developed by the equipment manufacturers themselves.

Ultrasonic testing A-, B-, and C-scans

UT techniques subdivide into three different scan 'types' known as: A-, B- and C-scans. Figure 6.24 shows the details.

- *A-scan*. This is the most common type used. The *x*-axis represents time of flight of the ultrasonic beam, indicating the depth of a defect and/or the location of the back wall of the component. The *y*-axis represents the *size* of the defect in comparison to a calibrated sample.
- *B-scan*. This scans the probe along one axis, thereby producing a cross-sectional view of the component under test. The *x*-axis therefore shows the *position* of a defect.
- *C-scan*. This shows a plan view of the component. The *x*- and y-axes form a system of coordinates that indicate the defect's position in the component.

The general principles of UT have been used to develop a range of advanced NDT techniques. These have been developed in response to particular engineering problems, experienced in in-service inspection.

UT corrosion mapping

Corrosion mapping is an ultrasonic testing technique developed specifically for in-service inspection of engineering components. Its main use is in the petrochemical and offshore industries where corrosion of pipework and vessels is a serious problem. Although the technique is straightforward in concept it requires a special equipment set-up, making it suitable for pre-planned rather than impromptu inspections.

How does it work?

Corrosion mapping uses a standard (normally 2 MHz) $0°$ ultrasonic probe, producing compression waves. This is the same type used for standard wall thickness measurement or lamination checks. The probe measures the thickness of the material being scanned using a simple

The basic A-scan pulse echo technique

Defect position (depth) is shown as a function of time

A 'pulsed' wave is used. It reflects off the back wall and any defects

Signal amplitude

Defect echo

Back wall echo

Time

The probe transmits and receives the waves

Defect

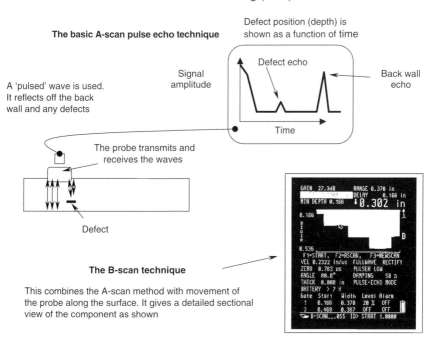

The B-scan technique

This combines the A-scan method with movement of the probe along the surface. It gives a detailed sectional view of the component as shown

The C-scan technique

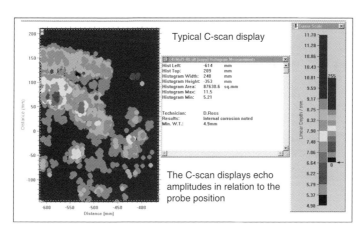

Typical C-scan display

The C-scan displays echo amplitudes in relation to the probe position

Fig. 6.24   Ultrasonic A, B, and C-scans

back-wall echo displayed on an A-scan screen. The main factors are (see Fig. 6.25):

- *Scanning.* The probe is scanned over 100 percent of the surface, thereby building up a complete picture of the wall thickness.
- *Display.* The display shows a colour representation of the wall thickness. The thickness readings that trigger each colour are pre-set to match the acceptance criteria for the specific job. For a 20 mm material wall thickness a typical display format would be:
  - <10 mm remaining wall thickness: Red
  - 10–15 mm remaining wall thickness: Yellow
  - 15–20 mm remaining wall thickness: Green
  - reading not yet recorded: Black

- *Resolution.* The resolution of the scan, in pixel size, can be chosen to suit the material, and the type of corrosion expected. A typical resolution set to detect pipe internal wall corrosion, including flow-accelerated corrosion and isolated oxygen pitting, would be 2 mm × 2 mm. Where corrosion is expected to be more widespread the pixels could be bigger, resulting in a quicker scan.
- *Location reference.* The scanned area is mapped out using a set of $x-y$ coordinates, referenced to a fixed origin point defined on the component – a light-emitting diode (LED) is normally used for this. Scan location is then plotted via a fixture-mounted video camera respective to the LED 'origin'. To help with location, grid-lines or datum points may be marked on the surface of the component itself.

What are the advantages of corrosion mapping?

The main advantage of corrosion mapping is that it guarantees 100 percent scan coverage of the area under examination. This gives a much-improved effectiveness over a standard 'random' UT wall thickness scan where it cannot be demonstrated whether a specific area has been fully examined or not. Tests with a corrosion mapping system will quickly show that, without the aid of a display confirming the unscanned areas (grey or black on the screen), even a competent technician doing a thorough technique will only cover about 60–70 percent of the scan area. This percentage reduces when the scan is complicated by poor surface finish or irregular geometry. Another practical advantage is that corrosion mapping produces a permanent record of corrosion measurements. This allows comparisons to be made between subsequent

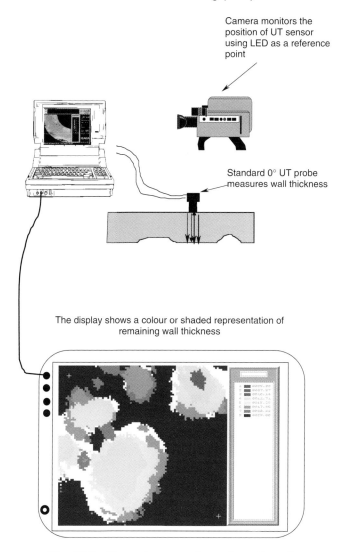

Camera monitors the
position of UT sensor
using LED as a reference
point

Standard 0° UT probe
measures wall thickness

The display shows a colour or shaded representation of
remaining wall thickness

Fig. 6.25   Corrosion mapping – how it works

in-service inspections to check the rate at which corrosion is progressing.

Are there any disadvantages?

Corrosion mapping suffers from the same disadvantages as standard UT wall thickness testing. These are:

- It can be difficult to get a clear back-wall echo from some corroded surfaces. This causes poor results. Internal surfaces that can be particularly difficult are:
  - jagged undercut pitting caused by acid attack;
  - oxidation pitting where the pits are covered by thin crusts or *tubercles.*
  - small isolated pits or pinholes that are surrounded by solid, uncorroded material;
  - areas near welds, e.g. weld attack or *tramline* corrosion around the internal weld toes of circumferential seam welds on stainless steel pipework and vessels;
  - areas where the external surface geometry restricts the application of the standard 0° probe (valve bodies, weld neck pipework flanges, eccentric reducers, etc.)
- The cost of the initial set-up.

Corrosion mapping is finding increasing use in those industries where high corrosion rates are a significant integrity issue. One driving force is the increased popularity of risk-based inspection (RBI), in which some internal examination during cold shutdown is replaced by in-service non-invasive inspection. Corrosion mapping fits this application well.

Remote UT thickness monitoring strips

This is a technique used for *in situ* testing of buried pipelines. There are several similar methods marketed under proprietary tradenames. Figure 6.26 shows the principles. A number (10+) of simple 0° UT mini-probes (fixed frequency) are built into a flexible plastic strip, each one linked by a flat copper conductor strip to wire connectors at the end of the strip. The strip is wrapped around the pipeline, under any wrapping or lagging and is connected by hard-wiring or radio link to a remote monitoring computer. The probes are triggered on a periodic basis, giving a crude measurement of the wall thickness at multiple radial points. The strips are fitted at key wall thinning points such as changes of section, flow restrictions, and bends. This technique is particularly useful for buried or long-distance desert/mountain terrain pipelines where access is difficult. Two limitations of this method are:

- It only gives a 'first estimate' of wall thickness – the probes are not as sensitive as full-scale UT testing.

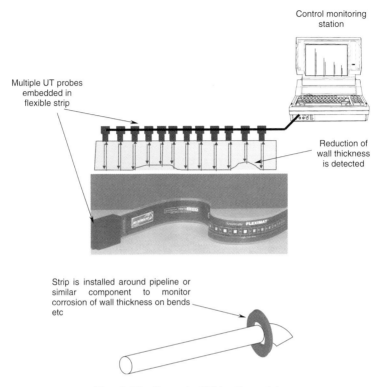

Control monitoring
station

Multiple UT probes
embedded in
flexible strip

Reduction of
wall thickness
is detected

Strip is installed around pipeline or
similar component to monitor
corrosion of wall thickness on bends
etc

Fig. 6.26   Remote UT testing strips

• Where low thicknesses are identified, the pipeline still has to be excavated for more detailed checks. Spurious results are not unknown.

From the viewpoint of the in-service inspector, the monitoring of remote thickness readings is a simple process. The difficulties come when results have to be interpreted and validated. In practice this can rarely be done without either a lot of historical experience for the pipeline, or 'check results' from full UT testing.

Time of flight diffraction (TOFD)

Time of flight diffraction (TOFD) is an advanced UT technique that has claimed advantages in the detection and sizing of defects. It has been in use for more than 20 years but has not yet achieved widespread recognition as a practical replacement for standard UT techniques.

The TOFD principle

TOFD works on the principle of ultrasonic beam *diffraction* rather than *reflection*, as in standard UT techniques. Simplistically, the amount of diffraction achieved is less dependent on ideal orientation of a defect than is reflection, with the result that TOFD is more sensitive than conventional UT techniques. Figures 6.27 to 6.29 show the principle – the ultrasonic wave diffraction from the tip of a defect is almost independent of the orientation of the defect relative to the beam path. The intensity of the diffraction is increased if the defect is sharp-edged; a useful characteristic as sharp-tipped defects pose a greater risk to integrity than blunt-tipped ones of the same length.

Unlike conventional UT compression and shear-wave techniques, TOFD uses separate probes for emitting the ultrasonic beam input and receiving the diffracted 'output'. Beam angles are also restricted, which

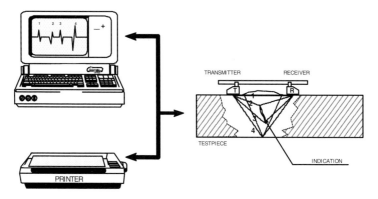

Fig. 6.27   Equipment set-up time of flight diffraction (TOFD)

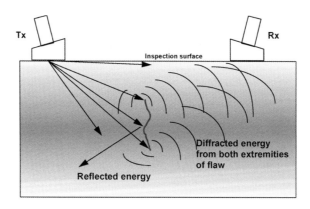

Fig. 6.28   TOFD diffraction display

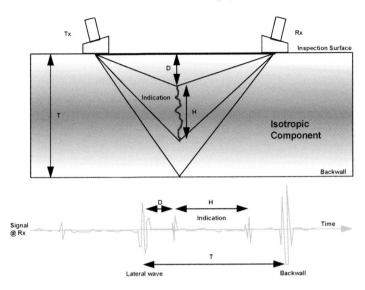

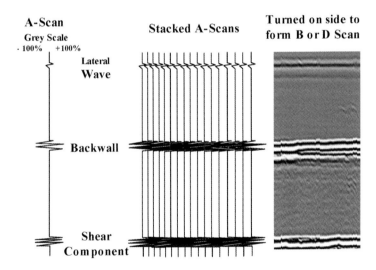

Fig. 6.29   TOFD scan patterns

means that the two probes must be separated by a minimum spacing distance. This is achieved by mounting the probes on a custom-made frame – the whole assembly being pushed along the weld. This constitutes the full scan, replacing the traditional longitudinal and transverse shear-wave scans used in conventional UT. Maximum scanning speed is about 150 mm/s, so, once set up, the technique allows

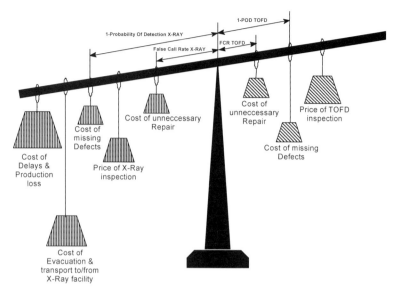

Fig. 6.30   TOFD comparisons

quick scanning of long, regular-geometry weld joints. Figure 6.29 shows a typical arrangement, and the way that various common defects appear on the display.

TOFD limitations

The nature of the TOFD technique results in some important limitations for on-site use. These are practical restrictions imposed by the practicality of doing in-service TOFD under site conditions rather than inherent technical problems with the technique itself. They are:

- *Materials require validation.* TOFD techniques must be carefully validated on a material before it can be reliably used on site. Common grades of low-carbon (and some low-alloy) steel have been success-fully validated but many other common ASTM and EN materials have not. Validation is particularly rare on high-strength/high-temperature creep-resistant materials – precisely the ones used in high-integrity defect-sensitive applications where TOFD *could* be of potential benefit.

- *Restriction on weld joint type.* TOFD is almost exclusively limited to butt welds because of the physical requirements of the probe frame set-up and the characteristics of the scanning pattern. Again, this is a practical restriction – many pressure equipment components do not have easy, uniform access around full-penetration nozzle and manway welds and nozzle compensation pad fillet welds, which are important in-service integrity areas. For pipework systems, the existence of valve bodies, reducers, and flange fittings means that multiple-probe frames set-ups would be needed to cater for all the different component geometries. This would be impractical (and expensive) for all but the most critical nuclear and safety-critical systems.

• *Defects do not always need to be sized accurately.* Most routine in-service inspections use techniques suitable for identifying and sizing defects in accordance with common pressure equipment codes and their supporting technical standards (ASME VIII/V/IX, EN 13445, BS PD 5500, and similar). The techniques need to be able to detect cracks that have propagated above a critical crack size for the material, but do not always need to be able to size the crack accurately.

TOFD cost and effectiveness comparisons

Various studies have been carried out to investigate the efficiency and overall cost effectiveness of the TOFD technique compared with other volumetric NDT techniques. Most of these have centred around the comparison between TOFD and radiographic testing (RT), and have been based mainly on pre-use site inspection rather than in-service inspections. Overall, TOFD shows several advantages over conventional RT – some of the main issues are outlined in Fig. 6.30 and in the references below.

TOFD bibliography

ASME VIII, Code case 2235: 1986 *Use of ultrasonic examination in lieu of radiography*, section VIII, Div. 1 and 2.

Bouma T., Halkes, C. J. van Leeuwen, W., and van Nisselroij, J. J. M. Non-destructive testing of thin plate (6–15 mm). Final report KINT project, April 1995.

BS 7706: 1993 *Guide to calibration and setting up of the ultrasonic time of flight (TOFD) technique for the detection, location and sizing of flaws.*

prEN 583 pt 6: 1999 *TOFD technique as a method for defect detection and sizing.*

Silk, M. G. Estimates of the probability of detection of flaws in TOFD data with varying levels of noise. *Birmingham Insight*, January 1996, **38** (1).

Verkooijen, J. TOFD used to replace radiography. *Birmingham Insight*, June 1995, **37** (6).

Wall, M. and Wedgwood, A. NDT value for money. *MTQ94 Birmingham Insight*, October 1994, **36** (10).

Radiographic testing

Radiographic testing (RT) is used regularly for in-service shutdown inspection as well as its more traditional use for 'works inspection' of new equipment during manufacture. RT techniques are well-proven and supported by a raft of European (EN) and American (ASME) published standards. As a strongly quantitative technique, RT is perhaps better controlled by published technical standards than almost any other area of inspection. The relative merits of RT versus UT change slightly when considered in the context of in-service rather than works inspection. Figure 6.31 shows some of the main considerations.

Radiography technique

A lot of the technology and standards of RT originates from pressure vessel practice. Hence you will find that the technical aspects of RT are very well covered by the various technical standards. The standards themselves contain definitive guidelines on defect names, and the techniques themselves, but for acceptance criteria relevant to in-service use, as discussed earlier, some interpretation and judgement is inevitably required. In-service RT techniques are almost identical to those used for works inspection of new equipment. In this section we will look at a typical radiographic test application, a full-penetration butt weld, and nozzle welds in steel vessels and pipework.

Figure 6.32 shows the position of four typical welds to be examined. Weld 1, the large-bore pipe butt weld can be accessed from both sides. It can be easily radiographed using a single-wall technique. Weld 2 is a small-bore full-penetration pipe butt weld. It can only be easily accessed from the outside, so it needs to be examined using a 'double-wall' method. In this case a double-wall, double-image technique is used. Weld 3 is a nozzle weld of the set-through type, as the small tube projects fully through the pipe wall. This weld could be radiographed but the technique would be difficult, mainly due to the problem of finding a good location for the film. Attempts to bend the film around the outside of the small-diameter nozzle weld would cause the film to crease, giving a poor image. As a general principle, nozzle to shell welds are more suited to being tested by ultrasonic means – radiography is difficult and impractical. Weld 4, the pipe-to-flange joint is of a common type often termed a 'weld neck' flange joint. Note that this is a single-sided weld with a root 'land'. It qualifies as a full-penetration weld and would require full NDT under most specifications and codes. This joint is capable of being radiographed using a single-wall technique similar to that shown for weld 1, as long as there is sufficient room for the film, i.e.

Safety and procedural issues

RT is much more trouble than UT for site use. RT requires planning for safety permits and evacuation zones; often, therefore, the only practical solution is to do the tests at night. UT is much easier and quicker.

Access

Both RT and UT require scaffolding access and cladding/lagging removal. On-site RT may be limited to gamma techniques (using a small portable radioactive source rather than an X-ray machine) if access is particularly awkward.

Suitability for in-service defect types

For crack-like defects (in welds)

- UT requires weld-cap grinding to obtain a full scan pattern. RT only requires cap grinding if the cap is of sufficiently poor profile that it would mask any defects located underneath it.
- UT crack detection is less dependent on the plane of the defect than RT, as long as several different scanning angles are used. RT can miss defects that are orientated in some planes.

UT is better at *sizing and classifying* crack-like defects than RT.

For corrosion-type defects

Both RT and UT techniques have limitations in finding and classifying particular corrosion and wall-thinning defects.

- RT can miss wall thinning altogether if the thinning is fairly uniform or hidden behind thick material – but it is good for use on welds.
- UT can miss degradation mechanisms such as weld root/HAZ attack and similar galvanic corrosion mechanisms unless weld caps are ground flat. This is particularly a problem on pipework.

Records

- RT has the advantage that it gives a permanent record, whereas standard UT techniques do not. In reality however progressive deterioration of components between inspections is *not easy* to determine from radiographs – UT is much better for defect classification and sizing purposes.
- UT has the *disadvantage* that technicians' written reports have to be taken by the in-service inspector at face value as there is no confirmatory record of defects found (or lack of them). This is a situation where *witnessing* of a sample of UT tests is important, to allow the inspector to build some confidence about the UT techniques used (and the technician who used them).

Fig. 6.31   RT-versus-UT techniques – advantages and disadvantages for in-service inspection

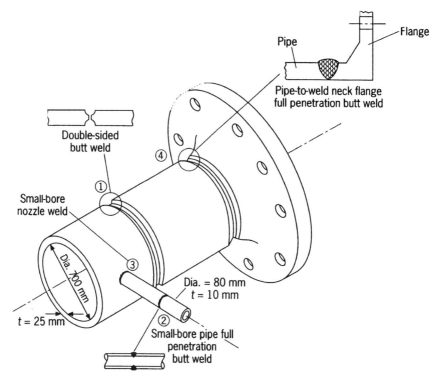

Fig. 6.32   Welds for radiographic testing

the width of the weld that can be examined should not be restricted by the position of the flange. It would also be possible to use an ultrasonic technique on this weld if required, scanning from the pipe outer surface and flange face.

We will look in detail at welds 1 and 2. Before reviewing the results of any radiographic examination it is wise to look at the details of how the examination was performed, i.e. the examination *procedure*. This may be a separate document, or included as part of the results sheet. Note the following points:

- *Single- or double-wall technique?* Weld 1 is a single-wall technique – the film shows a single weld image shot through a single-weld 'thickness'. Up to ten films are required to view this weld around its entire circumference. In contrast, weld 2 is a double-wall technique, suitable for smaller pipes. An offset double-image technique is used to avoid the two weld images being superimposed on the same area of film, and hence being difficult to interpret.

- *Position of the films.* For weld 1, ten films, each approximately 230 mm in length are required around the circumference. Starting from a datum, their position is indicated by lead numbers so the position of any defects seen in the radiograph can be identified on the weld itself.
- *The type of source used.* In-service inspectors only need do a broad check on the suitability of the source *type*. For all practical purposes this will be either X-ray or gamma ray.

  - X-rays produced by an X-ray tube are only effective on steel up to a material thickness of approximately 150 mm. Weld thickness of up to 10 mm requires about 140 kV. Thicknesses in excess of 50 mm need about 400–500 kV. A practical maximum is 1200 kV.
  - Gamma rays, produced by a radioactive isotope, can be used on similar thicknesses but definition is reduced. If a cobalt 60 gamma source is used, the best results are only obtained for material thicknesses between 50 and 150 mm. It does not give good results on thin-walled tube welds.
  - It is not possible to compare accurately results obtained by X-ray and gamma ray methods.

Checking RT films

Although 'real-time' viewing methods are technically feasible, most practical techniques use a photographic film which is developed after exposure. A visual record is therefore provided for review by all parties concerned. To check radiograph films properly consists of six essential steps. It is best to work through these in a structured way to ensure that the films are reviewed thoroughly.

1. *Check the film location.* Make sure you relate each film to its physical location on the examined weld or component. Use the procedure sketch and position markers but *also* look for recognizable features such as weld-tees, flanges, and surface marks to perform a double-check.
2. *Check sensitivity.* This is a check to determine whether the radiographic technique used is 'sensitive enough' to enable defects to be identified if they exist. It is expressed as a *percentage*, a lower percentage sensitivity indicating a better, more sensitive technique. A typical quoted sensitivity is 2 percent, indicating that, in principle, the technique will show defects with a minimum size of 2 percent of the thickness of material being examined.

The European standard for IQIs is <u>BS EN 462</u>

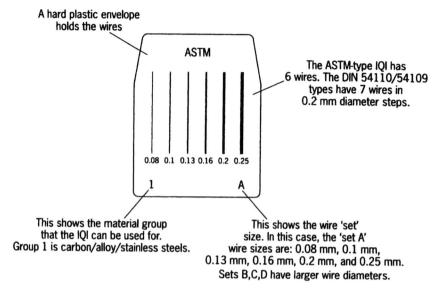

A hard plastic envelope
holds the wires

ASTM

The ASTM-type IQI has
6 wires. The DIN 54110/54109
types have 7 wires in
0.2 mm diameter steps.

0.08 0.1 0.13 0.16 0.2 0.25

1   A

This shows the material group
that the IQI can be used for.
Group 1 is carbon/alloy/stainless steels.

This shows the wire 'set'
size. In this case, the 'set A'
wire sizes are: 0.08 mm, 0.1 mm,
0.13 mm, 0.16 mm, 0.2 mm, and 0.25 mm.
Sets B,C,D have larger wire diameters.

Fig. 6.33   Using the wire-type penetrameter (IQI)

The actual value of sensitivity is determined using a penetrameter, also called an image quality indicator (IQI). The general principle is that one of the IQI 'wire' diameters is a given percentage of the material thickness being examined. If this wire is visible, it shows that the RT technique is being properly applied. There are several commonly used types that you will see, they are described by BS EN 462-1 and ASME/ASTM E142 standards. Figures 6.33 and 6.34 show two types and how each is used to calculate the sensitivity. Note that you cannot make an accurate comparison between different types – it is essential therefore to quote a sensitivity value against a specific IQI type for it to be capable of proper verification. If sensitivity is worse than that specified then the technique is unacceptable because it is not capable of finding relevant defects.

3. *Check unsharpness*, properly termed 'geometric unsharpness'. This is a measure of the contrast of the image, i.e. the difference between the dark and light areas. Although methods have been devised to measure the degree of contrast it still involves subjectivity. Check that the image is 'sharp and defined', without any obvious blurring – you can get a reasonable impression by looking at the IQI. If you do

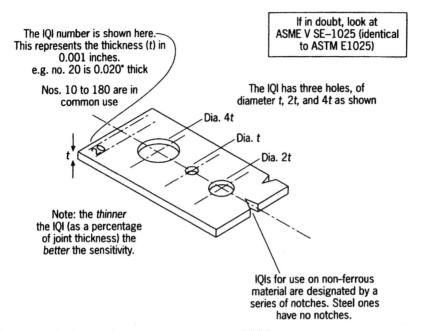

The IQI number is shown here.
This represents the thickness (t) in 0.001 inches.
e.g. no. 20 is 0.020" thick

Nos. 10 to 180 are in common use

If in doubt, look at ASME V SE–1025 (identical to ASTM E1025)

The IQI has three holes, of diameter t, 2t, and 4t as shown

Dia. 4t

Dia. t

Dia. 2t

t

Note: the *thinner* the IQI (as a percentage of joint thickness) the *better* the sensitivity.

IQIs for use on non–ferrous material are designated by a series of notches. Steel ones have no notches.

Image quality designation is expressed as

(X)–(Y)t:
(X) is the IQI thickness (t) expressed as a percentage of the joint thickness
(Y) (t) is the hole that must be visible

IQI designation	Sensitivity	Visible Hole*
1–2t	1	2t
2–1t	1.4	1t
2–2t	2.0	2t
2–4t	2.8	4t
4–2t	4.0	2t

* The hole that must be visible in order to ensure the sensitivity level shown

Fig. 6.34 How to read the ASTM penetrameter (IQI)

find problems with unsharpness, check the technique sheet for details of the source-to-film and object-to-film distances used; incorrect distances are one of the common causes of unsharpness. Radiographic standards such as BS 2910 (withdrawn) and EN 1435: 1997: *Radiographic examination of welded joints* give the minimum distances required.

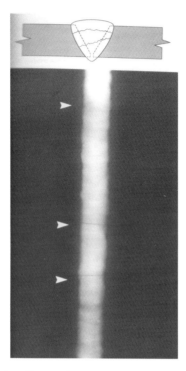

Fig. 6.35   Typical radiograph – this one shows a transverse crack

4. *Check the film density.* You can think of density as the 'degree of blackness' of the image. It is determined by an instrument known as a densitometer. Typical values should be between 3.5 and 4.5.

5. *View each film for indications.* The principle is that indications, (loosely termed 'defects') are identified and then classified according to type and size. Luckily, defect *classifications* are well accepted and defined – there is much less room for interpretation than with, for instance, the acceptance criteria for defects, once classified. Defects are broadly classified as either major or minor. Major defects are those that impinge directly on the *integrity* of the weld and form the core content of the acceptance criteria. Major defects often need repairs. Defects classified as 'minor' mainly appear on the surface of the weld and have generally less effect on the weld integrity.

6. *Marking up the films.* This is a key role for the in-service inspector. Work through the test report sheet viewing each radiograph film in turn, checking the reported indications carefully against what you see on the films. Mark important indications on the film with a chinagraph pencil, using the correct abbreviations. If a repair is

required mark the film 'R' in the top right-hand corner, making a corresponding annotation on the report sheet. Keep each packet of films and its report sheet together and check they are correctly cross-referenced in case they become separated. Figure 6.35 shows a typical radiograph.

These six steps are common to all radiograph review activities. Although routine, reviewing radiographs is a key aspect of checking for fitness for purpose.

NDT standards

EN 287–1: 1992	*Approval testing of welders – Fusion welding – Part 1: steels.*
EN 473: 2000	*Non-destructive testing – qualification and certification of NDT personnel – general principles.*
EN 571–1: 1997	*Non-destructive testing – Penetrant testing – Part 1: General principles.*
prEN 764–6: 2002	*Pressure equipment – Part 6: Operating instructions.*
EN 583–4: 1999	*Non-destructive testing – Ultrasonic examination – Part 4: Examination for discontinuities perpendicular to the surface.*
EN 970: 1997	*Non-destructive examination of fusion welds – Visual examination.*
EN 1289: 1998	*Non-destructive examination of welds – Penetrant testing of welds – Acceptance levels.*
EN 1290: 1998	*Non-destructive examination of welds – Magnetic particle examination of welds.*
EN 1291: 1998	*Non-destructive examination of welds – Magnetic particle testing of welds – Acceptance levels.*
EN 1418: 1997	*Welding personnel – Approval testing of welding operators for fusion welding and resistance weld setters for fully mechanized and automatic welding of metallic materials.*
EN 1435: 1997	*Non-destructive examination of welds – Radiographic examination of welded joints.*
EN 1712: 1997	*Non-destructive examination of welds – Ultrasonic examination of welded joints – Acceptance levels.*
EN 1713: 1998	*Non-destructive examination of welds – Ultrasonic examination – Characterization of indications in welds.*
EN 1714: 1997	*Non-destructive examination of welds – Ultrasonic examination of welded joints.*
EN 1779: 1999	*Non-destructive testing – Leak testing – Criteria for method and technique selection.*

EN 12062: 1997	*Non-destructive examination of welds – General rules for metallic materials.*
EN 12517: 1998	*Non-destructive examination of welds – Radiographic examination of welded joints – Acceptance levels.*
EN 13445–2: 2002	*Unfired pressure vessels – Part 2: Materials.*
EN 13445–3: 2002	*Unfired pressure vessels – Part 3: Design.*
EN 13445–4: 2002	*Unfired pressure vessels – Part 4: Fabrication.*
prEN ISO 5817: 2002	*Welding – Fusion welded joints in steel, nickel, titanium and their alloys (beam welding excluded) – Quality levels for imperfections (ISO/DIS 5817: 2002).*
EN ISO 6520–1: 1998	*Welding and allied processes – Classification of geometric imperfections in metallic materials – Part 1: Fusion welding (ISO 6520–1: 1998).*

Chapter 7

Risk-based inspection (RBI)

Few subjects in the world of inspection raise as much controversy as risk-based inspection (RBI). Over the past 10 years, some industries have started to use it extensively, with varying results, while others have ignored it. From its roots in the USA, Europe, and Japan, its use is also spreading internationally, to the point where it is now a permanent feature of the world of in-service inspection.

So what *is* RBI?

There is no formal definition – at least not one that everyone would agree with. RBI is an emotive subject in some parts of the in-service inspection business, but in reality, it is not that difficult to define. It is only when discussing the merits (or, conversely, *dangers*) of RBI that matters become controversial. Figure 7.1 shows the situation.

Simple though the definitions in Fig. 7.1 may sound, the whole concept of RBI is bound up tightly inside a situation that is much more complex. RBI, not unlike quality assurance (QA), is more of a way of

---

RBI is a method of:

- Ranking plant items according to their relative risk of failure. And then
- Designing the in-service inspection plan, i.e. the written scheme of examination (WSE), to suit.

RBI is therefore:

- a method of *optimizing* in-service inspection activities.
- an alternative to conventional calendar (or time)-based inspection.

Fig. 7.1   RBI: what exactly is it?

thinking (almost a type of thinking *model*) than a stand-alone engineering technique. The techniques of RBI only work when they are accompanied by a raft of other methods and assumptions about how things related to the world of engineering failure, analysis, and inspection actually work. Take away these supporting assumptions and the validity of RBI disappears with them. The objective of this chapter is to explain the techniques of RBI, and shed some light on the advantages and disadvantages, rather than simply to explain the techniques as a 'cure all' to the high cost and technical difficulties of in-service inspection.

Before we start – some important concepts

RBI is based on a perceived model of engineering reality. We can look at some of the concepts that make up the picture.

The Probability of Failure

RBI is based around the premise that engineering items can be allocated a *probability* that they will fail in a certain way. This puts RBI firmly in the category of a mathematical model. Note that this is not the same as saying that a failure can be *predicted* – merely that a number representing the probability of a failure occurring can be attached to it, without any guarantee that such failure will, or will not, eventually occur. This means that even a component that has been assessed as having a 99 percent probability of failure in the next year may not fail in that year, or even in the following 20, or even 50, years. Such a scenario sits perfectly comfortably in the world of mathematical models to which RBI belongs.

Probability should not therefore be confused with *reality*. Despite various other terms sometimes introduced to muddy the waters (*likelihood* and *possibility* are the common ones), there is absolutely no guarantee that the concept of probability will help predict the future of an engineering item (or anything else for that matter).

Risk

The concept of risk is central to RBI. It is common to hear RBI described as 'a technique to reduce the risk of failure'. Most RBI specialists would concur with the fact that it is only possible to assign *relative* risk to items of plant, i.e. to decide that one item has a lower or higher risk of failure than another, rather than an *absolute* risk.

Knowing the absolute risk brings us back to the idea of predicting accurately the future, which we know is impossible.

Even the idea of relative risk has serious weaknesses. No two people, anywhere, view or assess risk in the same way – what is a high risk to one person is seen as low risk, or justifiable risk, by another. The end result is that industry exhibits a complete spectrum of varying attitudes to risk – from the almost totally risk averse at one end to those that are apparently happy to live, day-to-day with very high levels of risk in almost all areas of their business.

The existence of different attitudes to risk is one of the main influences on the effectiveness of RBI as an engineering technique. The core value of RBI is that users are allowed to assess their own risk levels *themselves* – it is therefore a tool of *self-regulation* rather than one related to a scenario where a risk level is imposed by an external or regulatory body. This, coupled with the fact that most RBI programmes are kept 'internal' to the businesses that use them, guarantees that across industry the content of RBI programmes (and their resulting inspection periods, techniques, etc.) will all be different. Their results will vary across the risk spectrum from low-risk analyses and answers to high-risk analyses and answers. Hence there can be no true 'benchmark' for the validity of an RBI programme – none (however weak) can be adjudged to be 'totally wrong'. RBI is therefore self-regulation in a near-extreme form.

How does RBI affect in-service inspection?

RBI affects three aspects of inspection: the frequency or periodicity of inspection, the scope of inspection, and the techniques used to perform the inspections. We can look briefly at these in turn.

The effect on periodicity

Normally, the objective of RBI is to optimize the periodicity of inspection – often expressed as the maximum inspection period (MIP). In practice, it is rare that an RBI analysis is performed to try to decrease the MIP (i.e. that an inspection should be carried out more often) – the objective is normally to justify an increase in MIP, so that inspection can be carried out *less* often. In more complex cases the rationale of an RBI analysis may be double-edged: to decrease the MIP for shutdown inspections by replacing them with more frequent on-stream or non-invasive inspections performed under normal working conditions. Figure 7.2 shows the situation.

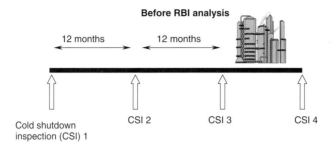

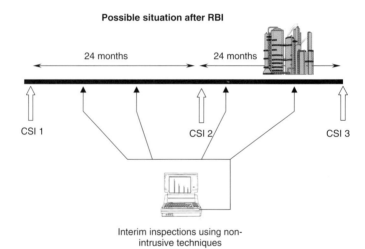

Fig. 7.2   RBI and inspection periodicity

The effect on scope of inspection

The result of increasing the MIP for a piece of plant is often accompanied by a matching increase in the scope of the (now less frequent) inspections. The inspection may look at more items, remove more lagging, etc. than was previously done. You can think of this (if you like) as the additional inspection activities compensating for the fact that they are carried out less often.

The effect on inspection techniques

The main effect of RBI tends to be an increase in the scope of non-intrusive inspection activities. Specialist techniques such as ultrasonic corrosion mapping, time of flight diffraction (TOFD), etc. (see Chapter 6) are used to provide an assessment of corrosion or other damage mechanisms without the need for isolation and strip-down of a pressure vessel or pipework system. Refineries and petrochemical plant are the main users of these types of techniques. As well as encouraging the use of advanced NDT techniques, RBI tends to encourage the use of more stringent 'conventional' NDT techniques. Typical changes are:

- The use of more stringent ultrasonic testing (UT) defect acceptance criteria, e.g. to look more carefully for damage mechanisms such as stress corrosion cracking (SCC) in stainless steel and reheat cracking on high-chromium alloys.
- Using replication to set a 'benchmark' for creep and cumulative creep fatigue damage early in the life of a high-temperature component.
- For surface crack detection, replacing basic magnetic particle inspection (MPI) examination with a more sensitive version such as fluorescent MPI, or extending a dye penetrant (DP) technique to use a better surface finish, developer dwell time, etc. (in excess of code requirements) obtain a more searching, reliable test.

RBI links with statutory regulations

Perhaps not surprisingly, most statutory regulations relating to the in-service inspection of plant worldwide do not delve too much into RBI techniques. There is an increasing trend for statutory regulations to *recognize* RBI as a valid technique used to define inspection scope and periodicities, but none address it in much detail. Many countries are following the trend of the UK and some other EU countries of not specifying rigid inspection periodicities for pressure equipment. Instead they rely on the decision of an industry 'Competent Person' (or sometimes the plant owner/operator themselves) to define the necessary periodicity and scope, on an item-by-item basis. The techniques of RBI fit neatly with this approach, providing a robust and documented technique on which to base these decisions.

RBI and the UK PSSRs

The UK Pressures System Safety Regulations (PSSRs, see Chapter 5) do not set rigid periodicities and scopes of inspection, leaving it up to the Competent Person to decide, and then document them in a written scheme of examination (WSE). There is no direct mention of RBI in the PSSRs themselves, or in the associated Approved Code of Practice (ACoP). There is, however, an *inference* in that decisions made by the Competent Person need to be justifiable if questioned by the enforcing authority (the Health and Safety Executive). RBI would be one way of demonstrating that a robust analysis had been done, in order to arrive at the scope and periodicity of inspection set out in the WSE.

RBI and American codes

American codes such as API 510 *In-service inspection of pressure vessels*, API 570 *In-service inspection of pipework systems*, and API 653 *In-service inspection of atmospheric storage tanks*, are a legal requirement in many parts of the USA. They are also in common use as technical guidelines in the refinery, petrochemical, and offshore industries in other countries. The general trend is now for these codes to recognize the validity of RBI as a technique for deciding inspection scope and periods. The concept is slow to be accepted fully (API 510 and API 570, for example, still quote maximum periods for on-stream inspections, thickness checks, etc.) but the trend definitely appears to be in that direction. The recent American RBI standards API 580/API 581 (see later in this chapter) are comprehensive and authoritative documents that are increasingly referenced in API and ASME codes, and published documents. It is doubtful whether RBI will ever totally replace code-based inspection periodicities, but it seems destined for increasing acceptance as a valid alternative.

RBI in different countries

Despite its focus on mechanical integrity and related safety issues, RBI does not fit equally well into all industry sectors. This is justified by historical experience over the past 5–10 years in Europe and the USA. Despite obvious synergy between the methodology of RBI and the characteristics of some industries, many have spent large sums on RBI programmes, only to find that they do not fit well with their business methods, or give inaccurate or misleading technical results, or both. Figures 7.3 and 7.4 give some insights into this situation. These figures are not absolute, and certainly not based on any validated data, but they probably contain more than an element of truth. Use them as a guideline, rather than as a prescription for RBI success. Note how Fig. 7.3 is based on the characteristics of industry while Fig. 7.4 takes a more targeted view on the types of plant and equipment used within a utility or plant operations business.

Resistance to RBI

Overall, few industries have adopted whole-heartedly the concept of RBI as a basis for managing the in-service inspection of their plant. Worldwide, perhaps 5–10 percent of plants have tried it in some limited form with only a handful operating, with any degree of success, a full RBI system on a continuous basis. The reasons for resistance to RBI are

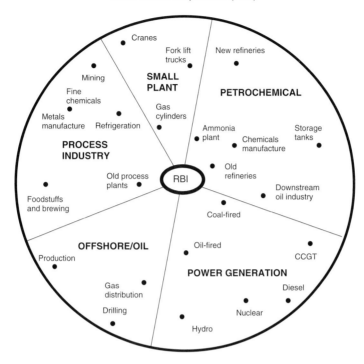

**The nearer an industry is to the centre of the
circle, the greater its acceptance of RBI
techniques**

Fig. 7.3   Industry acceptance of RBI

wide and varied. They include the following perceptions:

- RBI is of limited use on newer plant; you need a lot of old plant (mainly pipework systems) in poor condition before it can be of any use.

- RBI is petrochemical-industry orientated. The main technical documents such as API 580/581 are based almost entirely around degradation mechanisms found mainly in refinery and petrochemical plants.

- RBI is little more than a defensive technique, used to provide a raft of convincing technical excuses for increasing maximum periods between shutdown inspection.

- RBI is software-dominated, based on data and assumptions that are not representative of 'real plant' conditions.

- RBI is an academic, theoretical exercise submerged in jargon and acronyms.

RBI is safety- and mechanical-integrity-focused, *but* it works better under some conditions than others

Factors which *facilitate* an RBI-based approach are:	**Factors which *hinder* an RBI-based approach are:**
• A plant inventory containing large numbers of items of pressure equipment (e.g. pipes, PRVs, valves, and vessels).	• A plant inventory containing a small number of large, discrete items (e.g. rotating equipment, turbines, gas/steam turbines, storage tanks, etc.).
• Large variations in process conditions (e.g. temperature, pressure, fluid velocity, pH, corrosiveness, etc.).	• Most plant items operating under basically similar 'bands' of pressure and process conditions (with little real variation).
• A lot of old plant, of varying ages from about 5 years old to near-retirement.	• New plant, or all within an age range of about 2–3 years.

Fig. 7.4   Some plant characteristics that affect the use of RBI

- RBI absorbs unreasonable amounts of plant resources, involving key managers, operations staff and engineers sitting around in soul-destroying day-long meetings, diverting them from their proper work.

Whether justified or not, criticisms like this are absolutely valid in the minds of those that perceive them. Various different types and varieties of RBI programmes have been devised to contain such criticism, each one heralding its advantages over its shadowy predecessors by claiming to be simpler, more (or less) quantitative while producing 'ever increasing levels of risk optimization'. Some of this has engineering validity, some is pure jargon. Notwithstanding the differing views on RBI, it continues to exist, as an option, for plant owners and operators that choose to use it.

Technical limitations in scope

The clearest technical limitation of the effectiveness of RBI relates to its scope. In practice, RBI is limited almost exclusively to items of static mechanical equipment such as vessels, valves, and pipework. Critical plant items such as rotating equipment, electrical and control equipment, computer hardware and software rarely appear in an RBI analysis, even though they are the frequent cause of plant breakdown. In reality, RBI analysis is limited to those plant items which have, historically, been well served by manufacturing codes (ASME, API, BS, etc.) and have been the subject of statutory authority or insurance company-imposed inspections. Few industry bodies or individual plant operators feel the need to collect or distribute data on breakdowns and failures of non-statutory plant items, so this large proportion of plant inventory remains outside the RBI envelope. Viewed carefully, this shows up as a significant technical limitation on the effectiveness of RBI.

RBI – the basic techniques

There is no single unequivocal way of doing an RBI assessment – there are many different technical methods and multiple approaches for implementing them within the hierarchical levels of an organization. The level of detail can also vary considerably: from a quick risk-based 'look ahead' to a comprehensive programme designed to change the entire inspection regime of an operating plant. With the large variety of possibilities there are several 'lowest common denominator' require-ments that must be met, if an RBI analysis is to have credibility (see Fig. 7.5)

Warning: jargon dangers

The world of RBI is besieged with specialist terminology, jargon, and acronyms. Some of this is necessary to define essential concepts and some is either an overcomplex way to describe something simple, or belongs in the world of pure statistics and mathematics. The secret of dealing with RBI terminology is *selectivity*: there are a few that are really needed and the rest can be safely ignored.

Quantitative versus qualitative techniques

In the rarefied world of RBI specialists, RBI techniques divide into two categories: quantitative techniques and qualitative techniques. It is fair

---

To be credible, an RBI programme must:

- **Be systematic.** Any decision to analyse only the easy or non-controversial parts will devalue the whole exercise.
- **Have continuity.** This means it must have a *time context*. Any RBI programme that is performed as a 'snapshot' only can have limited effectiveness. Try to think of RBI as a strongly time-based exercise.
- **Incorporate the correct expert technical opinions.** This means expert technical opinion about the subject plant and its specific operation and degradation mechanisms, rather than expertise with numerical computer tools or commercial software packages.
- **Have an output.** Normally this involves changes to inspection periodicities and scope of examination in the written scheme of examination. An RBI programme with no output is absolutely worthless.

Fig. 7.5   Some 'lowest common denominator' requirements of an RBI programme

to say that these categories have been driven, at least in part, by the content of commercial software. Across the spectrum of RBI programmes the differences between quantitative and qualitative technique become blurred to the point where the different characteristics of each become clouded. Figure 7.6 is an attempt to explain the situation. Note the format of Fig. 7.6 – it presents the scope of RBI as a spectrum of techniques varying continuously between the quantitative and qualitative so, in reality, virtually all RBI analysis techniques end up being *semi-quantitative*. Because of this, it can be difficult to compare the methodology (and usefulness) of proprietary RBI software packages, each of which places different emphasis on the subtle combination of qualitative and quantitative methods on which it is based. Figure 7.7 might help you avoid some of the confusion.

The RBI programme itself

There is, perhaps, no such thing as a 'typical' RBI programme. They vary, as explained earlier, from the short and simplified to the long and complex, with many shades of grey in between. To look at the contents of a sample RBI programme, we need to see it from the viewpoint of the plant operator or owner: a role described as 'the Duty Holder'.

Quantitative Approach

- Based on 'pre-set' risk parameters
- Considers just about everything

Qualitative Approach

- Uses plant experience and feedback
- Decides priorities, based on risk

So the qualitative RBI approach is:

- More specific to an individual plant

- Uses mainly **your** experience (rather than the person who wrote the software)
  BUT
- It's different for each plant (so you have to start from scratch, each time)

Fig. 7.6   RBI – quantitative versus qualitative techniques

The Duty Holder role

In an operating plant, the final responsibility for the safety and integrity of the plant lies in virtually all countries of the world, with the plant user, operator, or owner. The exact party responsible is often difficult to define (because there are many possible variations of operation and ownership), so the term 'Duty Holder' is used. The Duty Holder is defined quite simply as the party who has the responsibility for meeting statutory and other requirements that have an influence on the way that a plant is inspected in service. The concept of 'Duty Holder' therefore transcends any contractual arrangements that may or may not exist.

The steps in Fig. 7.8 outline the basic programme for an RBI procedure as applied to large-scale industrial plant. Figure 7.9 shows the analysis in operation using a typical example.

Step 1. Define the 'system boundaries'

The first stage is to decide which systems and equipment are going to be within the scope of the RBI programme. These are not predetermined, it is necessary to choose them from the many possibilities available. Some possible approaches are:

- Choose those systems that suffer the most plant failures resulting in forced outages.
- Restrict the scope to discrete physical areas or systems of the plant.

- Some RBI software-package-based techniques either:
  - rely more on mathematical equations, models, and assumptions than you think, or
  - rely more on generic assumptions about corrosion rates, damage mechanisms, and the way that things fail (rather than actual site-specific conditions on your plant), than you think.
- RBI packages that claim to be fully qualitative must rely almost entirely on data provided by the plant itself. It they do not, (in order to make the analysis shorter and cheaper) they cannot really be fully qualitative.
- Many RBI programmes start in a flurry of activity, only to die slowly and quietly, so as not to embarrass anyone, within 12 months. The process may repeat itself (in a slightly different form perhaps) every few years.

Fig. 7.7   RBI realities?

- Concentrate on older plant in poor or unknown condition, or where previous inspection programmes have been random, or not always carried out as planned.
- Identify those areas of the plant that have the most influence on a specific plant operation objective. The main ones are:

  - safety;
  - production potential.

  Note that these two operating objectives, while both perfectly valid in themselves, are not *necessarily* complementary (although almost everyone will try to convince you that they are).

The choice of boundary of the system and plant areas that are to be included in an RBI programme has a big influence on whether or not an RBI programme will produce any useful results. There are few agreed precedents on where the boundaries *should* be, leading to a tendency for Duty Holders to define limits based on what is convenient, or easy, rather than what is *needed*.

Defining boundaries by 'corrosion loops'
One common way is to define system boundaries by process fluid, in particular those that experience similar identifiable corrosion mechan-

THIS IS THE **OUTPUT** OF THE RBI ANALYSIS

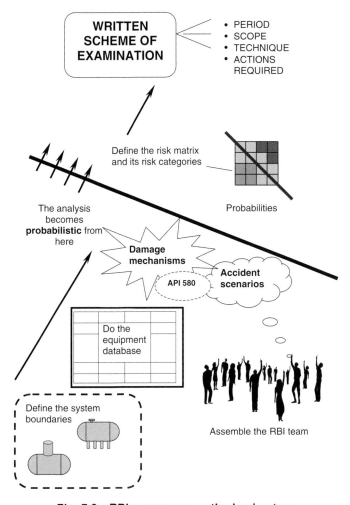

Fig. 7.8　**RBI programme – the basic steps**

isms (see Fig. 7.10). These are termed *corrosion loops*. This is a common approach in the petrochemical industry where process-induced corrosion is a major issue. It is less common in power plant and other installations where more widespread and unpredictable damage mechanisms such as corrosion under insulation (CUI), fatigue, and creep are more prevalent.

Links with statutory regulations
Do not expect statutory regulation such as the PSSRs to offer much guidance as to where to draw the system boundaries for an effective RBI

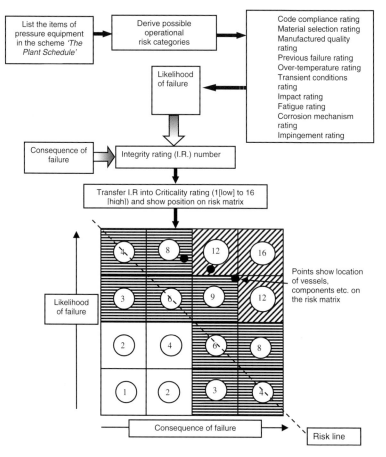

Fig. 7.9   A typical RBI matrix analysis

analysis. The PSSRs do define different types of system, but only by broad categories (major, minor, intermediate, etc.) and stop well short of specifying where the system boundaries might be. Although the PSSRs are based on a risk-based approach to in-service inspection, they do not offer guidance at the level of detail needed when designing an RBI programme. In addition, the PSSRs are only concerned with preventing failures involving the release of stored energy and most plant failures, particularly of pipework systems, do not fall into this neat category.

Perhaps the biggest problem with the boundaries of an RBI programme is caused by defining systems too widely. This results in too much analysis, producing too much information, leading to eventual confusion. Conversely, defining system boundaries too

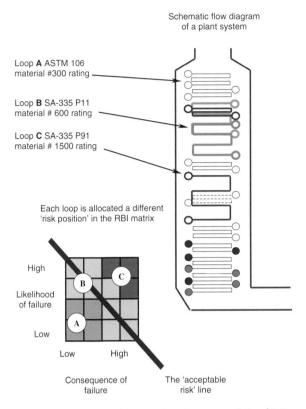

Schematic flow diagram
of a plant system

Loop **A** ASTM 106
material #300 rating

Loop **B** SA-335 P11
material # 600 rating

Loop **C** SA-335 P91
material # 1500 rating

Each loop is allocated a different
'risk position' in the RBI matrix

High

Likelihood
of failure

Low

Low   High

Consequence of   The 'acceptable
failure   risk' line

Fig. 7.10   An example of corrosion loops and the RBI matrix

narrowly has the opposite effect – the whole RBI process lacks
resolution and results in oversimplified output that does not reflect the
real engineering complexity of the plant to which it is meant to refer.
The message is that the definition of useful system boundaries for an
RBI analysis is *important*. Too many RBI programmes start off using
wrong, or poorly defined and unclear, ideas as to where their boundaries
are, and are doomed to fail from the start.

Step 2. Assemble the equipment database
Once the system boundaries have been defined, the next step is to list all
the equipment and their characteristics within the system boundaries.
The amount of information that has to be assembled is extensive; it
includes a lot of quantitative technical data, supplemented by empirical
data gathered from the plant during its previous operating life. Figures
7.11 and 7.12 show some characteristics (and probable difficulties) of
the type of data needed.

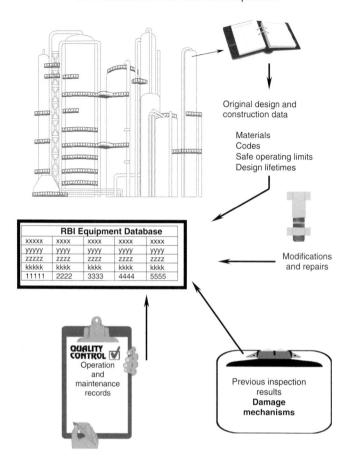

Fig. 7.11   Contents of the RBI equipment database

The equipment database contents

Equipment databases are divided into sections containing basic design data, materials, process conditions, and identification details such as P&ID references, equipment tag numbers and similar. Compilation of the equipment database is the starting point of the analysis phase of an RBI programme. In theory, all the information included in it should be 'validated' in some way (all specialist RBI literature will tell you this) in order to maintain the validity of the programme. In practice this is rarely possible (it would require massive manpower resources to do it properly), so there will always be a level of uncertainty.

Step 3. Assemble the RBI programme team

The idea of RBI is that the initial analysis and subsequent programme is performed *over time*, by a dedicated team of people containing a

For an RBI programme to be meaningful, it has to involve extensive gathering of essential data about the plant. Some difficulties are:

- If the plant is old, design data will not be available in a single location (database or similar). Each piece of information will have to be sourced from plant process and instrumentation, diagrams (P&IDs), general arrangement (GA) drawings, equipment nameplates, and datasheets. This is very time-consuming and may be practically impossible to complete with anything like 100 percent accuracy.
- If the plant is new, it should be possible to extract the necessary data from equipment databases already completed by the plant construction contractor. In reality this is not such an easy process as it sounds.
- Data gathering is frustrating, time-consuming work. It requires significant manpower input that has to be paid for.

Fig. 7.12   Initial data-gathering for an RBI programme

mixture of plant operators and maintenance staff, 'area' engineers (responsible for the plant items under analysis), corrosion engineers and metallurgists. These all work together with the Competent Person or external third-party representative and external 'RBI specialist' often sourced from one of the RBI consulting companies that sells RBI software.

Why a team approach?
A team approach is used for all but the smallest-scale RBI programmes. Conventional wisdom is that this allows for the inclusion of 'specialists' in the various areas considered essential to the RBI analysis, e.g. health and safety, plant operations, corrosion/damage mechanisms, inspections/NDT techniques, and similar. A further extension of the conventional wisdom is that the team should be managed (competently managed, presumably) by someone who has the authority to enable collection of the various plant and operations data, and then to get things done once the RBI programme has produced its results and recommendations. Figure 7.13 shows the situation.

The above rationale would, of course, sound equally valid applied to any activity carried out within an engineering organization (or any other organization for that matter). There is nothing new or RBI specific about such a recommendation. Some RBI documents go a little further in suggesting that outside parties such as the Competent Person

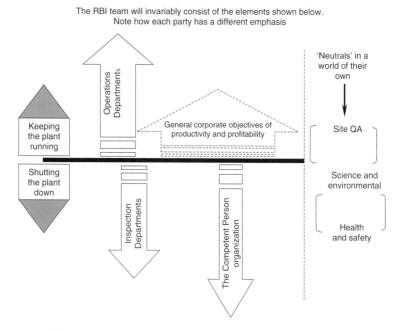

The RBI team will invariably consist of the elements shown below.
Note how each party has a different emphasis

Fig. 7.13   Initial data gathering for an RBI programme

inspection body should take a leading role in the RBI team without, perhaps, due consideration of the organizational realities of this – and its chance of success. It is sufficient for the moment to concede that a team approach will give desirable wide coverage of the multiple technical issues inherent in RBI.

Size of team?
The size of the team depends on the type of RBI that is planned. Opinions vary on whether it is the qualitative or quantitative approach that needs a larger team to do the job properly. These definitions themselves are not easy to agree upon so it is a source of fertile argument as to which needs the greater resource. In practice, all that can be sensibly concluded is that the wider and more detailed the study to be carried out, the greater the team size required. Inevitably, there will come a point where an RBI team becomes too large and unwieldy to decide anything – a later section in the chapter attempts to provide some basic suggestions and guidance to lessen the chances of this happening.

Relative team workloads?
Despite the wide variety of types of RBI programmes carried out across different industries, the relative workload of RBI team members is

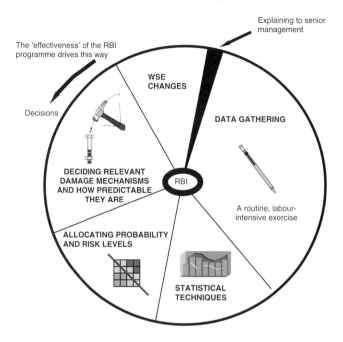

The 'effectiveness' of the RBI programme drives this way

Explaining to senior management

Decisions

WSE CHANGES

DATA GATHERING

DECIDING RELEVANT DAMAGE MECHANISMS AND HOW PREDICTABLE THEY ARE

RBI

A routine, labour-intensive exercise

ALLOCATING PROBABILITY AND RISK LEVELS

STATISTICAL TECHNIQUES

On balance, the time (and money) input requirements of each part of an RBI programme tend to distribute themselves like this

Fig. 7.14   Relative inputs of the RBI team

surprisingly consistent. A possible explanation for this is that the basic *activities* of RBI analysis are, themselves, much the same, even though the extent of statistical robustness, and the terminology, might differ. Figure 7.14 shows the general pattern of team workload input. Note one of the main conclusions from this figure i.e. if the RBI programme is to work, a large proportion of the resource is involved with routine *data gathering*. This information is rarely easy to assemble: it has to be taken from vendor manuals, P&IDs, and equipment nameplates with care taken to identify missing (and sometimes contradictory) data. A typical plant database spreadsheet can comprise 20 or 30 columns and hundreds or thousands of separate rows, each one dedicated to a discrete piece of equipment.

What about the decision-making content of Fig. 7.14? The key point is the fact that it is *distributed* around the RBI team – it does not reside with a single team member, or the department that they represent. Now we have a defining point for the character of the whole RBI process – it is an activity of *consensus*. It has to be, or it couldn't handle the scope of technical issues that it claims to cover. Figure 7.15 shows some possible

---

RBI is an activity of consensus, so, either:

- It produces results (technical outputs, assessments of risk, conclusions on inspection periods and techniques, etc.) made up of the *best parts* of the strengths of each of the Duty Holder's RBI team members. This means it is a hard, precise technique which cuts through technical problem (and plant politics) like a diamond. **Or**
- It ends up as an exercise in *lowest common denominator agreement* in which each RBI team member elects not to disagree with the others. Pride is taken in technical statements about risk, inspection periods etc. that sound *considered*, but do not actually mean much. This means that an RBI programme becomes a disguise, a defence against possible future criticism, real or imagined, about which the Duty Holder is frightened.

Fig. 7.15   RBI as an activity of consensus – some possible outcomes

characteristics of consensus activities; whichever ones are correct (decide for yourself) will govern the character of the RBI team, and the validity and usefulness of its technical output.

Step 4. Define accident scenarios and damage mechanisms

This is the core risk analysis part of an RBI programme. There is no single set formula or absolutely correct way to do it – each one must be tailored to the plant-specific situation. Specialist RBI published documents and standards provide a detailed breakdown of techniques used. API 581 *Risk-based-inspection – base resource document (BRD)* gives perhaps the most detailed coverage. Figure 7.16 shows, as a simplification, the basic elements of this part of the analysis.

Accident scenarios

You can *think* of this as the least technical part of an RBI analysis. The idea is to think of any possible accident scenario that could happen to the plant being analysed (within the pre-set boundaries set for the RBI programme). Owing to the definition of an 'accident', i.e. something which is *unexpected*, this is an activity that benefits from a wide imaginative approach. Semi-technical people responsible, for example, for health and safety, plant management, and procedural matters within a Duty Holder's plant, can provide input which is as valuable as specialist technical knowledge. Figure 7.17 shows some details.

- **Think about** failure and accident scenarios that could happen to the plant items and areas.
- **Identify** possible damage (or **deterioration** mechanisms) that could cause failure.
- **Confirm** the modes of failure that could result from the damage mechanisms identified.
- **Assess** (guess) the probability of each mode of failure actually happening.
- **Assess** (guess again) the consequences (i.e. the accident that will happen) resulting from equipment failure.
- **Decide** risk ranking 'categories' and actions to be taken for each category.

Fig. 7.16   RBI risk analysis – the basic elements of the process

Accident definitions

It is important to understand the relationship between the accident scenario and the damage mechanism element of an RBI analysis. An accident scenario is an undesirable final outcome (i.e. the accident) that results from a set of circumstances that *include* some kind of damage mechanism that has the ability of causing an engineering component or plant to fail. Look at Fig. 7.18 where this concept is shown graphically; note how there is a sequence of events that *cause* the accident. Note also how these events, although different in technical character, act together to produce the final outcome. There is also a *time context* to this – the events rarely happen simultaneously, they are much more likely to be spread over time. This means that, strictly, accidents do not happen immediately (as most people think they do).

Accident scenarios – unlikely but credible

This is a central issue to the idea of RBI, because most accidents, as proposed earlier, cannot be properly foreseen (otherwise how could they be classed as an accident?). There is no reason, however, why this should cloud your view when anticipating accident scenarios as part of an RBI analysis. The guideline is simple. You can think of an accident as:

- A chain of events which although *unlikely* is *credible*. Or
- Something which probably won't happen, but which *could*, if it felt like it.

Most of the techniques which are used to anticipate accident scenarios are based on this implicit way of looking at events. There

---

- Steam leak.
- Uncontrolled steam release (stored energy release).
- Hazardous release, e.g. hydrocarbon and flammable effects:
  - flash fire;
  - fireball;
  - jet flame;
  - pool fire;
  - explosion.
- Toxic release.
- Acid or hazardous release.
- Radioactive release.
- Flood.
- Fragmentation of metal parts.
- Environmental pollution.

Note how these are mainly orientated towards fluid and pressure equipment applications (remember that RBI programmes rarely consider plant, electrical equipment and control/instrumentation items). See API 581 further details, mainly for downstream hydrocarbon/petrochemical plant application.

Fig. 7.17. Typical 'accident scenarios' used in RBI analyses

are four basic techniques, with many variations of each. They are (along with their acronyms):

- HAZOPs (hazard and operability studies).
- FMEA (failure mode and effects analysis).
- FTA (fault tree analysis).
- ETA (event tree analysis).

HAZOPs
A HAZOP analysis looks at the operating conditions of a plant. The idea is to 'brainstorm' upset conditions of the plant (temperatures, pressures, process fluid pH, concentration, etc.) that could result in damage to its components. Not all HAZOP studies are necessarily safety related – many are concerned more with identifying credible production problem scenarios. This means that previous 'off-the-shelf' HAZOP analyses need careful checking for relevance before they can be taken as a valid part of an RBI programme, with its own terms of reference and physical plant boundary limits.

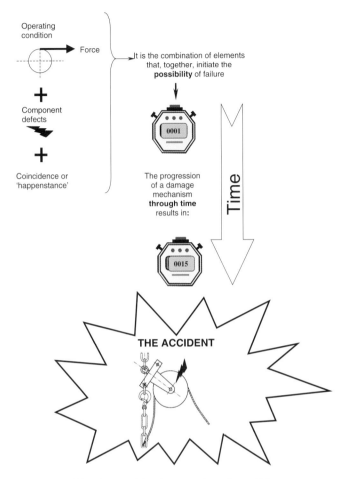

Fig. 7.18   An accident scenario – how it happens

FMEA

This is a more structured and step-by-step approach than the HAZOP analysis. Each plant component is considered in turn. Its failure modes and their effect on the rest of the system are listed. It uses greater input from technical specialists – the objective being to *analyse* each failure mode in significant depth, rather than simply *identify* them.

FTA

FTA is a technique that 'works backwards' from the answer, i.e. the assumption that a failure has occurred. The idea is to identify the chain (or *tree*) of events that could lead to the failure, and suggest intervention activities (extra inspection, process monitoring, or whatever), to prevent it occurring. The analysis involves producing formal logic diagrams in either tabular or graphical format. The power of FTA is its ability to identify secondary or contributory causes to failure, rather than a single simplified cause. This fits well with the real plant situation, where failures are nearly always caused by a combination of events rather than a single reason.

ETA

ETA is essentially FTA in reverse. It uses the same type of logic tables or diagrams but 'works forward' from the assumption that some *initiating* event has occurred. The objective is then to decide what series of events will follow. As with FTA, the analysis generally considers events that happen in quick sequence over a short, perhaps almost instantaneous, timescale. Again, this is representative of how most failures actually happen.

All the above techniques have their advantages and disadvantages in identifying and defining accident scenarios. A lot depends on the type, complexity, and age of the plant and how well its failure mechanisms are understood. The main practical restriction on their use as part of an RBI programme is *cost* – a full analysis of possible accident scenarios using any combination of these techniques needs large amounts of resources. Several man-months input is not unusual for an analysis of a few straightforward steam, process, or piping systems. A full analysis of an entire operating plant would be uneconomical to consider, except for some safety-critical plants such as those including highly toxic or nuclear materials. There are also problems of the knowledge level of the personnel involved – specialists at quantitative techniques such as HAZOP, FMEA, etc. sometimes lack real plant operating experience, being more used to relying on simulations and software models that they understand. This often causes 'mismatch' between members of the RBI team leading to big meetings and long discussions, increasing the cost. In extreme cases, the RBI team will fragment into several smaller ones, each following its own path, until eventually the whole programme drifts away and dies. This is the danger.

Defining damage mechanisms

Damage mechanisms (DMs), sometimes called *deterioration mechanisms*, are a feature of all RBI programmes. They comprise the various metallurgical mechanisms that cause material or components to deteriorate in a way that affects their fitness for purpose, i.e. corrosion, fatigue, creep, age-hardening, and others. DMs are therefore a 100 percent technical phenomenon – they are not involved with statistics or probability distributions. In a plant-operating organization, plant operators, design engineers, metallurgists, and corrosion engineers are all involved, in some way, with the consideration of DMs and their effect on the plant. Figure 7.19 gives a summary.

DMs – levels of definition
There are several different 'levels of definition' of DMs. They can all be used as valid input to the RBI process. The levels of definition are highly industry specific, and reflect the type of plant systems and equipment commonly seen in that industry and the content of published technical codes and documents relevant to the RBI process. Figure 7.20 shows the situation in simplified form.

In practice, most RBI programmes do not use a full Level 3 analysis, owing to the almost unrealistic manpower requirements necessary to do the job properly. External RBI consultants may claim to define DMs to this high level of resolution but, frankly, it is rare that they have sufficient real plant-specific knowledge to do so. Most use a traditional generic Level 2-based analysis, supplemented with whatever site-specific knowledge they can extract from plant operators, metallurgists, and corrosion engineers from the operating site.

Once the damage mechanisms are defined, using Levels 1, 2, or 3 as appropriate, they are then tabulated to sit alongside the accident scenarios to form the input to the risk-based analysis (look back at Fig. 7.16). This forms the end of the 'real world' deterministic part of an RBI analysis – the following stages are *probabilistic* – based on the concept of probability. There are several ways of expressing probabilities, either by statistical numbers in a table, or in a graphical diagram of some type. The most common is the use of base data held in the form of a table, then expressed in a graphical form (to make it more visual and easier to understand) as a *matrix*.

The risk matrix

The RBI risk matrix is used to display the results of the process of *risk ranking and categorization*. There are no fixed rules as to how large the matrix should be – practice varies from 3 × 3 to 6 × 6, depending on the depth and complexity of the RBI analysis being performed. Figure 7.9 showed the basic format of the risk matrix: the axes represent *consequence of failure* and *probability or 'likelihood' of failure*. The exact format will vary depending on whether the analysis is qualitative or quantitative, as described earlier.

For qualitative methods, the likelihood axis uses descriptive terms such as very unlikely, likely, probable, etc. coupled with a 'scoring'

- **Don't get confused by terminology**: you will hear damage mechanisms being referred to as:

  - failure modes or mechanisms;
  - deterioration modes or mechanisms;
  - modes of failure;
  - root causes of failure.

  Think of them as different ways of describing the same thing – it makes things easier.

- **Remember that DMs differ significantly** between types of plant – DMs for refinery, petrochemical, power boilers, offshore installations, etc. are different. Published documents such as API 580/581 contain lists of DMs but they are mainly applicable to downstream petrochemical plant.
- You don't *need* to be a metallurgist or corrosion engineer to discuss DMs – plant operators, welding/NDT engineers, etc. all have useful knowledge as well. Some DMs are well understood by metallurgists and some by corrosion engineers, while others are not very well understood at all.

Fig. 7.19 Damage mechanisms (DMs) – some guidance points

system. For quantitative schemes these are expressed as a probabilistic number, such as 6×10^5 per year. Remember two important facts:

- For both qualitative and quantitative RBI schemes, the scales of the matrix axes are only *indicative*, they do not carry any absolute message.
- The robustness of the matrix varies within the matrix itself. As a rule of thumb, the 'high-likelihood' end can be expected to be more representative of reality than the low-likelihood end. This is because the low-likelihood end is based on fewer data about rare occurrences, so there is a greater degree of guesswork involved.

In theory, each separate piece of equipment that lies within the system boundaries of the RBI programme should be allocated its own position in the risk matrix. Practically, the complexity of such an approach would make the programme unmanageable, so the solution is normally to amalgamate plant systems and equipment items into *groupings*. For petrochemical, refinery, offshore installations, and similar, where

- **Level 1 definition** is the use of *broad categories* of damage mechanism, without looking into too much detail. Typically, these are:
  - corrosion of all types;
  - high-temperature effects (mainly creep);
  - fatigue;
  - stress corrosion cracking (SCC);
  - embrittlement;
  - brittle fracture;
  - hydrogen damage.

Note how this categorization is very crude, each category being a convenient amalgamation of many different submechanisms.

- **Level 2 definition** uses a more complex breakdown of DM categories as given in published documents and codes. The main one is API 580/581 – this defines specific DMs in some detail, specifically relevant to the downstream air and petrochemical industry.

  Similar, but less detailed breakdown of DM categories given in published documents and codes are:
  - API 510 gives a (fairly broad) list of DMs expected in unfired pressure vessels (see Chapter 11).
  - API 570 gives DMs expected in pipework systems (see Chapter 13).
  - API 653 gives DMs expected in atmospheric storage tanks (see Chapter 14).
  - Other DMs are identified in API BP 572, 576, etc. and other API IRE guides. These tend to be in less well-structured form.

Level 2 definitions of DMs use more conventional wisdom than in Level 1, but are nevertheless still *generic*, rather than plant specific.

- **Level 3 definition** uses DMs derived exclusively from close experience of the *specific* plant being studied. It therefore forms the most robust basis for an RBI analysis and produces the most representative results. It requires huge resource input to investigate, rationalize, and document each individual damage mechanism.

Fig. 7.20   Damage mechanisms – 'levels of definition'

corrosion is the predominant damage mechanism, these groupings are often known as *corrosion loops*.

Corrosion loops

Corrosion loops are sections or systems of a plant that are subject to similar corrosive conditions and experience similar types and severity of corrosion damage mechanisms. Corrosion loops typically:

- *Consist of similar materials*: base material, piping and trim class, coating, etc.
- *Are exposed to similar corrosive conditions.*
- *Have similar operating conditions*: process fluid, percentage solids, and operating transients within the same 'step categories'.

If the above conditions remain true, then all equipment items within a corrosion loop can be considered as having the same criticality and so be allocated a single, collective position in the RBI matrix. Figure 7.10 shows the idea.

Risk categorization and mitigation

Once the RBI matrix is compiled, containing all the necessary plant items and 'loops', two steps follow: *categorizing risks* and then *deciding mitigation activities*.

Categorizing risks

Figures 7.9 and 7.21 give an example of how risks are categorized (i.e. apportioned into categories) after the matrix is completed. The matrix is divided into regions (categories) covering different ranges of risks; this is often done by colouring the categories red (for highest risk), blue, yellow, green, etc. to make it visually easier to understand. Again, there are no fixed rules on how many categories to use – most RBI programmes use either three or four, depending on the type and age of plant, and its previous inspection/operating history. The risk categories are allocated names or numbers (e.g. negligible risk, unacceptably high risk, low, R0, R1, etc.) reflecting the level of risk that they represent.

Deciding mitigation activity

The final stage is to make decisions on what actions are necessary to reduce (or *mitigate*) those risks that are deemed too high to be acceptable as they are. This is the core output activity of an RBI programme – to decide *how* to reduce overall risk exposure by making tangible changes to inspection scope, methods, and periodicity for

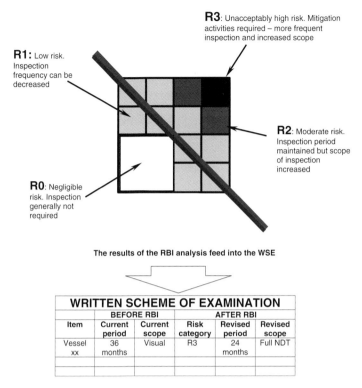

R3: Unacceptably high risk. Mitigation activities required – more frequent inspection and increased scope

R1: Low risk. Inspection frequency can be decreased

R2: Moderate risk. Inspection period maintained but scope of inspection increased

R0: Negligible risk. Inspection generally not required

The results of the RBI analysis feed into the WSE

WRITTEN SCHEME OF EXAMINATION					
	BEFORE RBI		AFTER RBI		
Item	Current period	Current scope	Risk category	Revised period	Revised scope
Vessel xx	36 months	Visual	R3	24 months	Full NDT

Fig. 7.21   Risk categorization, mitigation, and output into the WSE

individual items of equipment. This is the point at which the overall objectives and approach of the plant's management holds major influence; the approach taken to mitigation actions will reflect their attitude to risk. Many factors come into play, management's level of understanding of technical issues, corporate policy, history of previous accidents and failures, and their approach to risk and safety issues in general.

Revisions to the written scheme of examination (WSE)
The output from an RBI programme, if it is done properly, appears as changes to the WSE. Feasible changes to the WSE include:

- Increasing or decreasing inspection *periodicity*.
- Increasing or decreasing the *scope of inspection* (i.e. which systems and plant items actually appear in the WSE, irrespective of whether they are covered by statutory regulations or not).
- *Reallocation of priorities* between cold shutdown and on-stream inspections.

- Specifying more *non-intrusive techniques* such as thermography, corrosion mapping, and long-range UT techniques such as pulse-echo.

Figure 7.21 shows a simplified example of changes made to a WSE as a result of conclusions drawn from an RBI programme.

Changes to a plant WSE rarely happen without resistance from one or more of the involved partners employed by the duty holder (i.e. the plant management). Operations departments, maintenance department, corrosion engineers, metallurgists and the Competent Person organization are often reluctant to agree changes to a WSE, particularly if it involves a lessening of their role in some way. Plant managers can be difficult to convince that additional extensive inspection techniques are really necessary, when their need is based totally on the results of an RBI analysis based on subjective risk assessments. Third-party bodies also, despite their apparent approval of RBI analyses, are often reluctant to formally approve decreases in the frequency of inspections, particularly if these lie outside the scope of their past experience. It is an undeniable characteristic of third-party organizations (see Chapter 2) that they are highly risk averse and feel much more comfortable with previous 'proven' inspection periodicities than venturing into new, possibly risky territory.

The end result is that RBI-initiated changes to a plant WSE are often made 'bit-by-bit', rather than resulting in a totally step-change revision, however polished and convincing the results of an RBI programme may be. RBI programme recommendations frequently remain unimplemented (for no single reason in particular), awaiting the confirmatory results of some future review which never happens. This tends to reduce the validity of the concept of RBI as a lifetime tool for managing plant inspections. Once the process of implementing changes to the WSE is interrupted, the effectiveness of the whole RBI process falls rapidly. If the interruptions are *considered* (as they might be) the effectiveness falls rapidly to zero, and the whole thing crumbles, leaving only the arguments about apportionment of blame and whose fault it was. Unfortunately, this is a far from uncommon occurrence. RBI needs *continuity* if it is to work.

RBI standards and published documents

There are no *mandatory* 'codes' covering the mechanics or implementation of an RBI programme. The scope and inherent variety of the subject is simply too wide for agreement of a formal code and even if one were to be agreed within one sector of industry it would almost certainly be limited in application to the sector that produced it and have little relevance to plants in other industry sectors.

There are, however, several notable published documents which attempt to rationalize and explain the RBI process. Some of these are comprehensive and go into a lot of technical detail about the subject. They also contain detailed quantitative information that is actually *useful* in performing an RBI analysis. Others are more general and act as little more than a generic, sometimes confusing, introduction to the subject. All contain glossaries of the specialist terms and acronyms that abound in RBI, and lots of mathematical formulae that are difficult to relate to the day-to-day practicalities of inspecting and operating plant.

API 581 Risk-based inspection – Base resource document

This is part of a set of documents (API 581 and API 580) that claim to be a 'matching pair'. API 581 acts as a *Base Resource Document* while the more recent API RP 580 is the *Recommended Practice* for its implementation. In practice, their content overlaps. Perhaps the best description is that API 580 is a much simplified version of the essential principles and content of API 581.

API 581 consists of more than 500 pages of text, diagrams, worksheets, and tables of data. Figure 7.22 gives a summary. Its content is specifically orientated towards the downstream oil and petrochemical industry. It has little engineering relevance to other industries such as steam, power, or general process plant. The principles of the RBI analysis are similar but the mechanics of the process such as damage mechanisms, accident consequences, and numerical data are different. This does not prevent some industries from quoting API 581 as part of their approach to RBI, or adapting bits of it for use as necessary.

Some of the most useful parts of API 581 are the so-called 'technical modules' in the appendices to the document. These specify risk assessment methods (using questions and flowcharts) for common damage mechanisms such as thinning, SCC, H_2 attack, fatigue, brittle fracture, and external corrosion/corrosion under insulation (CUI). Table 7.1 outlines the most useful content. A full listing of DMs is given in Appendix A7 of this book.

ASME document CRTD-VOL 20–1

This was developed by the ASME Centre for Research and Technology Development (CRTD) from a 1991 document describing the general

---

API 581

- Is specific to the downstream oil and petrochemical industry.
- Uses the methodology of a **qualitative** analysis to identify critical plant items followed by a more detailed **quantitative** analysis.
- Proposes the use of a 5 × 5 matrix.
- Defines the **likelihood of failure** using six basic considerations:

  1. The number of equipment items in a plant or system.
  2. Expected damage mechanisms.
  3. The effectiveness of the current inspection regime.
  4. Condition of the equipment.
  5. Type of process.
  6. General safety considerations.

     Defines the **consequences of failure** using three basic considerations.

  1. The effect of fire.
  2. The effect of explosions.
  3. The effect of toxic releases.

- Suggests specific data values to use in various situations. These cross-reference standard reliability data sources.
- Contains numerous 'workbooks' that *could* be used to work through an RBI analysis for specific types of refinery and petrochemical plant.
- Can be dazzlingly confusing if you are not already very familiar with RBI techniques.

Fig. 7.22   API 581 Risk-based inspection – base resource document

principles of RBI. It follows similar principles to API 581 but is a little more theoretical, with increased emphasis on the mathematical aspects of RBI techniques.

Health and Safety Executive Research Report 363/2001

This is a 150-page document *Best practice for RBI as part of plant integrity management*. It provides a wide-ranging survey of RBI techniques and the mechanism of how an RBI programme works. It is easy to understand and does not rely on extensive jargon or acronyms. It is useful in relating RBI to specific in-service inspection legislation relating to pressure equipment, such as the UK PSSRs and

Table 7.1   API 581 – RBI Base resource document – some useful (plant-specific) content

Section	Subject	It is useful for:
10	Plant database structure	Useful as a guide to database structure for all plant types
App G	Thinning: technical modules	Components (tubes etc.) subject to pitting, erosion, or general electrochemical corrosion (mainly petrochemical industry applications)
App H	Stress corrosion cracking (SCC): technical module	Any plant subject to SCC/caustic/amine/hydrogen or chloride cracking of stainless steel or high-Ni-alloy components. Petrochemical, food, brewing, offshore industries, etc.
App I	High-temperature H_2 attack: technical module	Dry plant with carbon/low-alloy steels operating in hydrogen-rich environments
App J	Furnace tubes: technical module	Fired boilers (power/steam generation) and externally fired heaters (refinery applications)
App K	Mechanical fatigue of piping: technical module	Any pipework subject to fatigue loadings from flow effects, wind loadings, or imposed vibration from rotating equipment
App L	Brittle fracture: technical module	Low-temperature storage tanks, vessels, pipework systems or steel structures plus $1\frac{1}{4}$–$2\frac{1}{4}$% Cr materials subject to temper embrittlement
App N	External linings: technical module	Chemical, offshore plant, etc. subject to external corrosion or corrosion under insulation (CUI) (temperatures $<120\,°C$)

API 510. It also places RBI in context with fitness-for-service codes such as API 579 and BS 7910.

Other published documents about RBI tend to be either dated or highly mathematical. While of interest as theoretical background to the subject, there is little of practical use to the in-service inspector. In contrast, many power, petrochemical, and offshore plant owners have produced their own detailed in-house documents covering RBI programmes employed on their plants. These are highly relevant but are not generally available in published form.

THINKING OUTSIDE THE BOX –. 1 RBI AS AN 'INTEGRATED MANAGEMENT TOOL'

For		*Against*
• RBI certainly needs integration to work; plant operations, maintenance, corrosion engineers, as well as in-service inspectors and the 'Competent Person' all need to get involved.	INTEGRATED? ⟶	• Sorry, but an RBI programme does not bring integration with it. There is no free integration hiding at the bottom of the RBI box.
• An RBI programme is approved (and paid for) by plant management. Everybody chants the mantra that *without 'management buy-in' an RBI programme won't suceed.*	MANAGEMENT? ⟶	Do managers collect the data from P&IDs, fill in the spreadsheets or software fields, investigate the damage mechanisms, decide the risk categories then correct the written scheme of examination?
• Tools help change things for the better. A spreadsheet is a tool and computers are tools.	TOOL? ⟶	Plants are disassembled and inspected using spanners, hammers and wedges, scaffolding, chain blocks and huge multi-wheeled mobile cranes. These can not be tools because they do not have a keyboard and are not made out of paper.

Fig. 7.23   Thinking outside the box – 1. RBI as an 'integrated management tool'

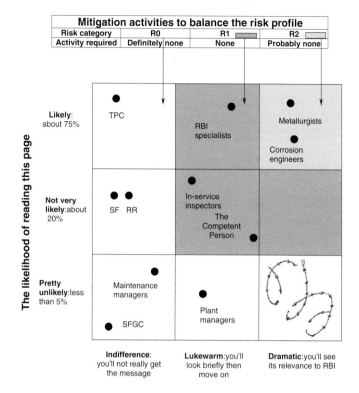

The consequence of reading this page

Fig. 7.24   Thinking outside the box (TOTB-2) – The 3 × 3 risk matrix

Chapter 8

Failure

In-service inspection is about the *avoidance of failure* in engineering components. The study of failure is a wide and complex subject ranging from the simple visual observation of corrosion and defects through to the complex numerical methods of fitness-for-purpose (FFP) studies to BS 7910 and API 579 (see Chapter 9). However robust the theories used to try to predict when an engineering component is going to fail, there is still a high level of uncertainty involved in the process. One reason for this is that there are many types of failure, and numerous subdivisions within each type, resulting in a wide variety of recognized ways in which even a simple engineering component can fail. These are called failure *mechanisms*.

Failure mechanisms

These are broadly divided into the categories shown in Fig. 8.1. They can act alone or, more commonly, in combination with each other. The combination may be simultaneous, i.e. both failure mechanisms working together, or fully or partially sequential, in which the failure mechanisms follow each other, in time, contributing eventually to the final failure of the component. Note how the failure categories shown in Fig. 8.1 are vastly different, in terms of their mechanisms, from each other. Despite their differences however, there are basic similarities based on the way materials actually fail.

Failure statistics

Various studies have been performed to collect data about the way equipment (mainly pressure vessels) fail and why they do it. The resulting statistics are interesting but of course are suitable for use as guidance only. Tables 8.1 to 8.3 show some recent data collected from a study extending over several years.

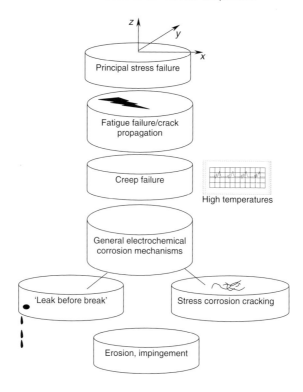

Fig. 8.1   Failure mechanisms

Table 8.1   Some typical failure rate statistics

	Failure rate (potential)		Failure rate (catastrophic)	
Vessel type	Average (per year)	95% CL* (per year)	Average (per year)	95% CL* (per year)
Boilers/steam receivers	1.5×10^{-4}	1.9×10^{-4}	2.2×10^{-5}	4.4×10^{-5}
All vessels	4.3×10^{-5}	5.2×10^{-5}	7.8×10^{-6}	1.2×10^{-5}

*CL, Confidence limit

Table 8.2   Average vessel population figures over 5-year period of survey

Vessel type	Number of vessels	%
Boilers	28 800	8.0
Steam receivers	24 840	6.9
Air receivers	260 280	72.3
Other pressure vessels	46 080	12.8
Total	360 000	100.0

Table 8.3   Failure classification system

Failure type	Definition
Disruptive	'Catastrophic' failure resulting in forcible release of contents
Deformation	'Catastrophic' failure resulting in gross distortion of vessel, with or without release of contents
Leakage	'Potential' failure resulting in release of small amount of contents
Inspection	'Potential' failure resulting in no release of contents

Note: 'Catastrophic' and 'Potential' failures as defined in test.

How materials fail

There is no single, universally accepted explanation covering the way that metallic materials fail. The study of pressure equipment has, however, led to several generally accepted theories. Figure 8.2 shows the generally accepted phases of failure. Elastic behaviour, up to yield point, is followed by increasing amounts of irreversible plastic flow. The fracture of the material starts from the point in time at which a crack initiation occurs and continues during the propagation phase until the material breaks or 'ruptures' (Fig. 8.3).

There are several approaches to both the characteristics of the original material and the way that the material behaves at a crack tip (see Fig. 8.4). Two of the more common ones applicable to pressure equipment and similar applications are:

- The linear elastic fracture mechanics (LEFM) approach with its related concept of the fracture toughness parameter, (K_{1c}) (a material property).
- Fully plastic behaviour at the crack tip, i.e. 'plastic collapse' approach.

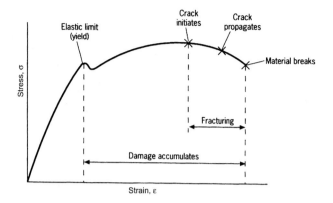

Fig. 8.2   Phases of failure

A useful standard is ASTM E399.

LEFM method
This is based on the 'fast fracture' equation:

$$K_{1c} = K_l = YS\sqrt{\pi a}$$

where

K_{1c} = plane strain fracture toughness

K_1 = stress intensity factor

a = crack length

Y = dimensionless factor based on geometry

S = stress level

Typical Y values used are shown in Fig. 8.5.

Multi-axis stress states
When stress is not uniaxial (as in many pressure-equipment components), yielding is governed by a combination of various stress
components acting together. There are several different 'approaches'
as to how this happens:

- *Von Mises criterion* (or 'distortion energy' theory). This states that
  yielding, S_y, will take place when

$$\tfrac{1}{2}^{1/2}[(S_1 - S_2)^2 + (S_2 - S_3)^2 + (S_3 - S_1)^2]^{1/2} = \pm S_y$$

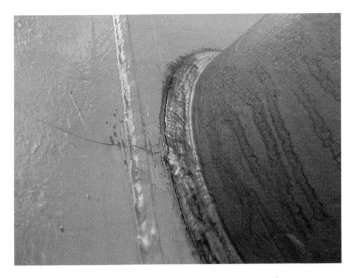

Fig. 8.3   A typical propagating crack

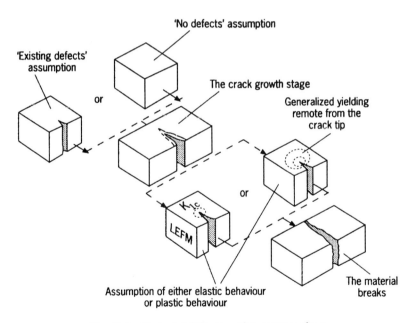

Fig. 8.4   Material failure – the approaches

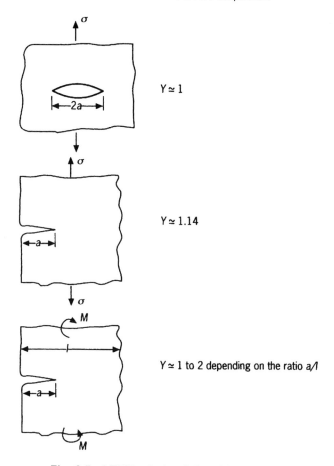

Fig. 8.5   LEFM – 'rule of thumb' Y factors

where S_1, S_2, S_3 are the principal stresses at a point in a component. It is a useful theory for ductile metals.

- *Tresca criterion* (or maximum shear stress theory).

$$\frac{(S_1 - S_2)}{2} \text{ or } \frac{(S_2 - S_3)}{2} \text{ or } \frac{(S_3 - S_1)}{2} = \pm \frac{S_y}{2}$$

This is also a useful theory for ductile materials.

- *Maximum principal stress theory*. This is a simpler theory which is a useful approximation for brittle metals. The material fails when

$$S_1 \text{ or } S_2 \text{ or } S_3 = \pm S_y$$

Principal stress failure

The basic failure assumptions outlined above are, simplistically, related to principal stress failures, i.e. failures resulting from excessive principal stress in a component. Principal stresses are assumed to be:

- Stresses at mutually perpendicular directions, i.e. the x-, y-, and z-axes: σ_x, σ_y, σ_z.
- Stresses that exist in the 'body' of a component, i.e. not involving stress concentrations.
- Stresses that are calculated using basic mechanical design and code calculations, e.g. for circumferential (hoop), axial, and radial stresses in cylinders.
- Stresses that arise from normal operation of a component i.e. not caused by upset conditions such as water hammer, excessive restraint, mechanical damage etc.

Design codes for pressure equipment and similar static mechanical components are constructed around the concept of principal stresses. The basic scantlings for vessel and pipework wall thicknesses are derived initially from principal stress consideration, with additional factors of safety added to cater for other possible failure mechanisms. In addition, those design codes that allow re-rating of pressure equipment on the basis of reduced wall thickness (due to corrosion) do so under the assumption that the main basis of the calculation is to ensure that the remaining wall thickness is adequate to resist principal stresses.

Failure – the reality

In reality, few mechanical failures are caused by principal stresses alone. Principal stresses eventually play a part in the *later stages* of failure but the initiation and propagation of the failure mechanism is more often caused by another mechanism. Typical contributing mechanisms are fatigue, creep, crack propagation, or any of many possible simple or composite corrosion mechanisms. For equipment operating at temperatures below about 400 °C, one of the most common failure mechanisms is fatigue.

Fatigue

Ductile materials can fail at stresses significantly less than their rated yield strength if they are subject to fatigue loadings. Fatigue data are displayed graphically on an *S–N* curve (see Fig. 8.6). Some pressure equipment materials exhibit a 'fatigue limit', representing the stress at which the material can be subjected to (in theory) an infinite number of

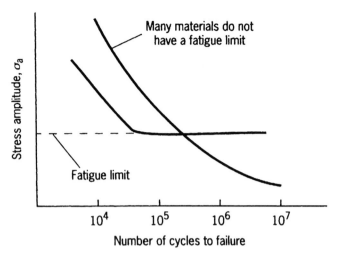

Fig. 8.6   A typical fatigue 'S-N' curve

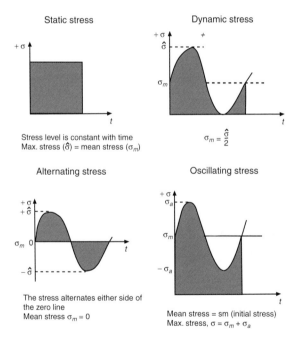

Fig. 8.7   Types of stress loading

Table 8.4 Fatigue strength (rules of thumb only)

	Bending			Tension			Torsion		
	$S_{w(b)}$	$S_{a(b)}$	$S_{y(b)}$	S_w	S_a	$S_{sw(t)}$	$S_{sa(t)}$	$S_{sy(t)}$
Steel (structural)	$0.5S_m$	$0.75S_m$	$1.5S_y$	$0.45S_m$	$0.59S_m$	$0.35S_m$	$0.38S_m$	$0.7S_m$
Steel (hardened and tempered)	$0.45S_m$	$0.77S_m$	$1.4S_y$	$0.4S_m$	$0.69S_m$	$0.3S_m$	$0.5S_m$	$0.7S_y$
Cast iron	$0.38S_m$	$0.68S_m$	—	$0.25S_m$	$0.4S_m$	$0.35S_m$	$0.56S_m$	—

$S_{w(b)}$ = fatigue strength under alternating stress (bending).
$S_{a(b)}$ = fatigue strength under fluctuating stress (bending).
$S_{y(b)}$ = yield point (bending).
S_w = fatigue strength under alternating stress (tension).
S_a = fatigue strength under fluctuating stress (tension).
S_y = yield point (tension).
$S_{sw(t)}$ = fatigue strength under alternating stress (torsion).
$S_{sa(t)}$ = fatigue strength under fluctuating stress (torsion).
$S_{sy(t)}$ = yield point (torsion).
S_m = ultimate stress value.

cycles without exhibiting any fatigue effects. This fatigue limit is influenced by the size and surface finish of the specimen, as well as the material's properties. Figure 8.7 shows types of stress loading.
   Characteristics of fatigue failures are:

• Visible crack-arrest and 'beach mark' lines on the fracture face.
• Striations (visible under magnification) – these are the result of deformation during individual stress cycles.
• An initiation point such as a crack, defect, or inclusion, normally on the surface of the material.

Fatigue strength – rules of thumb

The fatigue strength of a material varies significantly with the size and shape of section and the type of fatigue stresses to which it is subjected. Some 'rules of thumb' for pressure equipment materials are shown in Table 8.4. Note how they relate to yield, S_y, and ultimate, S_m, values in pure tension. European equivalent (SI) units are shown in Table 8.5.

Creep

Creep is one of the more complex degradation mechanisms found in pressure equipment. It is a high-temperature mechanism, rarely occurring in steels below about 390 °C. It *is* a specialist subject, but there are some straightforward guidelines (see below).

The Handbook of In-Service Inspection

Table 8.5   Strength units – European equivalents

	Yield strength	Ultimate tensile strength	Modulus
USCS (US imperial)	F_{ty} (ksi) or S_y (lb/in^2)	F_{tu} (ksi) or S_m (lb/in^2)	E (lb/in^2 × 10^6)
SI/European	R_e (MN/m^2)	R_m (MN/m^2)	E (N/m^2 × 10^9)

Conversions are: 1 ksi = 1000 lb/in^2 = 6.89 MPa = 6.89 MN/m^2 = 6.89 N/mm^2.
(ksi = kilopounds per square inch)

What is creep?

Creep is a degradation mechanism found in steels and other pressure-equipment materials.

What is it dependent on?

Creep is dependent on two parameters: *temperature* and *time*. For steels, creep rarely occurs below about 390 °C and can get progressively worse as the temperature increases above this. Creep damage also gets worse (more or less linearly) with the time that material is exposed to these high temperatures. Other factors such as corrosive atmospheres and fatigue conditions can act to make the situation worse.

What damage does creep cause?

The main effect is a permanent and progressively worsening reduction in tensile strength. The creep mechanism causes the metal structure to 'flow' (on a macro/microscopic scale) leaving holes or *voids* in the material matrix. Structures under pressure stress can therefore deform, and then fail.

How is creep detected during in-service inspections?

It is detected using metallographic replication (often simply termed a 'replica'). This is a non-destructive technique that enables a visual examination of the material's grain structure to look for the voids that are evidence of creep damage. The procedure comprises (see Chapter 15 for further details):

- Prepare the metal surface using mechanical cleaning/grinding and a chemical etch.
- Bond a special plastic tape to the surface and allow the adhesive to cure.
- Remove the tape, transferring a replica of the metal surface to the tape.

• Examine the replica under low magnification and compare with reference microstructure pictures.

Note that the results of replica tests are assessed in a *qualitative* way, comparing the number and size of voids with those on reference pictures for the material under consideration. If a material fails this test, the next step is to perform full (destructive) tests for tensile strength on a trepanned sample taken from the component. Although ultrasonic testing is often used to try to detect creep, it is only effective if the creep has progressed far enough to result in a significant crack size (above about 0.5 mm).

Where is creep probable in pressure equipment?

Common areas are:

• *Boilers*

  – Traditional boilers and heat recovery steam generators (HRSGs): superheater and reheater headers (see Chapter 15).
  – High-pressure boilers: platen superheater tube banks.

• *Valves*. Valve bodies and seats on high-temperature systems.
• *Turbine casings*. Stator blade casing diaphragms at the hot end of the casings.

What is a typical creep-resistant alloy?

Main steam pipelines for fossil or gas-fired power stations containing superheated steam at 520 °C + commonly use a DIN 14MoV63 alloy (or a comparable 0.5% Cr – 0.5% Mo – 0.25% V material). This has documented creep-resistant properties up to 100 000 h of exposure. High-chromium materials such as SA 335-P91 (see Chapter 15) are used as creep-resistant alloys in HRSGs. Figure 8.8 shows a glossary of creep-related terms.

Corrosion

Much of the job of the in-service inspector concerns the inspection of components that are corroded. Corrosion is perhaps the most common degradation mechanism to affect engineering materials and is responsible for more cases of retiral of equipment than all the other degradation mechanisms put together.

Creep

Creep is defined as deformation that occurs over a period of time when a material is subjected to constant stress at constant temperature. In metals, creep usually occurs only at elevated temperatures. Creep at room temperature is more common in plastic materials and is called cold flow or deformation under load. Data obtained in a creep test are usually presented as a plot of creep versus time with stress and temperature constant. The slope of the curve is creep rate and the end point of the curve is time for rupture. The creep of a material can be divided into three stages. The first stage, or primary creep, starts at a rapid rate and slows with time. The second stage (secondary) creep has a relatively uniform rate. The third stage (tertiary) creep has an accelerating creep rate and terminates by failure of material at time for rupture.

Creep limit

An alternative term for creep strength.

Creep rate

Creep rate is defined as the rate of deformation of a material subject to stress at a constant temperature. It is the slope of the creep versus time diagram obtained in a creep test. Units are mm/h or percent elongation/h. Minimum creep rate is the slope of the portion of the creep versus time diagram corresponding to secondary creep.

Creep recovery

Creep recovery is the rate of decrease in deformation that occurs when load is removed after prolonged application in a creep test. Constant temperature is maintained to eliminate effects of thermal expansion, and measurements are taken from a time load of zero to eliminate elastic effects.

Creep rupture strength

Stress required to cause fracture in a creep test within a specified time. An alternative term is *stress rupture strength*.

Creep strength

Maximum stress required to cause a specified amount of creep in a specified time. The term is also used to describe the maximum stress that can be generated in a material at constant temperature under which the creep rate decreases with time. An alternative term is *creep limit*.

Creep test

This is a method for determining creep or stress relaxation behaviour. To determine creep properties, material is subjected to prolonged constant tension or compression loading at constant temperature. Deformation is recorded at specified time intervals and a creep versus time diagram is plotted. The slope of the curve at any point is the creep rate. If failure occurs, it terminates the test and time for rupture is recorded. If the specimen does not fracture within a test period, creep recovery may be measured. To determine the *stress relaxation* of a material, a specimen is deformed a given amount and the decrease in stress over a prolonged period of exposure at constant temperature is recorded.

Fig. 8.8   Glossary of creep-related terms

Does corrosion cause failure?

Corrosion *does* cause failure, but not nearly as often as you might think. There are several reasons for this:

- *Leak before break*. Corrosion more often causes *leaks*, than it does catastrophic failure. This used in some plant to form the basis of a strategy of corrosion management.
- *Most corrosion is detectable*. Detection may need some inspection effort. Corrosion under insulation (CUI) and internal corrosion of pipework and vessels can remain hidden but, once looked for properly, are easily detectable by simple visual inspection.
- *Predictability*. Corrosion is, in general, well understood, so its occurrence and rate of corrosion can be predicted, within limits. The degree of predictability varies a lot between industries and process applications but it is still easier to predict than, for example, the existence of weld defects or fatigue damage.

Although corrosion is a pivotal issue in in-service inspection, it is far from the full story. In many cases, its existence poses little threat to FFP or integrity of the plant. Conversely, there are instances in which corrosion is an important FFP aspect, and some industries (e.g. the offshore oil/industry) where corrosion is the main driving force behind the in-service inspection process. Figure 8.9 shows the overall incidence (and hence the degree of risk) of corrosion in various industries. Treat this figure as a guideline only, but it is not too far from the truth. Against this background, one of the difficulties of the in-service inspection task is to report corrosion in a consistent and meaningful way.

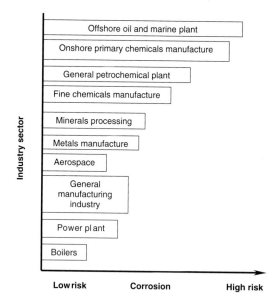

Fig. 8.9   The incidence (and risk) of corrosion in various industries

Inspection versus corrosion management

Corrosion management is a generic term given to a wide variety of different processes related to the discovery and prevention of corrosion in industrial plants. Unlike simpler engineering disciplines, it comprises a combination of three main disciplines: corrosion technology,

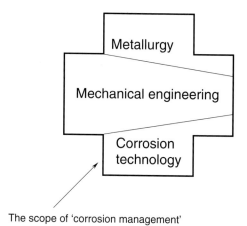

The scope of 'corrosion management'

Fig. 8.10   Inspection corrosion – the mix of knowledge involved in corrosion management

metallurgy, and mechanical engineering. Figure 8.10 shows the situation.

Do not confuse in-service inspection with corrosion management. Corrosion management *involves* in-service inspection, but it is only a fairly small part of the overall process. Corrosion management, as a process, is the remit of the plant owner or user, and is rarely delegated to independent inspectors. In addition to statutory inspection, under, for example the PSSRs (see Chapter 5), a good corrosion management programme includes the use of laboratory or pilot-plant tests, in-plant corrosion coupons, and advanced NDT procedures such as corrosion mapping.

Corrosion management versus RBI

Corrosion management is a subset of RBI rather than a replacement for it. Although the stated objective of most corrosion management programmes is to deal with all types of corrosion, they often ignore some of the more complex types that are closely linked with other types of degradation mechanism. Mechanisms such as corrosion fatigue and 'residual stress' corrosion cracking are examples of composite, complex degradation mechanisms that belong more to the wider scope of an RBI programme than a straightforward corrosion management approach.

Inspecting corrosion

Inspecting and reporting on in-service corrosion is not a simple process. There are few published technical standards that address the issue in any great depth and most procedures and acceptance criteria that do exist are highly company specific or even plant specific. The inspection task itself is not just linked to finding the corrosion – once found it has to be classified, and conclusions drawn on its criticality, and then a decision made on the way forward.

Finding corrosion

Most corrosion is found visually. The main limitation is that labour-intensive work (e.g. erection of scaffolding, removal of lagging, etc.) is often necessary to enable a proper visual inspection to be made. More advanced techniques that are used to detect corrosion tend to be highly application specific (see also Chapter 6) i.e.:

- UT sample wall thickness measurement using a normal 0° compression probe.

- Corrosion mapping – a UT scanning exercise which records 100 percent of the scanned area and provides a colour representation of wall-thinned areas caused by corrosion.
- UT using angle probes – useful for finding corrosion in the weld root area.
- Time of flight diffraction (TOFD) – also useful for finding weld root corrosion.
- Using a simple hammer or rapping test – an unsophisticated method that has limited use for high integrity and pressure plant.
- Indirect methods such as:

  - thermography – identifies areas of wet lagging where corrosion is more likely to be found;
  - corrosion products – finding corrosion product somewhere in a system or around areas suffering crevice corrosion, e.g. flange faces;
  - monitoring corrosion currents – in a system (pipeline etc.) fitted with cathodic protection.

Practically, the biggest limitation on finding corrosion in plants is self-inflicted, i.e. not removing sufficient lagging. Some industries, even those with a long history of corrosion under insulation (CUI), are reticent to remove enough lagging to enable an adequate visual inspection for corrosion – even during full plant cold shutdowns. Published standards such as API 570 *In-service inspection of pipework* rarely make full lagging removal mandatory. Most mention only areas of obviously damaged cladding, or leave the decision to the absolute discretion of the inspector. For statutory plant, the content of the written scheme of examination (WSE) is used as the guideline, but this again sometimes does not specify in detail exactly the extent of lagging removal. Figure 8.11 shows some broad guidelines to follow when looking for areas to inspect for CUI.

Classifying corrosion

Classifying means measuring the extent of corrosion, rather than deciding in chemical/metallurgical detail exactly what type it is. There are three main aspects to the assessment:

- *Extent.* How widespread is it, and how much of the component does it cover?
- *Depth.* How deep is it? Both as an absolute measurement and as a percentage of remaining wall thickness.

- Weathered mastic moisture barriers above supports, or lifting lugs on piping and vessel heads.
- Missing caulking at seams and connections.
- Split metal cladding sheets or missing screws.
- Areas around steam leaks.
- Areas of staining (red, brown, or white) of cladding sheets.
- Areas around unprotected lagging, where parts have been removed.
- Gaps around pipe hangers and lifting lugs.
- Gaps in cladding jackets at the top of vertical pipe runs.
- Any areas where the lagging is obviously wet.
- Plain carbon steel components operating at:
  - steady temperatures $\leq$ about 120 °C;
  - varying temperatures $\leq$ about 150 °C.

Fig. 8.11   Areas of inspection for high CUI risk

- *Is it active or passive?* Active corrosion is still in progress and is getting worse. Passive corrosion is evidence of some previous corrosion activity that has now stopped.

The way in which these three aspects are treated varies from site to site. There is no absolutely right or wrong method as long as the result is clear, and in harmony with the particular criteria and FFP assessment method used for the component in question.

Taken together, the three aspects are combined into an overall measure of criticality. This may be a quantitative number, e.g. on a scale of 1 to 12, or a qualitative ranking such as 'cosmetic', 'significant' or 'critical'. Again, the exact method used depends on the application. The allocation of a criticality rating is one of the main aspects of the in-service inspection process – it is not much use identifying corrosion if you cannot decide how critical it is, and what to do about it.

Diagnosing the corrosion mechanism

Arguably, the diagnosis of the type of corrosion, and its cause, is of less short-term importance than deciding its overall criticality. Diagnosis is important for the longer-term perspective, i.e. for preventing its reoccurrence. Accurate diagnosis is always difficult – there are hundreds of types of corrosion and the analysis process, if it is done properly, will involve metallurgists and corrosion engineers to add technical back-up to the inspector's observations. This means that inspectors can take a

---

- **Criticality.** Try to decide the *criticality* of all the corrosion you find.
- **Sketches.** Use sketches to show how extensive the corrosion is, rather than trying to describe it in words.
- **FFP.** Don't avoid the issue of fitness for purpose (FFP). FFP is a central issue in in-service inspection.
- **Qualitative terms.** Terms such as cosmetic, significantly, extensive, etc. mean vastly different things to different people – so they can be misinterpreted. Use quantitative terms such as '80% of the surface was corroded to a depth of 2 mm' instead.
- **Metallurgical terms.** Do not be frightened of using formal metallurgical terms, but make sure you use the correct ones.

Fig. 8.12   Corrosion reporting – some specific points

more relaxed view of it – perhaps concentrating more on the criticality assessment instead.

Corrosion reporting
Good corrosion reporting must include an assessment of criticality, as previously described. Some particularly useful points to note are shown in Fig. 8.12.

Types of corrosion

There are three basic types of simple corrosion relevant to mechanical and pressure-equipment applications:

- Chemical corrosion.
- Galvanic corrosion.
- Electrolytic corrosion.

To complicate matters there are a variety of subtypes, some hybrids, and some which are relevant mainly to boiler applications.

Chemical corrosion
This is caused by attack by chemical compounds in a material's environment. It is sometimes referred to as 'dry' corrosion or oxidation.

Galvanic corrosion
This is caused by two or more dissimilar metals or material structures in contact in the presence of a conducting electrolyte. One material becomes anodic to the other and corrodes (see Fig. 8.13). The tendency of a metal to become anodic or cathodic is governed by its position in

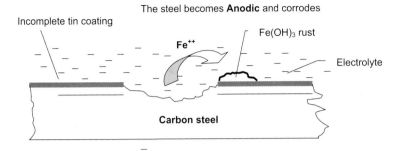

Fig. 8.13 Galvanic corrosion

the electrochemical series. This is, strictly, only accurate for pure metals rather than metallic compounds and alloys (see ASTM G135 and ASTM G102). A more general guide to the galvanic corrosion attack of common engineering materials is given in Fig. 8.14.

Electrolytic corrosion

This is sometimes referred to as 'wet' or 'electrolytic' corrosion. It is similar to galvanic corrosion in that it involves a potential difference

GALVANIC CORROSION ATTACK – GUIDELINES

Corrosion of the materials in each column is increased by contact with the materials in the row when the corresponding box is shaded.

Material	Steel and CI	Stainless steel 18% Cr	Stainless steel 11% Cr	Inconel/Ni alloys	Cu/Ni and bronzes	Cu and brass	PbSn and soft solder	Silver solder	Mg alloys	Chromium	Titanium	Al alloys	Zinc
Steel and CI		▓	▓	▓	▓	▓	▓		▓				
Stainless steel 18% Cr													
Stainless steel 11% Cr		▓						▨		▓	▓		
Inconel/Ni alloys		▓								▓			
Cu/Ni and bronzes		▓								▓			
Cu and brass		▓											
PbSn and soft solder		▓								▓			
Silver solder													
Mg alloys	▓	▓											
Chromium													
Titanium													
Al alloys	▓	▓	▓	▓	▓	▓	▓			▓			
Zinc	▓	▓	▓	▓						▓		▓	

Example: The corrosion rate of silver solder is increased when it is placed in contact with 11% Cr stainless steel.

Fig. 8.14 Galvanic corrosion attack – guidelines

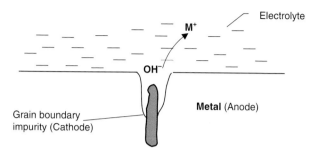

Fig. 8.15   Electrolytic corrosion

and an electrolyte but it does not need to have dissimilar materials. The galvanic action often happens on a microscopic scale. Examples are:

- Pitting of castings due to galvanic action between different parts of the crystals (which have different composition).
- Corrosion of castings due to grain boundary corrosion (Fig. 8.15).

Figure 8.16 shows an example.

Crevice corrosion
This occurs between close-fitting surfaces, crevice faces, or anywhere where a metal is restricted from forming a protective oxide layer (Fig. 8.17). Corrosion normally propagates in the form of pitting. Examples are:

- Corrosion in crevices in seal welds.

Fig. 8.16   General electrochemical corrosion of flange bolts

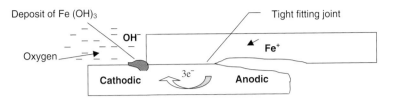

Fig. 8.17   Crevice corrosion

- Corrosion in lap-joints used in fabricated components and vessels.

Stress corrosion
This is caused by a combination of corrosive environment and tensile loading. Cracks in a material's brittle surface layer propagate into the material, resulting in multiple bifurcated (branching) cracks. Examples are:

- Failure in stainless steel pipes and bellows in a chlorate-rich environment.
- Corrosion of austenitic stainless steel pressure vessels.

Figure 8.18 shows stress corrosion cracking (SCC) in a stainless steel vessel.

Corrosion fatigue
This is a hybrid category in which the effect of a corrosion mechanism is increased by the existence of a fatigue condition. Sea water, fresh water and even air can reduce the fatigue life of a material.

Fig. 8.18   Stress corrosion cracking (SCC) in a stainless steel vessel

Intergranular corrosion

This is a form of local anodic attack at the grain boundaries of crystals due to microscopic differences in the metal structure and composition. An example is:

- 'Weld decay' in unstabilized austenitic stainless steels.

Erosion–corrosion

Almost any corrosion mechanism is made worse if the material is subject to simultaneous corrosion and abrasion. Abrasion removes the protective passive film that forms on the surface of many metals, exposing the underlying metal. An example is:

- The walls of pipes containing fast-flowing fluids and suspended solids.

Useful references

Websites

National Association of Corrosion Engineers (NACE), go to: www.nace.org.
    For a list of corrosion-related links, go to: www.nace.org/corlink/corplinkindex.htm.

Standards

ASTM G15: 1999   *Standard terminology relating to corrosion and corrosion testing.*
ASTM G102: 1999   *Standard practice for calculation of corrosion rates and related information from electrochemical measurements.*
ASTM G119: 1998   *Standard guide for determining synergism between wear and corrosion.*
ASTM G135: 1995   *Standard guide for computerized exchange of corrosion data for metals.*

Boiler failure modes

Steam boilers can exhibit various types of failure mechanism, caused by corrosion and other factors. Table 8.6 is a 'quick reference' of the main types. Figure 8.19 shows a list of formal failure-related terminology.

Table 8.6 Boiler failure modes

Failure mode	Mechanism	Appearance	Causes
Overheating	Short-term overheating of tubes leading to final failure by stress rupture	Longitudinal or 'fish-mouth' longitudinal rupture with thin edges	• Scale or debris blockage in the tube • Condensate locked in the tube owing to inadequate boil-out
High-temperature creep	High-temperature failure in the creep temperature range > 400 °C	Blister or larger 'fish-mouth' longitudinal rupture with thick edges Area around the rupture can have scaly appearance	• Excessive gas temperatures or overfiring during start-up • Hot gas flowing through an area of low circulation due to plugging or scaling, etc.

Caustic embrittlement	Intergranular attack along grain boundaries, causing cracking	Common on tubes rolled into vessel shells. General cracking with very little metal loss	• Excessive gas temperatures for the grade of steel used
			• A combination of:
			  – high stresses
			  – free caustic salts
			
			• A concentrating mechanism

Caustic gouging	Waterside corrosion that attacks the protective magnetite film on metal surfaces.	Common on the waterside of dirty boiler tubing as either small pinhole leaks or small bulges on thinned tubes, with thin-edged rupture failures	A combination of: • high heat flux • caustic contaminants • inconsistent boiler water chemistry
Pitting corrosion	Waterside corrosion	Generally localized, particularly on the inside of tubes. Sharp-edged craters surrounded by red/brown oxide deposits	Occurs in any areas where oxygen comes into contact with steel. Common causes are: • high O$_2$ levels in feedwater • condensate and air remaining in the boiler when shut down
Stress corrosion cracking	Waterside corrosion most commonly starting on the inside of tubes, resulting in cracks and leakage	Longitudinal or circumferential thick-edged cracks. The cracks often have a bifurcated (branched) appearance under magnification	Combination of the presence of: • stress • cyclic operation • caustic contaminants

Table 8.6 (Continued).

Failure mode	Mechanism	Appearance	Causes
Corrosion fatigue	Cracking failure of stressed components	Thick-edged parallel surface cracks with oxide coating and pits on the fracture surface	A combination of: • induced stress from a constrained joint or inaccurate assembly • a corrosive environment
Hydrogen attack	Waterside corrosion of dirty steam-generating tube-banks in high-pressure boilers	'Window' piece of tube falls out, leaving a thick-edged fracture, without any preliminary wastage or thinning of the tube	A combination of: • acidic boiler water • dirty tube surfaces • high heat flux
Waterwall corrosion	A general fire-side corrosion on the waterwall tubes in a boiler's combustion zone	General thinning of the tube material accompanied by deep longitudinal cracks and gouges. Hard slag deposits on the fire-side surface of the tube	Too little oxygen in the combustion. Excessive sulphides or chlorides in the fuel
Exfoliation	'Flaking' of tube and piping surfaces	Wastage or 'spalling' of the component from its inside surface	• Constituents of the metal oxide, resulting in differential expansion between the oxide layer and the metal surface • Quenching of tube internals during transient operations
Low-temperature corrosion	Common fire-side corrosion mechanism on boiler tubes	Gouging of the external surface and thin-edged ductile failure, often in the form of a hole	Acidic contaminants (such as sulphur or ash) in the furnace gas

Acid. A compound that yields hydrogen ions (H^+) when dissolved in water.

Alkaline. (a) Having properties of an alkali; (b) having a pH greater than 7.

Alloy steel. Steel containing specified quantities of alloying elements added to effect changes in mechanical or physical properties.

Amphoteric. Capable of reacting chemically either as an acid or a base. In reference to certain metals, signifies their tendency to corrode at both high and low pH.

Anode. In a corrosion cell, the area over which corrosion occurs and metal ions enter solution; oxidation is the principal reaction.

Austenite. A face-centred cubic solid solution of carbon or other elements in non-magnetic iron.

Austenitic stainless steel. A non-magnetic stainless steel possessing a microstructure of austenite. In addition to chromium, these steels commonly contain at least 8 percent nickel.

Blowdown.. In connection with boilers, the process of discharging a significant portion of the aqueous solution in order to remove accumulated salts, deposits, and other impurities.

Brittle fracture. Separation of a solid accompanied by little or no macroscopic plastic deformation.

Cathode. In a corrosion cell, the area over which reduction is the principal reaction. It is usually an area that is not attacked.

Caustic cracking. A form of stress corrosion cracking affecting carbon steels and austenitic stainless steels when exposed to concentrated caustic, i.e. high-alkaline solutions.

Caustic embrittlement. Term denoting a form of stress corrosion cracking most frequently encountered in carbon steels or iron–chromium–nickel alloys that are exposed to concentrated hydroxide solutions at temperatures of 200–250 °C (400–480 °F).

Cavitation. The formation and instantaneous collapse of innumerable tiny voids or cavities within a liquid subjected to rapid and intense pressure changes.

Cavitation damage. The degradation of a solid body resulting from its exposure to cavitation. This may include loss of material, surface deformation, or changes in properties or appearance.

Cementite. A compound of iron and carbon, known chemically as iron carbide and having the approximate chemical formula FE_3C.

Cold work. Permanent deformation of a metal produced by an external force.

Corrosion fatigue. The process in which a metal fractures prematurely under conditions of simultaneous corrosion and repeated cyclic

Fig. 8.19   Failure-related terminology

loading – fracture occurs at lower stress levels or fewer cycles than would be required in the absence of the corrosive environment.

Corrosion product. Substance formed as a result of corrosion.

Creep. Time-dependent deformation occurring under stress and high temperature.

Creep rupture. See **Stress rupture.**

Dealloying. (see also **Selective leaching.**) The selective corrosion of one or more components of a solid solution alloy. Also called 'parting' or 'selective leaching'.

Denickelification. Corrosion in which nickel is selectively leached from nickel-containing alloys. Most commonly observed in copper–nickel alloys after extended service in fresh water.

Dezincification. Corrosion in which zinc is selectively leached from zinc-containing alloys. Most commonly found in copper–zinc alloys containing less than 85 percent copper after extended service in water containing dissolved oxygen.

Downcomer. Boiler tubes in which fluid flow is away from the steam drum.

Ductile fracture. Fracture characterized by tearing of metal accompanied by appreciable gross plastic deformation and expenditure of considerable energy.

Ductility. The ability of a material to deform plastically without fracturing.

Erosion. Destruction of metals or other materials by the abrasive action of moving fluids, usually accelerated by the presence of solid particles or matter in suspension. When corrosion occurs simultaneously, the term 'erosion–corrosion' is often used.

Eutectic structure. The microstructure resulting from the freezing of liquid metal such that two or more distinct, solid phases are formed.

Exfoliation. A type of corrosion that progresses approximately parallel to the outer surface of the metal, causing layers of the metal or its oxide to be elevated by the formation of corrosion products.

Failure. A general term used to imply that a part in service: (a) has become completely inoperable, (b) is still operable but is incapable of satisfactorily performing its intended function, or (c) has deteriorated seriously, to the point that it has become unreliable or unsafe for continued use.

Fatigue. The phenomenon leading to fracture under repeated or fluctuating mechanical stresses having a maximum value less than the tensile strength of the material.

Fig. 8.19 (Continued)

Ferritic stainless steel. A magnetic stainless steel possessing a microstructure of alpha ferrite. Its chromium content varies from 11.5 to 27 percent but it contains no nickel.

Fish-mouth rupture. A thin or thick-lipped burst in a boiler tube that resembles the open mouth of a fish.

Gas porosity. Fine holes or pores within a metal that are caused by entrapped gas or by evolution of dissolved gas during solidification.

Grain boundary. A narrow zone in a metal corresponding to the transition from one crystallographic orientation to another, thus separating one grain from another.

Graphitic corrosion. Corrosion of grey iron in which the iron matrix is selectively leached away, leaving a porous mass of graphite behind. Graphitic corrosion occurs in relatively mild aqueous solutions and on buried pipe fittings.

Graphitization. A metallurgical term describing the formation of graphite in iron or steel, usually from decomposition of iron carbide at elevated temperatures.

Heat-affected zone. In welding, that portion of the base metal that was not melted during welding, but whose microstructure and mechanical properties were altered by the heat.

Inclusions. Particles of foreign material in a metallic matrix. The particles are usually compounds but may be any substance that is foreign to (and essentially insoluble in) the matrix.

Intergranular corrosion. Corrosion occurring preferentially at grain boundaries, usually with a slight or negligible attack on the adjacent grains.

Lap. A surface imperfection having the appearance of a seam, and caused by hot metal, fins, or sharp corners being folded over and then being rolled or forged into the surface without being welded.

Magnetite. A magnetic form of iron oxide, Fe_3O_4. Magnetite is dark grey to black, and forms a protective film on iron surfaces.

Microstructure. The structure of a metal as revealed by microscopic examination of the etched surface of a polished specimen.

Overheating. Heating of a metal or alloy to such a high temperature that its properties are impaired.

Pitting. The formation of small, sharp cavities in a metal surface by corrosion.

Residual stress. Stresses that remain within a body as a result of plastic deformation.

Root crack. A crack in either a weld or the heat-affected zone at the root of a weld.

Fig. 8.19 (Continued)

Scaling. The formation at high temperatures of thick layers of corrosion product on a metal surface.

Scaling temperature. A temperature or range of temperatures at which the resistance of a metal to thermal corrosion breaks down.

Selective leaching. Corrosion in which one element is preferentially removed from an alloy, leaving a residue (often porous) of the elements that are more resistant to the particular environment.

Spalling. The cracking and flaking of particles out of a surface.

Stress–corrosion cracking. Failure by cracking under combined action of corrosion and stress, either external (applied) stress or internal (residual) stress. Cracking may be either intergranular or transgranular, depending on the metal and the corrosive medium.

Stress raisers. Changes in contour or discontinuities in structure that cause local increases in stress.

Stress rupture (creep rupture). A fracture that results from creep.

Tuberculation. The formation of localized corrosion products in the form of knob-like mounds called 'tubercles'.

Underbead crack. A subsurface crack in the base metal near a weld.

Fig. 8.19   (Continued)

Chapter 9

Fitness-for-purpose/service assessment

Fitness-for-purpose (FFP) assessments, termed fitness-for-service (FFS) assessments in the USA, play a prominent role in the world of in-service inspection. Their most common use is for pressurized equipment and structural components where *integrity* is an important issue.

FFP assessments – what are they?

An FFP assessment is a set of calculations performed on a component that contains some kind of feature that is suspected to affect its integrity. It is pure calculation so, although the calculation routines are based on engineering reality (supposedly), it is a purely theoretical exercise rather than a practical one. The calculations themselves are based predominantly on the principles of mechanics. The whole concept of FFP assessment carries with it a specific set of terminology, which, while perfectly consistent in itself, does not necessarily coincide with common, or even general, engineering usage of the words – so it can be a bit confusing. Figure 9.1 shows some of the most important 'foundation' usage of the terms – there are hundreds of others, with whole sections of the technical standards covering FFP assessments set aside for their explanation.

FFP assessments – the approach

The central concept of an FFP assessment is that it looks at the effect on FFP of an existing flaw (specific terminology – see Fig. 9.1). It tries to predict how and when the flaw will cause the component to fail; it does this by considering the various failure mechanisms (ways of failure) that could occur. There is a graded approach to assessment with three possible levels of investigation. The higher the level of investigation, the more the failure mechanisms are considered in depth, and the greater the mathematical rigour that is used.

- **FFP or FFS?** These mean absolutely the same. European practice is to use FFP whereas American practice tends to refer to FFS. You can consider them *interchangeable* for most purposes.
- **Flaws.** The word *flaw* is used in FFP assessments as a lowest common denominator term for a feature that has a possible effect on integrity. It encompasses other more common terms such as 'defect', 'discontinuity', 'crack', and suchlike, but is a bit less controversial.
- **Assessment.** Used alone, this has at least three meanings because both major families of FFP assessment have three levels (1, 2, and 3) which can produce different answers.
- **Bounds (upper and lower) and shelves (top and bottom).** These are terms used to *qualify* the methodologies and results of FFP assessments. They are evidence that such assessments are more qualitative than they admit to, and almost never give a single, unequivocal, quantitative *answer* about which everyone will agree.

Remember – FFP assessment terminology is complex and not necessarily the same as 'common usage' engineering terms. Some terms are specifically included to address the high complexity of the subject, and the inevitable need to give qualified answers.

Fig. 9.1   FFP assessment – some terminology

FFP approach – two key points

A key point of the approach of an FFP assessment is its relationship to code compliance, e.g. the compliance of a pressure vessel with the construction requirements of ASME VIII, EN 13445, or another of the recognized published technical standards. Note two key points:

- Most FFP assessments are done on components that are outside code compliance limits in some way, i.e. they are *rejectable* under the code. This means that FFP assessments are, fundamentally, based on different acceptance criteria than design codes. This introduces the apparent paradox of there being more than one valid set of flaw acceptance criteria for the same engineering component. Figure 9.2 provides the explanation and exposes the real character of design codes themselves.
- The defect acceptance criteria given in technical codes (ASME VIII, EN 13445, etc.) are *not* arbiters of FFP. They are best thought of as *indicative* levels that can be reached if all the code requirements on materials, QA/QC, design features, etc. are met.

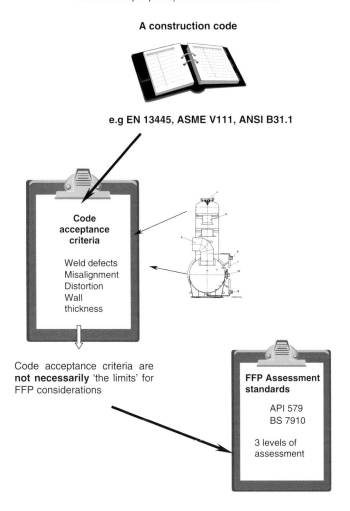

A construction code

e.g EN 13445, ASME V111, ANSI B31.1

Code
acceptance
criteria

Weld defects
Misalignment
Distortion
Wall
thickness

Code acceptance criteria are
not necessarily 'the limits' for
FFP considerations

**FFP Assessment
standards**

API 579
BS 7910

3 levels of
assessment

FFP assessment standards take
a *more detailed view* of defect
acceptance criteria

Fig. 9.2   Code compliance versus FFP assessments

The inference here is that the existence of out-of-code flaws (using the FFP assessment term) does not mean that a component is not fit for purpose or unsafe. It means, merely that it does not meet the full requirements of the code, some of which may be FFP related but many of which are chosen for practical reasons, or simply convenience. All of this reinforces the need for in-service inspectors to appreciate the basic principles of FFP assessment and to recognize the different approach to that of simple code compliance.

Application to in-service defects

FFP assessments can be used to analyse either out-of-code construction defects or those that have occurred or worsened during a component's service lifetime. The mechanics of the assessment are equally applicable to both situations, albeit with different emphasis. In practice their frequency of use in each of these two applications is probably about equal. Figure 9.3 shows the main reasons why an FFP assessment would be commissioned following an in-service inspection activity. The methodology of the FFP assessment is much the same in each situation although the form of the output, and the way in which conclusions are presented, will differ.

Note how Fig. 9.3 incorporates a wide variety of in-service findings. All of these can be found, singly or in combination, during shutdown inspection of pressure equipment and structural components. The in-service conditions that cause them are equally varied, covering different types of corrosion, low- or high-temperature effects, process upsets, and similar.

FFP assessments – pro-active or defensive?

Nominally, FFP assessments take their rightful place in the complex process of engineering design, i.e. they are an integral part of it. As such, they are sold as a pro-active activity, used for the purpose of improving the structural integrity of critical engineering components. Opinions vary as to whether this is true. The following points are valid to the argument:

- FFP assessments do not feature heavily in 'code-cases' that qualify design codes, or indeed even in the codes themselves. Codes prefer to rely on more simplistic defect acceptance criteria.
- Most FFP assessments are carried out *after* a defect is found and sized rather than before (e.g. to suggest the sizes of defect everybody *should be* looking for).
- Because FFP assessments can be performed at several levels of analysis, it is possible to predetermine the result (within limits) by deciding which level of analysis to use.
- Many more FFP assessments end up concluding that the flaw is *not* a threat to integrity rather than suggesting it *is* a threat, hence requiring repair or replacement.

You can see from these points that, although factual, they do not fit well with FFP assessments being a 'front-end', largely pro-active part of the design process. They look more *defensive* in character – something

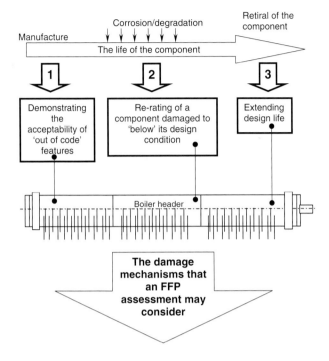

Fig. 9.3   The purpose of FFP assessments

that can be used as a justification for continued use of a component rather than to repair it, downrate it, or scrap it altogether. They certainly have all the necessary features – the assessments are complex, can be governed by individuals' technical preferences rather than prescribed rules, and are sufficiently theoretical to confuse even experienced engineers that are not dealing with them on a regular basis.

So, you can view FFP assessments in several ways: as a panacea for dealing with defects, as a secure justification for balanced technical decisions, or as an excuse for doing nothing when defects are found. You decide.

FFP assessment standards

There are two main sets of published standards used for assessing the acceptability of flaws in metallic equipment. These apply mainly to pressure equipment and structural items but can also be applied, with some limitation and adaptations, to some types of rotating plant. Both standards have been developed via several revisions over the past 20 years and are well accepted. There are also various supporting standards, and a few industry-specific guidelines and methodologies which have gained sufficient credibility to be accepted as authoritative.

The two main standards, BS 7910 and API 579, were developed in the UK/Europe and the USA. While they use similar concepts (because the laws of mechanics do not change), the technical approach of each is influenced by the technical methodologies and preferences of their host country. This means that they differ in *character*, as well as content.

BS 7910: 1999

The full title of this standard is BS 7910: 1999 *Guide for assessing the acceptability of flaws in metallic structures*. It evolved from the British Standard Published Document PD 6493 first published in 1980. This document was intended more 'for information only' rather than as a technical standard, and included several alternative methods of fracture assessment rather than a single recommended way of doing things. BS 7910: 1999 was issued as a more comprehensive version of BS PD 6493, covering both high- and low-temperature failure modes resulting from flaws in welds and some types of corrosion.

Figure 9.4 and 9.5 show the overall structure of the BS 7910 document. The various parts outline methods for dealing with the main degradation methods that can feasibly affect a component in service. Note how there are three progressive levels of fracture assessment:

- *Level 1.* A broad screening assessment using very conservative assumptions. This is normally the first level of assessment used when a flaws are found in a component.
- *Level 2.* A more in-depth assessment with lower margins of error.
- *Level 3.* The highest-integrity assessment involving detailed mathematical analysis and requiring the most accurate material properties data.

Note that the three levels apply only to the *fracture* part of an FFP assessment. All the other (equally valid) parts: creep, fatigue, buckling, etc. have only a single level. This has important implications for the assessment of engineering components that are subject to these effects – the priority of the different elements of the analysis must be chosen carefully, if the results are to have any real meaning.

For defects discovered during in-service inspection the use of these levels normally follows the pattern shown in Fig. 9.5. The Level 1 assessment is well within the capabilities of a plant Competent Person organization or, in many cases, the in-service inspection engineers themselves. Only a small number of cases (perhaps 15–20 percent) are completed at this Level 1 stage – most Level 1 results conclude that further Level 2 investigation is required. Level 2 assessment is a much more specialized task and not all Competent Person inspection organizations feel comfortable with it (see Chapters 3 and 4 for the strengths and weaknesses of these organizations). Level 3 assessments are even more complex (and expensive), and can require specific material tests to validate the material property parameters to be used in the calculation. This can be a daunting process, almost attaining the status of an on-going research programme.

BS 7910 methodology
BS 7910 is divided into discrete sections. Figure 9.4 shows the layout. Note the spread of different methodologies; the standard is much more complex than simply providing a single set of calculations which can be used in all situations. The choice of which methodology to use, or which takes priority, is one of the most difficult parts of performing an FFP assessment.

Analysing fracture – the failure assessment diagram (FAD)

Fracture is a failure mechanism that involves the propagation of a crack. The way in which a crack is predicted to propagate depends on the application, the applied stress, the properties of the steel, and temperature (Fig. 9.6). At high temperatures (and for specific materials such as austenitic stainless steel) the fracture behaviour is likely to be ductile. At low temperatures, particularly in ferrite steels, it is more likely to be brittle. The BS 7910 methodology uses the three parameters shown in Fig. 9.6 and expresses them in mathematical form. The basic steps are as described below (See Fig. 9.7).

BS 7910: 1999

Section	Content	How it works
1–6	Explains the form of input data required for FFP assessment calculations.	Gives a general (complicated) explanation of the format of an FFP assessment
7	Fracture assessment procedures.	Uses the principles of a fracture assessment diagram (FAD). A failure line on the diagram shows whether failure is likely or unlikely. Figure 9.5 shows how it works. Remember there are three levels. 1, 2 and 3. Level 1 is the simplest and most conservative.
8	Fatigue assessment procedures.	These use theoretical crack growth laws (the Paris law and similar) to plot a logarithmic graph. A simplified 'approximate' method is available using the traditional S–N curve.
9	High-temperature failure assessment.	This is based on theories about the way that cracks behave at high temperatures. Annex U of BS for 7910 gives a worked example.
10	Assessment of: ● yielding ● buckling ● corrosion ● SCC ● other modes	These use a selection of simple, broader-based techniques to cover different damage mechanisms. They all use the crack growth methodology as a base.
Annexes (1–21)	Reference information needed to perform FFP assessments.	Typical annex data includes: **Annex M–P**: Gives intensity factors used for various crack geometries. **Annex Q**: Recommendations on residual stress distribution to be used for as-welded joints in plates and pipes. **Annex J**: Gives correlations between fracture toughness (difficult to measure) and charpy impact values (easy to measure). **Annex K**: Outlines the concept of using worst-case scenarios to reduce overall risks of failure. **Other annex topics include**: Effects of weld misalignment, failure characteristics of tubular joints, corrosion assessment procedures, leak-before-break analyses.

Fig. 9.4   BS 7910: – contents

The FAD vertical axis

- *The crack driving force*, K_1 (known as the stress intensity factor), is calculated using the known geometry of the component, the applied stress, and the dimension of the flaw (crack). Defining the applied stress is the most difficult part as it must take into consideration normal principal stresses, design stress concentrations, and then make allowance for realistic out-of-code features such as weld misalignment, shell out-of-roundness, and profile distortion.
- *The material fracture toughness*, K_{mat}. This is a material property related to the Charpy impact strength. K_{mat} has to be found by doing specific test from similar material, with samples taken from the same area in which the crack is located (weld metal, HAZ, etc.) as it is highly affected by microstructure, heat condition, and material thickness. The precise way to determine K_{mat} is set out in published industry standards such as BS 7448: 1991 *Fracture mechanics toughness tests*. In some simplified cases it is possible to take K_{mat} values from published tables but this gives simplified, often overconservative results.
- *The ratio of* K_1/K_{mat} is plotted on the vertical axis of the FAD (see Fig. 9.7) and designated as K_r. If the ratio K_r is near to 1, then the component is predicted to fail by brittle fracture.

The FAD horizontal axis

The horizontal axis of the FAD is there to represent the possibility of the component failing by *plastic collapse*, i.e. not strictly by crack propagation. In most pressure equipment and structural steels, the material will yield and deform plastically (stretch) before actual failure, even if the failure process has been started by the propagation of a crack.

The horizontal axis is a ratio designated S_r. This is defined as follows

$$S_r = \sigma_n/\sigma_f$$

where

σ_n = 'net section stress', i.e. the stress (in MPa) that the component is actually seeing in use.

σ_f = 'flow stress', a strangely named term meaning the stress that the material needs to 'flow' or stretch. In many cases this is taken simplistically as the yield stress R_e.

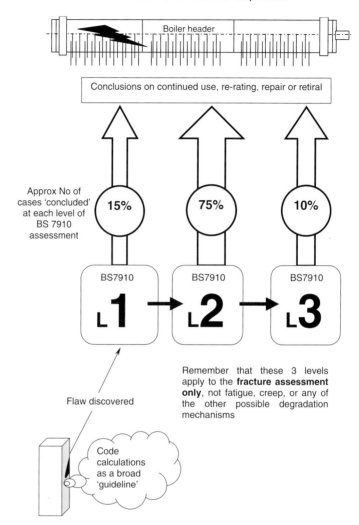

Fig. 9.5   BS 7910 – levels of fracture assessment

Again the S_r is a ratio, so if S_r is close to 1 then failure is predicted to occur by pure plastic collapse rather than the direct effects of crack propagation.

The connection between the vertical and horizontal axes of the FAD (i.e. between K_r and S_r) is defined by the failure *locus*. An individual case is assessed against the locus. If the assessment point lies on or below the locus, the flaw is stable and is deemed not to present a significant risk of failure. If the point is above the locus then there is a significant risk of failure. For any combination of crack propagation and plastic collapse

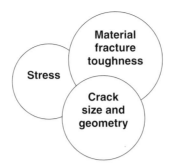

The principle is that crack propagation (i.e whether a crack will grow or not) is governed by a combination of the above three interlinked factors

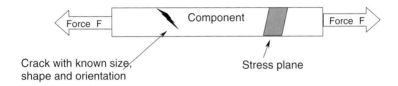

Fig. 9.6   The principle of a fracture assessment

(depending on where the point actually lies in relation to the K_r–S_r axes) the component could fail at any time when these criteria are met.

BS 7910 FAD levels

The main difference between the Level 1, 2, and 3 fracture assessments in BS 7910 is the derivation of the FAD and the way that it is interpreted. Level 1 assumes a simple rectangular locus while Level 2 uses a more complex curve *but* requires more accurate data to validate the curve. Level 2 is subdivided into Levels 2a and 2b; Level 2a is used for HAZs where full data may be difficult to get. Figure 9.7 shows a comparison of FADs for Levels 1 and 2 for a simple structural steel component.

A Level 3 fracture assessment under BS 7910 carries a more complex raft of technical assumptions. The routine is formally termed a *ductile tearing instability assessment* and has connections with the so-called R6 method (see later in the chapter). This type of assessment may include detailed elastic-plastic finite element analysis to give more accurate predictions of structural behaviour.

FAD – the basic idea

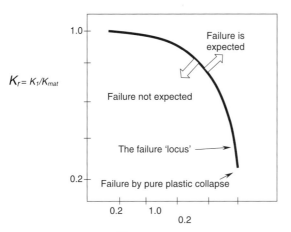

Stress ratio S_r = Net section stress σ_n / Flow stress σ_f

FADs used in BS 7910 (simplified)

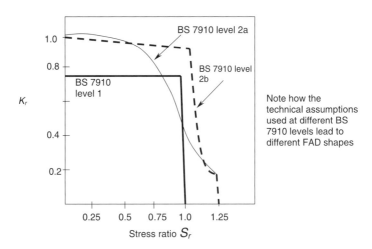

Note how the technical assumptions used at different BS 7910 levels lead to different FAD shapes

Fig. 9.7   How to use failure assessment diagrams (FADs)

Fracture assessments – summary

So, exactly how realistic are BS 7910 fracture assessments? Seen from the in-service inspectors viewpoint, the following points are worth bearing in mind:

- *The theory.* The fracture theory is well recognized and generally accepted as being robust.

- *Numerical values.* It is *very* difficult to define the 'correct' values to use for net section stress, (σ_n), on a complex-geometry component because of the complicating effect of changes in section and stress concentrations. It is equally difficult to know exactly what values to use for K_1 and K_{mat} on the vertical axis of the FAD. It is easy to use assumptions for these parameter values without actually knowing if they are correct for the case in question (because they are all different).

- *The proof?* This is the most dangerous aspect of the whole FAD affair. Once the best possible parameters have been chosen, the FAD drawn, and the specific assessment made, there is *absolutely no guarantee* that the results are correct. There are many examples of fracture assessments where components precisely concluded as being 'safe' have failed in service – and vice versa.

The difficulty about the three points above is that they are not always sufficiently well highlighted as an uncertainty in the conclusions of an FFP assessment. The desperate uncertainty of the results are well disguised behind a mass of obscure text and symbols, or simply left to be inferred, under the assumption that those who read the assessment reports do not need to be told.

Analysing fatigue

BS 7910 appears to take a compromise approach on the analysis of fatigue conditions. Fatigue introduces the added complication of trying to define the way that cracked materials respond to multiple crack-propagating stress cycles, rather than a stress which, although possibly difficult to determine, is at least consistent over time. BS 7910 gives two possible ways of doing an assessment: the simplified *quality categories* approach and the *crack growth laws* approach.

Fatigue assessment – the quality category approach

This predicts what will happen to a flaw by comparing it with information contained on a grid of *S–N* curves. These plot the stress, *S*, on a component against the number, *N*, of cycles to expected failure – Fig. 9.8 shows a simplified example. In practice the grid contains a family of *S–N* curves acting, in effect, as quality *categories*. The location of an individual case within the family of curves gives a prediction as to whether it is:

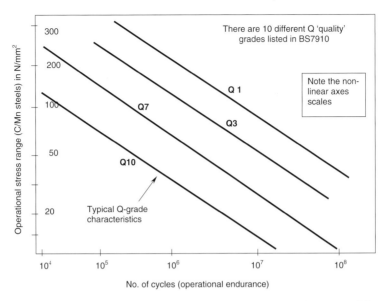

Fig. 9.8   **Quality category 'fatigue curves' (ref. BS 7910, Fig. 16)**

- Low risk, i.e. 'high quality', because it falls into the high cycles-to-failure region of the grid.
- High risk, i.e. or 'low quality' because it falls in an area where fewer predicted stress cycles are required before the component fails.

Figure 9.8 shows the basic form of the BS 7910 quality category curves. This type of procedure is used for preliminary screening assessments for simple components or applications that are not so critical. It is a straightforward method, but can be heavily subjective and is not difficult to discredit.

Fatigue assessment – the crack growth laws approach

This is a more accurate but complex method. It uses as a basis the fundamental assumption that the increment of crack extension, per cycle, is a factor of the applied stress intensity range. This is expressed in mathematical terms as

$$\mathrm{d}a/\mathrm{d}N = A\Delta K_1^m$$

where

a = crack length
$\mathrm{d}a$ = increment of crack extension per stress cycle
$\mathrm{d}N$ = each incremental stress cycle

K_1 = stress intensity ratio using applied stress ranges rather than the absolute magnitude of the applied stresses

A, m = constants which can only be determined empirically by testing the material in question

The most common crack growth laws that are used are known as the 'Paris laws'. BS 7910 uses these, applying various 'bound' criteria to account for residual uncertainties owing to the significant spread of data that are available. Figure 9.9 shows how the results are displayed graphically.

The output of the fatigue analysis part of a BS 7910 FFP assessment is an estimate of the number of stress cycles that it will take for a flaw to extend to a size where it will cause failure. The initial flaw size is usually the crack length or height in a critical direction (often the through-wall direction for pressure equipment). The final limiting size is that assessed with respect to the limiting failure condition, e.g. through-wall ligament cracking or, in structural components, reduction of an important area of 'ruling section'.

Final failure of a fatigued component happens by fracture once a flaw has reached a critical length. In such cases, the state reached just as the crack reaches its critical length corresponds to a point exactly on the FAD locus. This infers that although fracture and fatigue are separate parts of a BS 7910 FFP assessment they are *linked* by the similarities in the stress ratios that they use.

Analysing creep

BS 7910 Section 9 contains methods for investigating the effect of creep on the life of a component. The general concept is similar to those for fracture and fatigue – a mathematical routine is used to estimate how a creep-induced defect will progress, so giving a projected remnant life of the component before its integrity is threatened. Creep is a high-temperature mechanism rarely occurring below about 380 °C in plain carbon steels. It manifests itself as creep voids which act as crack-like flaws, propagating until they eventually result in failure by ductile tearing. The creep mechanism is dependent on three main parameters: stress, time, and temperature.

The creep assessment procedures in BS 7910 Section 9 are an updated version of those used in the superseded document, BS PD 6493 – most of the principles are unchanged. Creep assessment is one of the areas of BS 7910 which relies on a single model of the way that a material

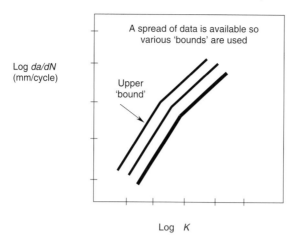

Log da/dN (mm/cycle)

A spread of data is available so various 'bounds' are used

Upper 'bound'

Log K

Where K = stress ranges corresponding to K_1 / K_{mat}

Units of K are MPa $m^{1/2}$

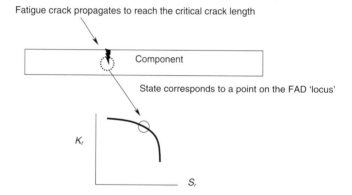

Fatigue crack propagates to reach the critical crack length

Component

State corresponds to a point on the FAD 'locus'

K_r

S_r

Fig. 9.9   Fatigue assessment 'crack laws'

behaves when under stress at high temperature. The methodology is broadly as follows (see Fig. 9.10).

- The analysis centres around what happens at a crack tip.
- It is assumed that the elastic stress existing around the crack tip will be highest at the actual tip itself, reducing with distance from it. The magnitude of this stress is represented by an elastic stress intensity factor K.

- The combined effect of time and temperature (i.e. the creep condition) has the effect of redistributing the stress up to the point where a steady state is reached. This state can be described by a creep fracture mechanism parameter C^*.
- This parameter C^* can be expressed in the following mathematical relationship.

$$a = AC^{*q}$$

where

   a = creep crack growth rate (in m/h)
   A, q = material constants determined by empirical tests
   C^* = creep fracture mechanics parameter (in $MJ/M^2 h$)

- A further parameter ε_f^* is introduced representing the *creep ductility*. This can be thought of as a measure of how ductile the material at the crack tip is.
- It is assumed that the creep tip progresses (actual damage takes place) when the limiting value of ε_f^* is exceeded. It is this crack progression which constitutes the creep damage.

Is this a realistic procedure?

The realism of this methodology is more reliant on the validity of the material data and mathematical constants than the fracture or fatigue evaluation parts of BS 7910. The mathematics is fully robust in itself but the realism of the results is highly dependent on the accuracy of the material property data fed in. The quality of individual data varies widely – information is available on some steels but there are hundreds where none is available or, what is, is poorly validated. Even the well-validated data suffer from a wide 'spread', making it necessary to define upper and lower 'bounds' or 'shelves' to qualify its use.

  A further uncertainty in creep assessments comes from the way that material crack propagation responds to high temperatures. Small differences in temperature can result in large changes in the material's creep properties, and therefore creep life. It is fairly widely accepted that a difference of 5 °C between estimated and actual temperature can totally invalidate the rationale of a creep life assessment in some alloys.

  From an in-service inspection viewpoint, the apparent predictability of creep suggested by a BS 7910 assessment must be treated with some caution. The greatest uncertainty is produced when creep and fatigue conditions act *together* on a component. The cumulative effects can exceed the sum of the individual mechanisms, resulting in a high-risk

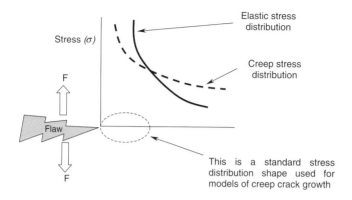

Assumed stress distribution around a flaw

Stress (σ)

Elastic stress
distribution

Creep stress
distribution

F

Flaw

This is a standard stress
distribution shape used for
models of creep crack growth

F

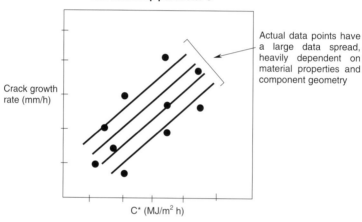

**Relationship between crack growth
rate and creep parameter C***

Crack growth
rate (mm/h)

Actual data points have
a large data spread,
heavily dependent on
material properties and
component geometry

C* (MJ/m² h)

Fig. 9.10   The principles of creep assessment

situation in which integrity is threatened. High-temperature reactor
vessels and power HRSG boilers (see Chapter 15) are classic examples
of this – they are subject to a complex combination of fatigue and creep
conditions in use, only some of which can be quantified fully.

Analysing corrosion

Analysing the effects of corrosion is relevant to many types of pressure
equipment and structural components. The breadth of the subject

means that there is no single assessment method that is considered valid for all cases – different methods are used for different applications. The multiple possibilities of size, shape, and location of corrosion 'defects' adds additional complexity to the problem.

Corrosion assessment is addressed in Annex G of BS 7910: 1999. This is fairly new material but is biased towards pipeline corrosion rather than general pressure equipment or structural applications. The principles, however, can be considered similar. Annex G contains some of the best-validated methods of all of BS 7910 – it also uses a very practical approach with straightforward mathematics that is not too difficult to follow.

The general approach

The general approach looks at components that possess corrosion damage in the form of:

- General corrosion (wall thinning).
- Corrosion pits.
- Combinations of the above defects which are located close enough to *interact* with each other.

An important assumption is that the component is not subject to a brittle fracture risk, i.e. it is operating well above the brittle–ductile fracture transition temperature.

The effect of loss of wall thickness caused by overall or pitting corrosion is easy to calculate using standard membrane stress theory so the emphasis of the assessment concentrates instead on the way that such defects *interact*. Figure 9.11 shows the principle – defects that are located within a certain distance from each other will add to each other's effects, producing a risk of failure greater than that of each defect taken separately. This contradicts conclusions inferred by less FFP-orientated published in-service inspection standards such as API 653/API 510/API 570 which allow isolated corrosion pits to be almost ignored for integrity purposes.

As with fracture assessment, corrosion assessment can take place at three levels starting with the simplest Level 1 'screening' assessment. Figure 9.12 shows the basics of each method – note the differences in approach.

API 579: 1999

API 579: 1999 *Recommended practice for fitness for service (FFS)* is the American equivalent to BS 7910. It has its roots in the refinery and petrochemical industry but has extended to deal with all mainstream types of pressure equipment. API 579 fits with the other three American codes covering in-service inspection issues: API 510 (vessels), API 570 (pipework), and API 653 (storage tanks) but provides much more technical information and detailed analysis methods. There is also a development relationship, of sorts, with the corresponding construction codes (see Fig. 9.13).

API 579 is a large document containing over 1000 pages. It provides very detailed coverage of different flaw and damage types. In line with the methodology of BS 7910, three levels of assessment (Levels 1, 2, and 3) are provided for each type. Although there are some methodological differences, the levels follow a similar philosophy to those in BS 7910, i.e.:

- API 579 Level 1 is a quick, conservative screening assessment that requires a minimum amount of real material or component data.
- API 579 Level 2 is a more detailed assessment which gives more accurate, less conservative results. The amount of material and component data required is still low but the calculations are more involved.
- API 579 Level 3 is the most detailed assessment, typically involving finite element techniques. Detailed component and material data are needed, often requiring specific material tests to be performed. The technical level is comparable with a BS 7910 Level 3 assessment.

API 579 – the content

Figure 9.14 shows the way that the content of API 579 is structured. Sections 3 to 10 deal with specific types of defects and 'flaws' (similar terminology to BS 7910) while Appendices A to J cover specific technical issues, including various data tables and reference information. Figure 9.15 gives further information on the damage mechanisms covered.

In line with the general philosophy of API technical codes, API 579: 1999 is heavily influenced by the calculation methodologies recognized by ASME I/VIII. There is also less emphasis on *fatigue* as a predominant damage mechanism affecting pressurized plant – again, this is a common ASME/API approach. Further development of

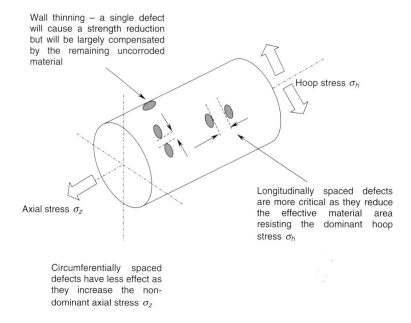

Wall thinning – a single defect will cause a strength reduction but will be largely compensated by the remaining uncorroded material

Hoop stress σ_h

Axial stress σ_z

Longitudinally spaced defects are more critical as they reduce the effective material area resisting the dominant hoop stress σ_h

Circumferentially spaced defects have less effect as they increase the non-dominant axial stress σ_z

Fig. 9.11   Corrosion assessment – the effect of interacting defects

API 579: 1999 is in progress – future editions should include technical enhancements and revised assessment procedures for creep, fatigue, dents and profile distortions, hydrogen-induced cracking, laminations, and similar. It is fair to predict that many of these will have direct relevance to refinery/petrochemical plant and be linked to the approach of API 580/581 (risk-based inspection).

Other FFP standards

There are a handful of other published standards and documents which contain information on FFP assessments. Most have been developed for particular specialized applications but, over time, have become used in a wider context. Technically, they use similar approaches to the mainstream FFP standards, with a few differences in methodology (see Table 9.1).

ASME B31G pipeline assessment codes

This is an FFP methodology published as a supplement to the parts of ANSI B31 code covering various types of process pipelines (e.g. B31.8 etc.). It has the following features:

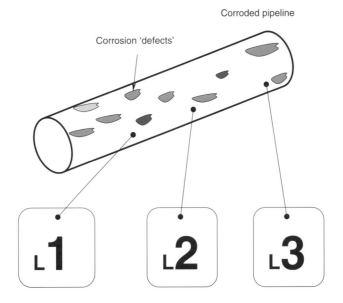

Fig. 9.12   Levels of corrosion assessment

- It covers only weldable carbon steels such as ASTM A106, A381, etc. and API 5L.
- It is applicable only for pipeline defects that have smooth, continuous loss of wall thickness due to electrolyte/galvanic corrosion or erosion. It does not cover mechanical damage, gouges, and similar.
- It does not cover fracture, creep, fatigue, or other failure mechanisms.
- Similar to other FFP methods, the B31G methodology is concerned only with structural integrity under pressure rather than the effect of any fluid leaks.

The method has similarities with Annex G of BS 7910. A corrosion defect is described in terms of length, depth, and proximity to other

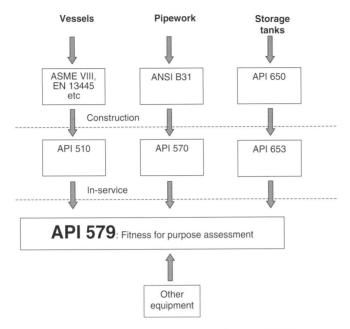

Fig. 9.13   Codes for manufacture, inspection, and FFP assessment

defects, then calculations are performed to predict its effect on the strength of that section of pipeline.

Nuclear Electric R6 assessment method

The R6 method is a specific assessment methodology designed by Nuclear Electric plc in the UK. It was devised to fit a need for a validated defect assessment procedure to suit nuclear plant for the purpose of life extension, and provide a justified technical justification for run-or-repair situations. The R6 method is a calculation-based routine along similar lines to BS 7910 and its predecessor BS PD 6493. The R6 rationale was itself preceded by the R5 method: *Assessment procedure for the high-temperature response of structures.* R6 is concerned with components at lower temperatures, i.e. below the creep range, and concentrates on fatigue crack growth. It does not cover the more straightforward damage mechanisms such as wall thinning due to corrosion/erosion or 'construction-stage' defects such as weld misalignment. It is rare to find the R6 used for everyday FFP assessments on pressure equipment or structural components. It is a comprehensive complex method that finds use mainly in nuclear and high-integrity chemical plant application, or in material research

Section 1	Introduction
Section 2	FFS Engineering Evaluation Procedure
Section 3	Assessment of Equipment for Brittle Fracture
Section 4	Assessment of General Metal Loss (when $t_{available} < t_{min}$ over a large area)
Section 5	Assessment of Localized Metal Loss (when $t_{available} < t_{min}$ over a local area)
Section 6	Assessment of Pitting Corrosion
Section 7	Assessment of Blisters and Laminations
Section 8	Assessment of Weld Misalignment and Shell Distortions
Section 9	Assessment of Crack-like Flaws
Section 10	Assessment of Fire Damage
Appendix A	Thickness, MAWP, and Membrane Stress Equations for an FFS Assessment
Appendix B	Stress Analysis Overview for FFS Assessment
Appendix C	Compendium of Stress Intensity Factor Solutions
Appendix D	Compendium of Reference Stress Solutions
Appendix E	Residual Stresses for an FFS Assessment of Crack-like Flaws
Appendix F	Material Properties
Appendix G	Deterioration and Failure Modes
Appendix H	Validation
Appendix I	Glossary of Terms and Definitions

Fig. 9.14   API 579: 1999 – structure

projects where it is valued for its robust approach. It is certainly not a technique that an in-service inspector could use 'on-site'.

Summary – FFP assessments

From the discussion in this chapter, you can see that there is no such thing as 'the correct' FFP assessment to use in an individual situation. FFP assessments exist as a family of techniques to be used when the situation demands. Different techniques, or levels within those techniques, *will* give different results so the main skill is one of *interpretation* of assessment results rather than blind reliance on mathematics. If in-service inspection is about reducing risk (which it is) then you have to treat mathematical answers with care – the numerical answer to an engineering problem will never be absolutely correct, it will always be wrong – it is simply a matter of *how wrong*.

Damage mechanism	API 579 section	Summary
Brittle fracture	3.0	Assessment procedures are provided for evaluating the resistance to brittle fracture of existing carbon and low-alloy steel pressure vessels, piping, and storage tanks. Criteria are provided to evaluate normal operating, start-up, upset, and shutdown conditions.
General metal loss	4.0	Assessment procedures are provided to evaluate general corrosion. Thickness data used for the assessment can be either point thickness readings or detailed thickness profiles. A methodology is provided to utilize the assessment procedures of Section 5.0 when the thickness data indicate that the metal loss can be treated as localized.
Local metal loss	5.0	Assessment techniques are provided to evaluate single and networks of local thin areas and groove-like flaws in pressurized components. Detailed thickness profiles are required for the assessment. The assessment procedures can also be utilized to evaluate blisters as provided for in Section 7.0.
Pitting corrosion	6.0	Assessment procedures are provided to evaluate widely scattered pitting, localized pitting, pitting which occurs within a region of local metal loss, and a region of localized metal loss located within a region of widely scattered pitting.
Blisters and laminations	7.0	Assessment procedures are provided to evaluate isolated and networks of blisters and laminations. The assessment guidelines include provisions for blisters located at weld joints and structural discontinuities such as shell transitions, stiffening rings, and nozzles.

Fig. 9.15   API 579: 1999 – damage mechanisms covered

Weld misalignment and shell distortions	8.0	Assessment procedures are provided to evaluate stresses resulting from geometric discontinuities in shell – type structures including weld misalignment and shell distortions (e.g. out of roundness, bulges, and dents).
Crack-like flaws	9.0	Assessment procedures are provided to evaluate crack-like flaws. Solutions for stress intensity factors and reference stress are included in Appendices C and D, respectively. Methods to evaluate residual stress as required by the assessment procedure are described in Appendix E. Material properties required for the assessment are provided in Appendix F. Recommendations for evaluating crack growth including environmental concerns are also covered.
High-temperature operation and creep	10.0	Assessment procedures are provided to determine the remaining life of a component operating in the creep regime. Material properties required for the assessment are provided in Appendix F.
Fire damage	11.0	Assessment procedures are provided to evaluate equipment subject to fire damage. A methodology is provided to rank and screen components for evaluation based on the heat exposure experienced during the fire. The assessment procedures of the other sections of this publication are utilized to evaluate component damage.

Fig. 9.15   (Continued)

Table 9.1 FFP assessment standards – a summary

	Fracture	Fatigue	Thermal fatigue	Creep	Metal loss	Pitting	Blisters and gouges	Misalignment	Fire damage
BS 7910	Yes	Yes	Yes	Yes	Yes	No	Partial	Partial	No
API 579	Yes	Partial	No	No	Yes	Yes	Yes	Yes	Yes
B31.G	No	No	No	No	Yes	No	No	No	No
R6/R5 (combination)	Yes	Yes	Yes	Yes	No	No	No	No	No

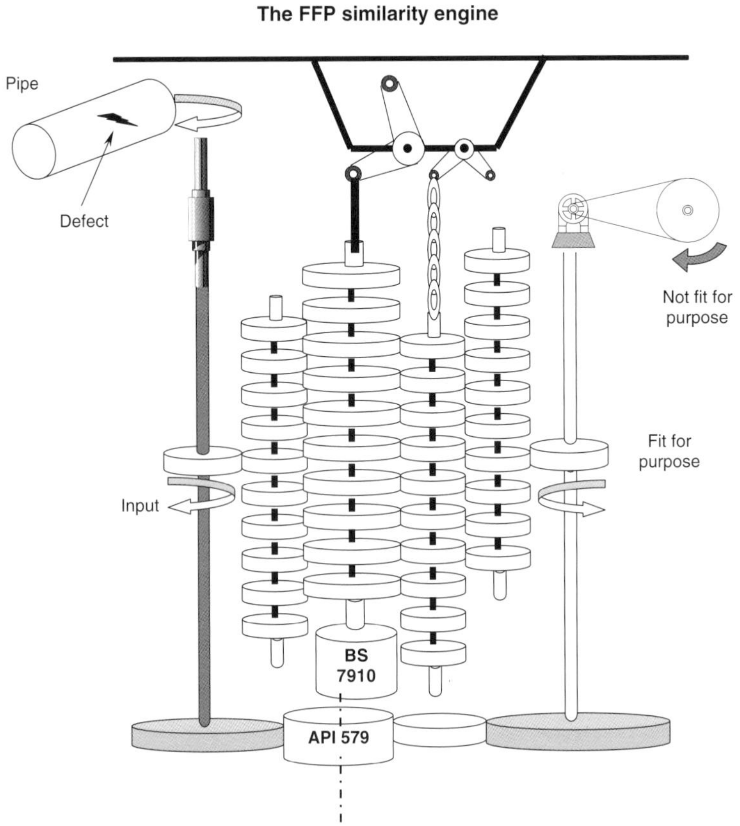

The FFP similarity engine

Fig. 9.16   Thinking outside the box. FFP studies – the FFP similarity engine

Sample BS 7910 calculation procedure

The following shows a typical example of the use of BS 7910 to analyse a header wall defect on a high-temperature boiler application.

Scenario

A plant began operation in April 1985 and a crack was detected during an inspection in July 1990. An outline calculation is required to decide whether the plant can continue to be operated until June 2005. The operating schedule of the plant is as shown in Table 9.2.

Fracture and crack growth properties

From text data, the following material properties are known:

Table 9.2 Operating conditions

Month	Pressure (bar)	Temperature (°C)
April, May, June, August, September, October, January, February, March	40	575
November, December	60	550
July	Shut down	

- Fracture toughness K_{IC}
  Lower bound $= 105$ MPa m$^{1/2}$

- Fatigue crack growth, upper bound

$(da/dN)_f = C(\Delta K)^m$

with

$C = 8 \times 10^{-11}$ and $m = 3$

for crack propagation in metres/cycle

- Creep crack growth

$\dot{a} = A(C^*)^q$
$A = 0.023$ upper bound
$A = 0.003$ mean
$q = 0.81$
for $\dot{a}$ in m/h, and C^* in MPa m/h

Hence for the worst loading conditions

$\Delta K_a = 19.29$ MPa m$^{1/2}$
$\Delta K_l = 13.60$ MPa m$^{1/2}$

Substitution in the fatigue law gives

$(da/dN)_f = 5.74 \times 10^{-4}$ mm/cycle
$(dl/dN)_f = 2.01 \times 10^{-4}$ mm/cycle

This crack growth during start-up is negligible and can be ignored.

Fatigue crack growth rate assessment for high-pressure start-up (U.3.7)

Margin against fast fracture

- Assume 2 blocks/month, 15 min start-up period, and 672 h steady operation.
- Need to check margin against fast fracture for most severe loading (high-pressure start-up).
- Level 2 procedure used.

$$K_r = \frac{K^{(p+s)}}{K_{\text{mat}}} + \rho$$

$$L_r = \frac{\sigma_{\text{ref}}}{\sigma_y}$$

$$K_{\text{mat}} = K_{\text{IC}} = 105 \, \text{MPa} \, \text{m}^{1/2}$$

$$\sigma_y = 113 \, \text{MPa}$$

Figure 9.17 and Table 9.3 show the situation.

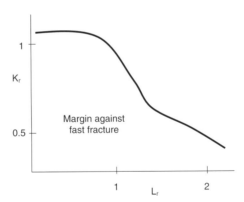

Fig. 9.17   Analysis

Table 9.3   Data summary

	$K^{(p+s)}$ (MPa m$^{1/2}$)	ρ	K_r	L_r
At surface	13.60	0.0212	0.151	0.96
Deepest point	19.29	0.08184	0.202	0.96

Determine margin against creep rupture in July 1990

Assume crack was present but not detected in April 1985.
Calculate creep damage D_c:

$$D_c = \frac{10 \times 672}{t_R(\sigma_{\mathrm{ref1}})} + \frac{48 \times 672}{t_R(\sigma_{\mathrm{ref2}})}$$

for 10 months at high pressure and 48 months at low pressure

At 550 °C $t_R\ (\sigma_{\mathrm{ref}\ 1}) = 3.87 \cdot 10^6\,\mathrm{h}$
At 600 °C $t_R\ (\sigma_{\mathrm{ref}\ 2}) = 4.40 \cdot 10^6\,\mathrm{h}$
$D_c = 0.009\ 07$
$D_c = \, << 1$

Hence there is adequate margin against creep rupture (U.3.8.3).
Units of C^*

$$1\,\mathrm{MJ\,m^2 h} = 1\,\mathrm{MPa\,m/h} = 1000\,\mathrm{N/mm\,h}$$

To convert crack growth law

$$\dot{a} = AC^{*q}$$

with $\dot{a}$ in mm/h and C^* in $\mathrm{MJ/m^2 h}$ to

$$\dot{a} = CC^{*q}$$

with $\dot{a}$ in mm/h and C^* in N/mm h

$$C = \frac{A}{(1000)^q}$$

Calculate performance from August 1990

It is necessary to calculate ε_c at July 1990 and increment of damage ΔD_c and creep strain $\Delta \varepsilon_c$ for each successive month using updated values of σ_{ref} for each new crack length. Table 9.4 shows the typical situation for August 1990 (U.3.8.4).

Table 9.4   August 1990 (U.3.8.4)

σ_{ref}	72.3 MPa
ΔD_c	1.53×10^{-4}
$\Delta \varepsilon_c$	9.17×10^{-6}
$K_a{}^p$	$9.05\,\mathrm{MPa\,m^{1/2}}$
$K_l{}^p$	$5.96\,\mathrm{MPa\,m^{1/2}}$
$C^*{}_a$	$1.54 \times 10^{-8}\,\mathrm{MPa\,m/h}$
$C^*{}_l$	$6.69 \times 10^{-9}\,\mathrm{MPa\,m/h}$

Calculate

$$C^* = \sigma_{\mathrm{ref}} \frac{\Delta_{\varepsilon_c}}{\Delta_t} \left(\frac{K^p}{\sigma_{\mathrm{ref}}} \right)^2$$

For

$\Delta t = 672\,\mathrm{h}$
$C_a^* = 1.54 \times 10^{-8}\,\mathrm{MPa\,m/h}$
$C_1^* = 6.69 \times 10^{-9}\,\mathrm{MPa\,m/h}$

Creep crack growth in August 1990
$\Delta a = AC^{*q}\Delta t$
$\quad = 7.3 \times 10^{-3}\,\mathrm{mm}$
$\Delta l = 3.7 \times 10^{-3}\,\mathrm{mm}$

Figure 9.18 shows the estimated creep crack growth characteristic.

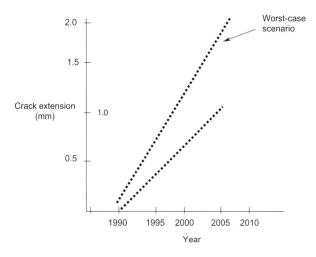

Fig. 9.18   Estimated creep crack growth characteristic

Bibliography

ASME B31G-1991 *Manual for determining the remaining strength of corroded pipelines – A supplement to ASME B31 code for pressure piping* (The American Society of Mechanical Engineers, New York).

Bjornoy, O.H., Fu, B., Sigurdsson, G., Cramer, E.H., and Ritchie, D. Introduction and background to DNV recommended practice, corroded pipelines (DNV RP-F101). In Proceedings of the 10th International Offshore and Polar Engineering Conference (ISOP-99), Honolulu, USA, 1999.

BS 7910: 1999 *Guide on methods for assessing the acceptability of flaws in fusion welded structures – The assessment of corrosion in pipes and pressure vessels* (British Standard Institution, London).

DNV recommended practice – corroded pipelines (RP-F101): 1999 (Det Norske Veritas, Oslo).

Fu, B. Advanced engineering methods for assessing the remaining strength of corroded pipelines. In Ageing Pipelines Conference on *Optimising the Management and Operation: Low Pressure and High pressure*, 11–13 October 1999, Newcastle upon Tyne, UK (The Institution of Mechanical Engineers, London).

Fu, B. and Batte, A.D. *Advanced methods for the assessment of corrosion in linepipe* OTO 97065, 1998, (UK Health and Safety Executive, London).

Fu, B., Jones, C.L., Stephens, D., and Ritchie, D. *Methods for assessing corroded pipeline – review, validation and recommendations, GRTC R3281.* Draft report to PRC International (BG Technology, Loughborough).

Hopkins, P. and Jones, D.G. A study of the behaviour of long and complex-shaped corrosion in transmission pipelines. In Proceedings of the 11th International Conference on *Offshore Mechanics and Arctic Engineering* (OMAE'9), Vol 5 (Pipeline Technology), Calgary, 1991, pp. 211–218 (The American Society of Mechanical Engineers).

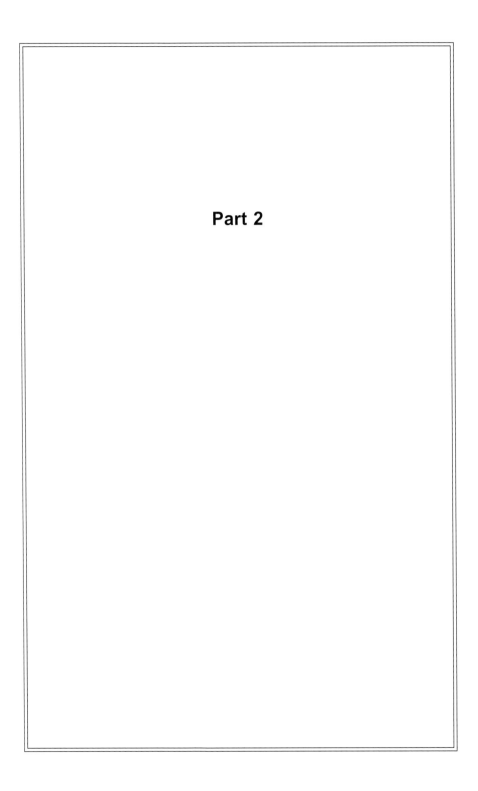

Part 2

Chapter 10

Pressure testing

The activity of pressure testing plays an important part in both pre-use 'works' inspection and in-service inspection of engineering equipment. It is mainly (although not exclusively) used for pressure equipment. Several types of test procedure including hydrostatic testing, proof testing, vacuum leak testing, pneumatic/leak testing and a few other specialist types all come under the generic name of *pressure testing*.

Pressure testing has two characteristics that are sometimes misunderstood:

- Pressure testing rarely has a *single objective*. Most types of test have two or more complementary objectives.
- Pressure testing is not a foolproof test of fitness for purpose (FFP). Neither is it a complete test for proving *integrity*.

These apparent weaknesses are not absolute – they vary with the type of test and are strongly linked to the technical aspects of an individual test procedure. This overall view of pressure testing is important – we will look at its component parts later in the chapter.

Pressure testing – safety first

Pressure testing can be a dangerous activity. Many types of pressure testing involve large quantities of stored energy with the ever-present danger of uncontrolled release. The risk is increased when the test involves in-service pressure equipment that is not new and which may contain defects, corrosion, repairs, etc. that add uncertainty to the situation. The individual hazards of pressure testing are well known and are shown in Fig. 10.1. All have been responsible for fatal accidents.

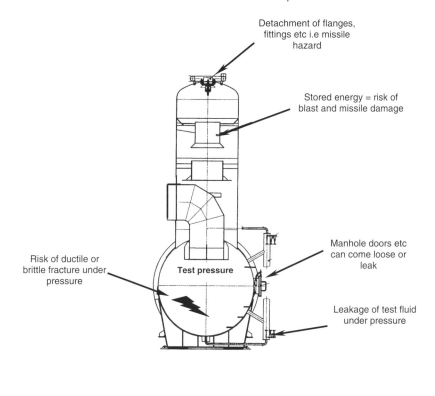

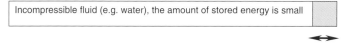

Fig. 10.1   Pressure test hazards

Stored energy

Stored energy constitutes the main danger associated with pressure testing (and the use of pressure equipment in normal service). Hazards are caused when stored energy is unintentionally released. The amount of energy stored depends on the phase of the test fluid:

- Compressed *liquid* has low stored energy. Most liquids (including water) are only compressible by a few percent (<4 percent) of their volume, even at high pressures. The amount of energy stored is therefore low (see Fig. 10.1).

- Compressed *gas or vapour*, either alone or as part of a multi-phase fluid mixture, is very compressible (up to more than 200 times that of water). This means that if a compressed gas or vapour (e.g. air or steam) is unintentionally released, large amounts of stored energy are released also. This results in the hazards of blast and missile projection.

Unintentional release of stored energy = blast damage + missile damage

The prime safety objective of pressure testing is therefore to reduce, as far as possible, the amount of stored energy held in the system under test. The ideal situation is for the test medium to be entirely liquid, with all air, gas, and vapour excluded. Even a small percentage of contained gas can increase significantly the amount of stored energy in a system. This method of testing, known as a *hydraulic test*, is not without risk but is by far the safest method and is used in preference to pneumatic testing wherever possible.

Pressure tests – practical dangers

Even when the amount of stored energy has been reduced to a minimum practical level, a pressure testing procedure involves many practical dangers. Figure 10.2 shows the main dangers in diagrammatic form, and the necessary precautions that need to be taken. Further details are given in two reference documents covering pressure test safety:

- HSE guidance note GS4 *Safety in pressure testing*, (3rd edition, 1998).
- HSE contract research report CRR 168 HSE.

Types of pressure test

Pressure testing is carried out on a wide variety of engineering equipment. Many items that are pressure tested during manufacture in the works are subject to periodic 'in-service' pressure tests during their operational life. The most common items are:

- pressure vessels (coded and uncoded),
- pipework systems,
- pipelines,
- isolation, control, and safety/relief valves,
- turbine casings,

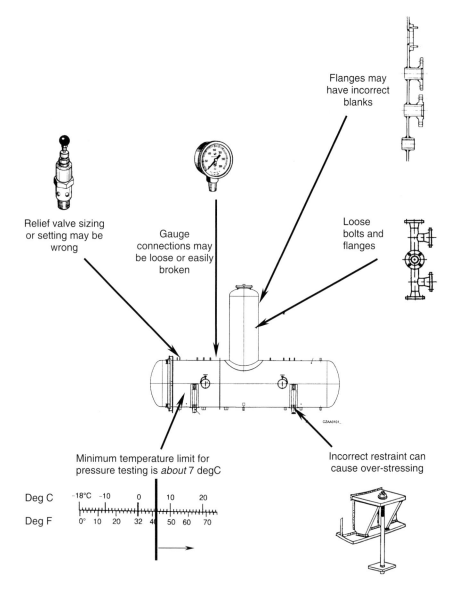

Fig. 10.2   Pressure testing – some practical dangers

- pump casings,
- heat exchangers,
- storage tanks (atmospheric and refrigerated),
- boilers,
- gas cylinders [termed transportable pressure receptacles (TRPs)],
- engine cylinders.

Some of these items are constructed to well-established design codes that specify pressure testing as a mandatory requirement while others are built to manufacturers' own technical standards, with pressure testing simply used as a technical option. In general, published codes and standards that deal with in-service inspection tend to be *less specific* about pressure testing than those that deal with new manufacture of the same equipment. The precise objectives of a pressure test depend both on the nature of the test and the type of equipment on which it is done. The most common types of pressure test are described below.

The standard hydraulic test

This is sometimes referred to as a *hydrostatic test*. Strictly, a hydrostatic test should involve a static head only (as in a test on a storage tank), i.e. without external pressure applied from a pump, but, in practice, the terms are often used interchangeably. It is performed on equipment in which the design criteria are well understood (in published design codes etc.) and hence the required pressure envelope wall thickness and stresses are predictable. A standard hydraulic test is carried out at a set multiple of design pressure. It is the most common pressure test.

The proof hydraulic test

This is a 'proving' hydraulic test used for new equipment or that which has been subjected to major design changes in service. The most common application of a proof test is on equipment in which the required wall thicknesses of all parts of the pressure envelope have not (or *cannot*) be accurately calculated. Pressure vessels of an unusual shape that do not fit comfortably into the scope of accepted codes (ASME VIII, PD 5500, EN 13445, etc.) are tested like this. A proof test is therefore part of the *design* process. There are few instances where in-service plant has its design amended sufficiently to warrant a true proof test. Note that a hydraulic test necessary because of a reduction in wall thickness of a vessel due to in-service corrosion is *not* considered a proof test – it is only a proof test if the remaining wall thickness *cannot be calculated*.

The full pneumatic test

This is similar to the standard hydraulic test except that it is carried out using air or nitrogen, rather than a liquid. A pneumatic test may be required if:

- The vessel or its supporting structure/foundations cannot physically support the weight of water necessary for a hydraulic test. Typical examples are large air-cooled condensers or 'fin-fan' coolers, large diameter (3–4 m+) horizontal low-pressure/vacuum vessels and ductwork, and large vessels on offshore platforms.
- The presence of liquid would contaminate the process, e.g. refrigeration, nuclear and specialist chemical process applications.

The full pneumatic test is very dangerous owing to the large amount of stored energy, and resulting blast/missile damage hazard in the event of an uncontrolled release or failure. Evacuation of persons from the immediate area is required and the test is often carried out in an underground pit or within composite steel and concrete blast walls. Specialist documents such as HSE GS4 contain data on calculating the amount of stored energy and blast wall requirements for various gaseous substances.

Pressure equipment design codes specify mandatory requirements before a pneumatic test can be carried out. As a rule, these apply to both new and in-service testing and are primarily concerned with doing non-destructive testing (NDT) before the test, and limiting the test pressure to give additional assurances on integrity. Figure 10.3 shows typical requirements.

Underwater full pneumatic test

The main application for this is for batch-manufactured gas cylinders (see Chapter 17). It combines a full pneumatic test with a leak test. The cylinder is immersed to a predetermined depth in a reinforced test tank full of water and then pressurized with air or inert gas. The test comprises two stages:

- *The full pneumatic test.* This requires similar safety precautions to a normal pneumatic test, with the concession that the personnel evacuation distance may be less (as long as the water tank is strong enough to withstand the hydraulic shock of an uncontrolled pressure release).
- *The leak test.* The pressure is reduced to <10 percent design of pressure before personnel can safely approach the tank and look for leaks, shown by bubbles. This leak test must only be done *after* the full pneumatic test.

- PD 5500 requires that a design review be carried out to quantify the factors of safety inherent in the vessel design. NDT requirements are those specified for the relevant Category 1 or Category 2 application *plus* 100 percent surface crack detection (MPI or DP) on all other welds.
- ASME VIII (part UW-50) specifies that all welds near openings and all attachment welds should be subject to 100 percent surface crack detection (MPI or DP).
- It is *good practice* to carry out 100 percent volumetric NDT and surface crack detection of all welding prior to a pneumatic test – even if the vessel code does not specifically require it.

The test procedure

- The vessel should be in a pit, or surrounded by concrete blast walls.
- Ambient temperature should be well above the brittle fracture transition temperature.
- Air can be used but inert gas (such as nitrogen) is better.
- Pressure should be increased very slowly in steps of 5–10 percent – allow stabilization between each step.
- PD 5500 specifies a maximum test pressure of 150 percent of design pressure.
- ASME specifies a maximum test pressure of 125 percent of design pressure, but consult the code carefully – there are conditions attached.
- EN 13445 and other harmonized standards specify requirements in accordance with the Pressure Equipment Directive (PED).
- When test pressure is reached, isolate the vessel and watch for pressure drops. Remember that the temperature rise caused by the compression can affect the pressure reading (the gas laws).

Fig. 10.3   Precautionary measures before a pneumatic test

The underwater test provides a very stringent test for leakage. Air will leak easily from small defects or areas of porosity, etc. that would not be shown by a standard hydraulic test.

The low-pressure pneumatic leak test

This is a standard test used for many types of heat exchangers, condensers and vessel/pipework systems. Test pressure is restricted to a few percent of design pressure (see Fig. 10.4) so that the extensive safety precautions associated with a full-pressure pneumatic test are not required.

Vacuum leak testing

Vacuum testing is less common during in-service inspection than during works inspection of new vessels.

Vacuum leak testing

Vacuum tests (more correctly termed *vacuum leak tests*) are different to the standard hydrostatic and pneumatic tests. The main applications are for condensers and their associated air ejection plant. This is known as 'coarse vacuum' equipment, designed to operate only down to a pressure of about 1 mmHg absolute. Most general power and process engineering vacuum plant falls into this category but there are other industrial and laboratory applications where a much higher 'fine' vacuum is specified.

The objective of a coarse vacuum test is normally as a proving test on the vacuum system rather than just the vessel itself. A vacuum test is much more searching than a hydrostatic test. It will register even the smallest of leaks that would not show during a hydrostatic test, even if a higher test pressure was used. Because of this the purpose of a vacuum leak test is not to try and verify whether leakage exists – rather it is to determine the *leak rate* from the system and compare it with a specified acceptance level.

Leak rate and its units

The most common test used is the isolation and pressure-drop method. The vessel system is evacuated to the specified coarse vacuum level using a rotary or vapour-type vacuum pump and then isolated.

- The acceptable leak rate is generally expressed in the form of an allowable pressure rise, p. This has been obtained by the designer from consideration of the leak rate in Torr litres per second.

  Leak rate $= dp \times$ volume of the vessel system/time, t, in seconds

  Note the units are Torr litres per second (Torr l/s); 1 Torr can be considered as being effectively equal to 1 mmHg. Vacuum pressures are traditionally expressed in absolute terms, so a vacuum of 759 mmHg below atmospheric is shown as +1 mmHg.

- It is a test for leakage only. It does not test the strength of the item in any way.
- It is useful for complex systems such as refrigeration or condensing plant.
- Gas leakage (normally nitrogen or similar inert gas) is detected using a hand-held 'sniffer' unit that can detect minute concentrations of escaping gas in an 'open-air' environment.
- It is a practical technique for use on existing plant, without major preparation work.
- Areas of likely leakage such as flange faces can be surrounded by adhesive tape, and the sniffer probe inserted through the tape, to capture minute leakages. This technique is used on equipment containing highly flammable or toxic fluid, where even the smallest leak would be a serious safety hazard.

Fig. 10.4   Some features of a low-pressure leak test

- It is also acceptable to express leak rate in other units (such as $1 \, \mu mHg/s$; known as a 'lusec') and other combinations. These are mainly used for fine vacuum systems.
- Because the leak rate is a function of volume, the volumes of all the system components: vessels, pipes, traps, bypasses, and valves need to be calculated accurately. It is not sufficient just to use the approximate 'design' volume of the vessels.

The test procedure

The procedure for the 'isolation and pressure drop' test is simple. Evacuate the system, close the valve, and then monitor pressure rise over time. The main effort however, needs to be directed towards the preparation for the test – it is surprisingly easy to waste time obtaining meaningless results if the preparation is not done properly.

Leaks are often difficult to locate. If the observed leak rate is above the specified levels but still relatively small make a double check on the tightness of the pressure gauge and instrumentation fittings – they are a common source of air ingress, particularly if they are well used and have slightly worn union connections. The next step is to isolate the various parts of the system from each other to identify the leaking area. The system can then be pressurized using low-pressure air and soap/water mixture brushed onto suspect areas. Concentrate on joints and connections – leaks will show up as bubbles on the surface.

Volumetric expansion testing

This is an adaptation of the standard hydraulic test used for gas cylinders. It is a destructive 'burst' test performed on a sample of cylinders from each batch. The cylinder is tested to destruction and the volumetric expansion attained before failure is measured (it is calculated from the additional volume of water added). The objective is to verify if the cylinder meets a specified *minimum* acceptable volumetric expansion – thereby giving an indication of the inherent ductility of the component design and the materials and fabrication techniques used. See Chapter 17 for further details on gas cylinder volumetric testing.

Pressure tests – technical aspects

The technical aspects of pressure tests are more complex than they first appear. It is also a fact that a variety of technical opinion exists about certain matters surrounding pressure testing. These differences manifest themselves as technical differences in the various codes and standards covering design and in-service inspection of pressure equipment.

Objectives of the hydraulic test

Figure 10.5 summarizes the main technical objectives of the standard hydraulic test. The principles are much the same for vessels or pipework, although they differ slightly in effect, owing to the different technical assumptions and factors of safety inherent in their design. Note the major point:

- A hydraulic test is *not* a complete test for fitness for purpose (FFP).

This applies equally to hydraulic tests during shutdown of in-service plant as it does to new plant tested during manufacture – these tests can only ever be a partial test for FFP.

A check for leakage under pressure
This is the main purpose. A hydraulic test will show areas of incomplete welding, material or weld porosity, or through-wall corrosion that might not be evident from casual visual observation. For assembled skids, pipework systems, and the like, it will reveal leaking flange joints, screwed connections, expansion bellows, etc. The use of a test pressure greater than design pressure helps exaggerate any leakage paths that may exist.

A standard hydraulic test is **not** a complete
test for FFP integrity of an item of pressure
equipment

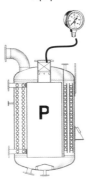

It has 3 main objectives

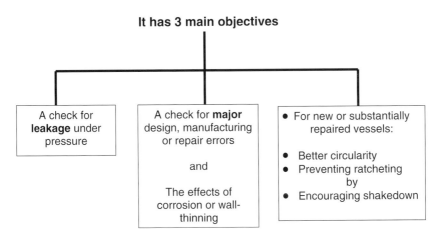

| A check for **leakage** under pressure | A check for **major** design, manufacturing or repair errors

and

The effects of corrosion or wall-thinning | • For new or substantially repaired vessels:

• Better circularity
• Preventing ratcheting by
• Encouraging shakedown |

Fig. 10.5   A standard hydraulic test – the technical objectives

A check for major design or construction errors

One way to think of this is that a hydraulic test provides a 'first estimate' of the ability of a component to resist static pressure stresses that will be imposed on it by the process fluid. It will reveal major errors in design (original and repaired parts) by plastic deformation (bulging or buckling) or general failure (*rupture*). There are several weaknesses in the true validity of this test:

• The test applies *static 'principal' stresses only*. It does not simulate any cyclic or pulsation effects that components will experience in service.
• The test is at *ambient temperature*. Although most design codes apply a factor to the test pressure to simulate the reduction of material yield

strength at elevated temperatures, this does not simulate stresses caused by thermal expansions or restraint that may occur during actual use. It also does not simulate any time-dependent reduction in yield or tensile strength that may result from operation in the creep range (400 °C+).

- It does not test the component to the limit of its 'factor of safety'. All well-designed and manufactured pressure equipment contains factors of safety to allow for uncertainty in the design process and material properties. These factors are well understood and allowed for when calculating the ratio between design pressure and hydraulic test pressure. In practice however these factors are frequently increased by:

    - 'overtolerance' of purchased material;
    - the use of a corrosion allowance on wall thickness;
    - the use of oversize stock material for convenience.

The net result is that the stress induced in the component during a hydraulic test is frequently *less* than intended. This means that the test is less stringent than perhaps was intended by the equipment designer – not necessarily a bad thing, but unintended nevertheless. In extreme cases, thicker material sections can actually *reduce* the strength of the material and produce an increased risk of defects, which will propagate over time (rather than instantaneously, during the hydraulic test).

An increase in design life

Some technical opinion maintains that the stressing of a component during a standard hydraulic test helps increase the design lifetime. The main mechanism is local yielding (plastic deformation) of highly stressed areas. This results in:

- Improved 'circularity' of the component. This leads to a more uniform stress distribution and a better approximation of the pure membrane stress assumptions used for the design of thin-walled pressure envelopes.
- Improved resistance to fatigue via a process known as *shakedown* (see later in this chapter).

These objectives of pressure testing are complementary rather than discrete. Their validity to the testing and re-testing of in-service equipment varies with each individual application. Even taken together, however, they do not form a foolproof test of FFP or integrity. Paradoxically, most in-service failure of pressure equipment results

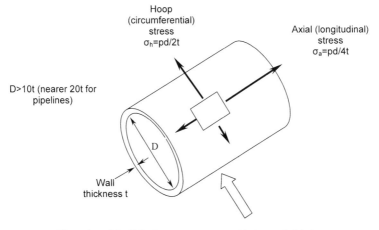

Hoop
(circumferential)
stress
$\sigma_h = pd/2t$

Axial (longitudinal)
stress
$\sigma_a = pd/4t$

D>10t (nearer 20t for
pipelines)

D

Wall
thickness t

Through-wall (radial) stresses are assumed to be negligible i.e.
it acts like a cylinder with no effective thickness. Under these
conditions, the hoop and axial stresses are known as
membrane stresses

In practice, more than 99% of pressure equipment items follow these 'thin-walled'
assumptions closely enough to give valid calculations of the stresses

**Fig. 10.6a   Thin-walled shell stresses in shells under internal pressure
(d is nominal diameter; p is internal pressure)**

from failure mechanisms, acting either singly or in combination, that
would *not* have necessarily been predicted, or prevented, by a prior
hydraulic test.

Stresses during hydraulic testing

The theory of hydraulic testing is based around the premise that most
items of pressure equipment are designed using *thin-walled shells*
assumptions (Fig. 10.6). This fits well with the actual design dimensions
of most pressure vessels and pipework and, to a partial extent, to
components with thicker walls such as headers, pump casings, turbine
casings, and valves. The thin-walled shell theory assumes that all shell
stresses are two-dimensional and tensile. The resultant circumferential
(hoop) and longitudinal (axial) stresses in cylindrical shells are known as
membrane stresses.

When a piece of pressure equipment is subject to internal pressure, the
stresses in the material and welds approximate to the general form of the
stress–strain curve for a tensile test. This assumption is not *exact*, but is
sufficiently accurate for equipment design and testing purposes. The
maximum stress is the circumferential (or hoop) stress, σ_h, i.e. that

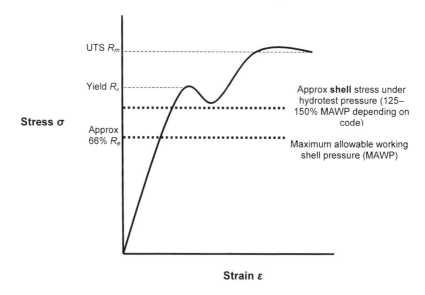

Fig. 10.6b   Shell stresses during the standard hydraulic test

Typical hydraulic test pressures table contents:

Code	Hydrotest pressure multiplier (x MAWP)	Notes
EN13445 unfired vessels	1.43	See section 10.2.3.3 of the code
EN 13480 piping	1.43	See section 9.3.2.2 of the code
ASME V111-1	1.3–1.5 depending on code	

which is trying to split the vessel open along a longitudinal plane. The design procedure therefore uses σ_h as the governing stress on which to base design calculations. As a general design principle, vessels are designed to experience a governing stress of about two-thirds of yield (66 percent of R_e in European notation and 66 percent of F_{ty} in US notation) when under their maximum allowable working pressure (MAWP). This varies numerically between design codes (and is subject to a range of complicating factors) but the general principles are much the same.

Figure 10.6 shows what happens during the standard hydraulic test. As the pressure is raised to 150 percent of MAWP or similar, the governing stress rises to > 90 percent of R_e, i.e. near the yield point in the parent material of the shell (sometimes called 'body stress'). This varies, depending on which design code is used as the basis of the calculation.

Temperature correction

Most hydraulic tests are carried out at ambient temperature, whereas the operating temperature of the vessel in service may be much higher; up to 560–600 °C in the case of boiler or gas turbine applications. A correction is therefore required to allow for the fact that the tensile yield properties vary with temperature. Three guidelines apply.

- Tensile yield strength, R_e, *reduces* as temperature *increases*. It is not an exact linear relationship but can be *approximated* to one.
- As a 'rule of thumb' for most materials, R_e does not reduce significantly up to about 200 °C. Above this, the material starts to soften.
- A general (ASME) calculation used for temperature correction is:

Hydraulic test pressure $1.5 \times \text{MAWP}$

$$\times \left(\frac{\text{allowable stress at test temperature}}{\text{allowable stress at design temperature}} \right)$$

Figure 10.7 shows a simple example. Note how the temperature correction *increases* the test pressure at ambient temperature.

Ratcheting and shakedown

Ratcheting and shakedown are two opposing characteristics of pressure vessel technology. They can also be controversial – not all design codes mention them, and those that do, do not go into much technical detail. They are, however, relevant to the testing of most types of new and used pressure equipment.

Ratcheting

Ratcheting is an *undesirable* characteristic in pressure equipment. It consists of a small but progressive plastic deformation of the pressure envelope material each time it is pressurized to its working pressure (see Fig. 10.8). Note that the material has not reached its *general* yield point at any time but is nevertheless still undergoing some plastic deformation. This deformation is caused by the physical design of the component producing local stress concentrations (i.e. it is not a simple idealized cylindrical shell without nozzles or fittings). If it is not

---

Hydraulic test pressure 1.5 × MAWP

$$\times\left(\frac{\text{allowable stress at test temperature}}{\text{allowable stress at design temperature}}\right)$$

So for a typical low-alloy steel material vessel designed to operate at 100 barg, 400 °C:

Allowable stress at 20°C = 250 MPa ⎫

Allowable stress at 200°C = 210 MPa ⎬   from ASME VIII – I

Allowable stress at 400°C = 155 MPa ⎭

Required test pressure at 20°C = 1.5 × 100$\left[\frac{250}{155}\right]$ = 242 ∗ barg

∗Note how this is *higher* than a normal 1.5 × MAWP (e.g. 150 barg). A test at 1.5 × MAWP would not result in a high enough stress to give a valid test.

Fig. 10.7   Hydraulic test temperature correction (ASME VIII-1)

addressed, ratcheting will continue through the life of the vessel resulting in progressive weakening of the material, bulging (strain), and eventually, the risk of failure at high-stress points. The term used for this is *incremental collapse*. The overall effect is therefore to shorten the life, or at least introduce a degree of uncertainty into the design life calculation.

Shakedown

Shakedown is a mechanism used to minimize ratcheting of a pressurized component during multiple pressurization cycles. Shakedown is therefore *desirable*. The idea is that the hydraulic test pressure chosen is sufficiently high to encourage significant local yielding at high-stress points. This causes a small degree of permanent strain, resulting in some work hardening, which *increases* the local effective tensile yield strength. This yielding happens at stress concentrations only, not in the 'body' of the component material (e.g. the shell or nozzle walls). When the test pressure is released, the 'shakedown' yielded areas are restrained by the surrounding bulk (unyielded) material. Figures 10.8 and 10.9 show the situation.

For standard pressure vessels, high-stress points such as set-through nozzles and compensation pads (particularly weld toes on these components) experience this desirable *shakedown* during a correctly

Ratcheting is:

Progressive plastic deformation of a vessel or pipe each time
it is pressurized to working pressure

The effects are:

Ratcheting – this is what happens

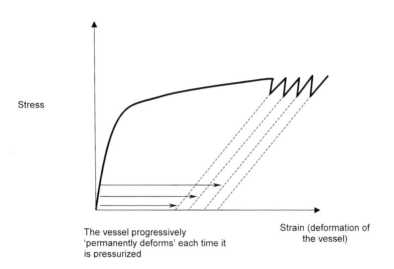

Fig. 10.8   Ratcheting

chosen hydraulic test. Such design features cause high local stress concentrations, so shakedown is achievable while the overall body stress in the shell is less than 100 percent R_e. Shakedown is the main reason why the hydraulic test pressure should not be too low – if a test results in shell stresses of less than about 90 percent R_e, there is no guarantee that high-stress weld toes will have achieved shakedown. This requirement is often ignored during a hydraulic test, particularly on used or heavily repaired plant, where the natural tendency is to reduce the test pressure from code requirements, 'so as not to overstress it'.

Shakedown:

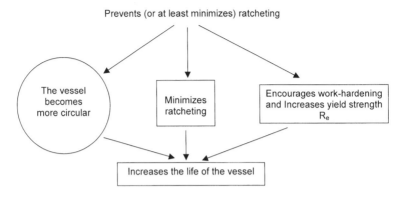

Shakedown – this is what happens

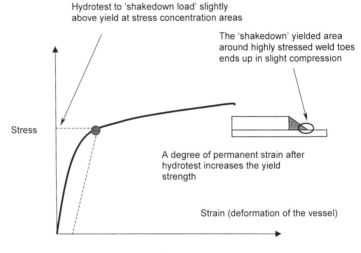

Fig. 10.9   Shakedown

In theory, all types of pressure equipment can benefit from shakedown. In practice, however, not all achieve it, and there is the added difficulty that it cannot be easily measured or its benefits verified.

Stresses in thick-walled components

An entirely different stress regime exists in pressure-equipment components that are 'thick walled', i.e. the wall thickness is greater

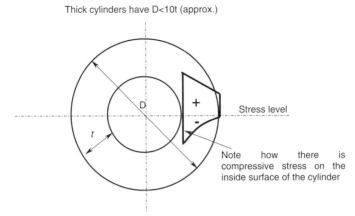

Thick cylinders have D<10t (approx.)

Stress level

Note how there is compressive stress on the inside surface of the cylinder

For thick-walled components, radial stresses become important. The stress regime is 3-dimensional and no longer follows simple 2-dimensional 'membrane stress 'assumptions. Complex 'Lamé' equations are required to calculate the stresses.

In practice, only a few items of pressure equipment (specialized headers, reactor vessels, etc.) require thick walled assumptions

Fig. 10.10   Stresses in thick-walled pressure equipment components

than about one-tenth of the effective diameter. Stresses do not follow the simple two-dimensional assumption of membrane stresses – instead they follow the so-called Lamé equations. Stresses are three-dimensional, i.e. with the addition of radial stress which may be compressive at the inner wall (see Fig. 10.10). This situation is outside the normal scope of most design codes. There are few types of pressure equipment that are truly 'thick walled' so in most cases, the simplified two-dimensional stress thin-walled assumptions are all that are needed.

In-service test procedures for pressure-equipment types

The term 'in-service pressure testing' refers to a situation where an item of pressure equipment is subject to a pressure test during a plant shutdown, at some interim point in its working life, i.e. it is not *new*. The requirement may be indicated on the written scheme of examination (WSE) for the equipment or may be required following major repair or discovery of defects or corrosion. In all cases, the need for in-service pressure testing has to be tempered by practicality – there are many pieces of equipment that, because of their physical geometry, are impossible to retest. There are also situations where a test is *possible* but may be deemed *undesirable* (too high risk or whatever). We can look at individual types of equipment in turn.

Unfired pressure vessels

Leak tests

These are common in vacuum vessels such as heat exchangers, ductwork, and vacuum condensers which need to be leak-tight for process reasons. Vessels which operate at a vacuum are normally *designed* to resist 1 barg positive pressure so they can be leak tested at a low positive pressure without problems. Common methods used are:

- Pressurize to 0.5 barg using air and detect leaks by covering in soap solution and looking for bubbles.
- Pressurize using nitrogen and detect leaks by gas sniffer.

Many large air-cooled condensers and similar will not support the physical weight of a full charge of water needed for a hydraulic test so leak testing becomes the only viable option. Leak tests are of less use on more standard vessels that contain water or other liquid under pressure in normal use. A leak test will only indicate whether there is any full through-wall defect and does not test the resistance to pressure stresses in any way.

Heat exchanger leak tests

Heat exchangers are often leak tested by filling one side (either the tube nest or the shell) with water dosed with a flourescent tracer chemical (normally hydrostatic head only). Leakage from tube-to-tubesheet expansions or cracked tubes can be detected using ultraviolet light. Figure 10.11 shows the procedure. Points to note are:

- This is a good in-service test for exchangers that use tubes made from cupro-nickel or titanium. These can suffer from microcracking and/or corrosion in service which is difficult to spot by visual observation alone even if the tube nest is pulled out of the heater during a shutdown inspection.
- It will *not always* detect leaking expanded tube-to-tubesheet joints. These have a frustrating tendency to leak at operational temperatures owing to differential expansion but not during a test at ambient temperature. This is a common occurrence in seawater cooling systems.
- The test can be particularly difficult for floating-head designs of heat exchangers (see Chapter 16). Unlike fixed-head types, floating head

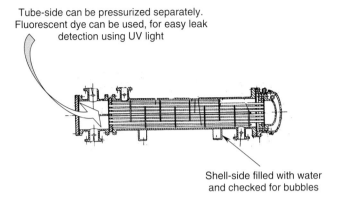

Tube-side can be pressurized separately.
Fluorescent dye can be used, for easy leak
detection using UV light

Shell-side filled with water
and checked for bubbles

Tube expansions sometimes leak at
operating temperatures but not at the
ambient test temperature

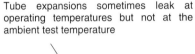

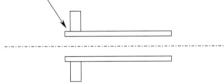

Fig. 10.11   Heat exchanger leak testing

designs do not always allow access to both ends of the tube nest
during a shell-side leak test. Special test rigs may be needed.

Hydraulic testing after repair

This is the most common reason for hydraulic testing of in-service
vessels and pipework. Vessel construction codes specify the conditions
under which hydraulic testing is required after a repair has been carried
out. The situation is not straightforward – some codes require specific
mandatory testing on certain thicknesses of material and when the
extent of repair exceeds a certain level while others leave the decision to
the judgement of an 'Authorized Inspector'. In all cases, there are
exemption clauses which can be implemented if it is technically
impossible to hydraulically test the item because of access, loading,
isolation, or similar restrictions.

Opinions vary about whether the benefits of 'shakedown' are
achieved properly after a repair to an in-service vessel. In theory, as

long as the vessel is hydraulically tested after repair to the original
design code requirements (and the repairs themselves are compliant with
original code requirements), then the repair will undergo shakedown in
the same way it would if the vessel was new. In practice, however,
vessels requiring major repair in-service are often heavily corroded, with
reduced wall thickness in other unrepaired areas. This means that the
hydraulic test pressure has to be recalculated (reduced) based on the
minimum remaining wall thickness. The result is that the new repaired
areas are not tested to a stress at which local yielding of stress
concentration areas happens and so shakedown does not occur. On
balance, shakedown is *not* normally the main objective of a hydraulic
test performed after a repair – leak tightness and general strength
considerations are seen as much more important.

Vessel test procedures
Design codes differ on the exact procedure specified for a hydraulic test.
Figure 10.12 shows the most common type – the pressure is raised
gradually to the code test pressure and held for a minimum time. It is
then reduced (gradually again) to MAWP, when the close visual
examination is carried out. The residence time at full test pressure is an
important aspect of a hydraulic test. There have been many cases where
leaks caused by defective weld material or porosity have not appeared
immediately – the test fluid takes time to find its way to the surface.
Figure 10.12 also shows some useful procedural 'good practice' points
to be observed during a hydraulic test.

Hydraulic testing of lined vessels

Pressure equipment may be lined with rubber, glass-reinforced plastic
(GRP), or various types of metal cladding for resistance to corrosion or
erosion. In most cases pressure testing is performed to code requirements *before* the application of the lining. The lining is then itself tested
after application using spark tests, adhesion tests, etc. Two exceptions
where the hydraulic test is carried out after cladding are:

- Vessels with internal coating applied as stainless steel or other
  corrosion resistant weld metal (known as *buttering*).
- Components with internal cladding applied as loose sheets and
  secured to the backing material by metal inert gas (*MIG*) or electrical
  resistance spot welds (known as *wallpapering*).

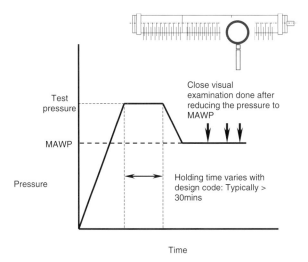

Test
pressure

Close visual
examination done after
reducing the pressure to
MAWP

MAWP

Pressure

Holding time varies with
design code: Typically >
30mins

Time

Witnessing hydraulic tests: some useful guidelines

• Surfaces should be blown dry with an air-line so that leakage can be detected accurately

• Tests need a minimum of two pressure gauges with the test pressure approx 75% full-scale deflection

• On large vessels (boilers, condensers etc) containing large volumes of test fluid some leaks can only be detected by a visual examination in dead spaces etc. It is not sufficient just to check the gauges for pressure drops

• For tests on austenitic stainless steel components, potable-grade or chemically dosed water is used to reduce the risk of stress corrosion cracking (SCC) initiated by chloride contamination

Fig. 10.12   A typical hydraulic test procedure

These two types are tested after lining because the lining application includes a welding process that, if done incorrectly, can affect the integrity of the parent material.

Pipelines

Pipelines are subjected to full hydraulic testing when new, and after major repairs. They may also be subject to (infrequent) periodic in-service testing depending on their contents and location. Hydraulic testing of pipelines follows a different procedure to that of vessels because of the difficulties of excluding all the air. Although pipeline design includes multiple bleed points, in a long pipeline with numerous bends, fittings, etc. it is almost impossible to exclude all of the air. This raises serious issues about the safety of this type of testing – barriers and exclusion areas are needed to keep personnel at a safe distance from above-ground parts of a pipeline.

The test itself typically lasts for up to 48 h. The pipeline is filled with water, pressurized to test pressure, and then calculations made of the volume of water added and the volume of entrapped air. The system is held at pressure for 24 h to allow temperatures to stabilize. A further measured volume of water is added over this period to maintain the pressure (otherwise it will slowly release owing to the entrapped air). Pipeline codes such as BS 8010 give acceptance criteria.

It is *normal* for pipelines to leak during the hydraulic test. For a pipeline to be fully leak-tight, it must be specified and constructed as a 'zero-leakage' design. An acceptance leakage rate for a non-hazardous process fluid would be

Maximum leakage rate = 0.1 litres per mm pipe diameter per km of pipeline per 24 hours for each 30 m$H_2$0 of pressure head.

So, for a 1 km pipeline of diameter 500 mm under a test pressure of 9 bar

Maximum leakage rate = $0.1 \times 500 \times 1 \times 3 = 150$ litres per 24 hours

Another relevant standard is ASTM E1003. This allows pressure to be pulsed, rather than maintained steady. In the UK this would normally need to be preceded by a full risk assessment to identify any dangers that might result.

Valves

Most pressure tests on valves are carried out in the manufacturer's works rather than during in-service shutdown inspections on site. Works tests may need to be witnessed by site inspectors, however, to make sure that they are done correctly and properly documented. The most common types of pressure tests are outlined below.

Valve shop tests

- Pressure relief valves (PRVs), control valves, and most types of general isolation valves are subject to a full hydraulic test on the valve body (the *body test*) and a leak test using air on the seat, to check its tightness. Tables 10.1 and 10.2 show typical data used for these tests.
- PRVs are 'pop-tested' to verify their lift pressure. See Chapter 12 for full details.

Table 10.1   Valve Hydraulic test pressures (ANSI B16.34 basis)

Material group no.	SHK1.L test by class – all pressures are gauge													
	150		300		400		600		900		1500		2500	
	lb/in²	bar	lb/in²	bar	lb/in²	bar	lb/in²	bar	lb/in²	bar	lb/in²	bar	lb/in²	bar
1.1	450	30	1125	77	1500	103	2225	154	3350	230	5575	383	9275	639
1.2	450	30	1125	78	1500	104	2250	156	3375	233	5625	388	9375	647
1.4	375	25	950	64	1250	86	1875	128	2775	192	4650	320	7725	532
1.5	400	28	1050	72	1400	96	2100	144	3150	216	5225	360	8700	599
1.7	450	30	1125	78	1500	104	2250	156	3375	233	5625	388	9375	647
1.9	450	30	1125	78	1500	104	2250	156	3375	233	5625	388	9375	647
1.10	450	30	1125	78	1500	104	2250	156	3375	233	5625	388	9375	647
1.13	450	30	1125	78	1500	104	2250	156	3375	233	5625	388	9375	647
1.14	450	30	1125	78	1500	104	2250	156	3375	233	5625	388	9375	647
2.1	425	29	1100	75	1450	100	2175	149	3250	224	5400	373	9000	621
2.2	425	29	1100	75	1450	100	2175	149	3250	224	5400	373	9000	621
2.3	350	24	900	63	1200	83	1800	125	2700	187	4500	311	7500	517
2.4	425	29	1100	75	1450	100	2175	149	3250	224	5400	373	9000	621
2.5	425	29	1100	75	1450	100	2175	149	3250	224	5400	373	9000	621
2.6	400	27	1025	70	1350	93	2025	140	3025	209	5050	348	8400	580
2.7	400	27	1025	70	1350	93	2025	140	3025	209	5050	348	8400	580

Valve 'in-service' tests

- PRVs are periodically 'hot-floated' in service to verify the lift pressure and blowdown settings in the operational hot condition.
- Valves and associated pipework systems are leak tested in 'test packs'. A test pack is a system or part of a system that is isolated using a closed valve or blank flanges so that it can be hydraulically tested in a single operation. Individual valves can be checked for leak tightness by subdividing the test pack into separate sections. Figure 10.13 shows a typical arrangement.

Table 10.2   Valve hydraulic test duration

Nominal size DN (NPS)	Minimum duration (min)	
	≤PN100 (≤Class 600)	>PN100 (>Class 600)
≤50 mm (2″)	1	3
65 mm (2½″) – 200 mm (8″)	3	8
≤250 mm (10″)	6	10

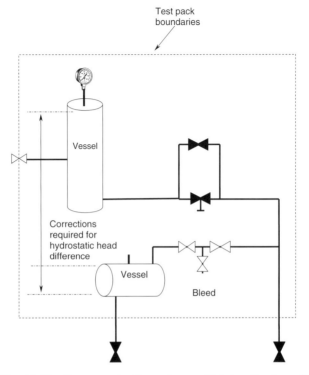

Fig. 10.13   Pressure testing of vessel/piping 'test pack'

Valve leak testing

Valve leak testing is applicable mainly to control valves and is covered by the American Fluid Controls Institute standard FCI 70-2 and API 598. Tables 10.3 and 10.4 show commonly used test data.

Table 10.3   *Control valve leakage classes (ANSI 70-2 basis)*

Class	Permittable leakage
I	No leakage permitted
II	0.5% of valve capacity
III	0.1% of valve capacity
IV	0.01% of valve capacity
V	$3 \times 10^{-7} \times D \times \Delta p$ (1/min)

D, seat diameter (in mm); Δp, maximum
pressure drop across the valve plug in bar.

Table 10.4   Valve leak testing (API 598 basis)

Drops per time unit	Amount (ml/min)
1 drop/min	0.0625
2 drops/min	0.125
3 drops/min	0.1875
4 drops/min	0.25
5 drops/min	0.3125
1 drop/2 min	0.03125
1 drop/3 min	0.0208
2 drops/3 min	0.0417
1 drop/4 min	0.0156
2 drops/4 min	0.0312
3 drops/4 min	0.0469
1 drop/5 min	0.0125

A rule of thumb conversion is: 1 ml $\cong$ 16 drops;
1 drop $\cong$ 0.0625 ml.

Storage tanks

Atmospheric storage tanks constructed to API 650 and BS 2654 are subject to a full hydrostatic test (no external pump, hydrostatic head only) after construction and after major repairs involving the floor annulus plates or lower shell strakes. The requirement for hydrostatic test after repairs, however, is optional in many cases (see the exemption chart in Chapter 14). As most tanks are designed to contain petroleum products with a specific gravity less than that of water (typically 0.8–0.9), the hydrostatic test using water acts effectively to simulate an overpressure test. Chapter 14 gives further details on the principles of testing storage tanks.

Rotating equipment casings

Casings for turbines, pumps, and some types of compressors are subject to hydraulic test during manufacture. They are generally cast components (and so are at risk of large-volumetric casting defects such as porosity) designed to the basis of pressure vessel codes (ASME-intent or similar) and so are tested to the relevant code hydraulic test requirements. Repeat hydraulic tests in-service are rare, unless there have been major repairs or a change in operating conditions

Pressure testing – the risk assessment

Integrity issues apart, the issue of personnel safety remains the most important aspect of pressure testing. While the number of accidents during pressure testing is small, the risks remain high. The risk is increased during in-service shutdown pressure tests which may be carried out on old and corroded plant, possibly with a poorly documented history of operating conditions, process transients, inspections, and repairs. For this reason, most responsible operating sites specify that pressure testing of all types should be covered by a permit-to-work (PTW) system.

The main part of the pressure test PTW is a detailed risk assessment. This includes all staff (in-house and external inspectors and others) that intend to be involved in it in any way.

Chapter 11

Pressure vessel inspections: API 510

The formal title for this code is API 510 *Pressure vessel inspection code: maintenance, inspection, rating, repair and alteration.* It is one of the few codes available that actually addresses the needs of in-service inspection and is kept frequently updated. Before using API 510 it is important to understand its relationship to other codes and published documents, and its formal status in the country in which it is applied. This is best explained by reference to Table 11.1 and Fig. 11.1. Table 11.1 shows the generalized situation for US-based code and documents while Fig. 11.1 attempts to explain the situation for API 510 in particular.

How is API 510 used in in-service inspection?

Outside the USA, API 510 is used (mainly in the refinery/petrochemical industries) for four main reasons:

- There is no equivalent European standard.
- It is recognized by regulatory bodies as representing valid technical practice within the industry.
- In terms of content, it matches reasonably well with European practice (or at least it does not directly *conflict* with it).
- It is concise and not too difficult to follow.

On close analysis, however, some of the methods in API 510 (and the cross-referenced assumptions of ASME VIII) differ significantly from the vessel design methods and tenets used in the European harmonized standard for unfired pressure vessels, EN 13445, and other industry documents such as BS PD 5500. It also has limitations because of its roots in the refinery industry, i.e. it does not match well with some accepted methods of defect detection and analysis used in other industries. In the UK situation, there is little or no cross-reference to the

Table 11.1 The US-based 'families' of codes and published documents

Components	Storage tanks	Power boilers	Unfired vessels	Piping systems	Valves	Pressure relief valves
Construction codes (i.e. manufacture only)	API 620 API 650	ASME 1	ASME VIII-1	ASME B31-1 ASME B31.3	ASME B16.34 API 600-609	ASME 1 ASME VIII API 2000
'In-service' codes (covering inspection, repair, alternation and re-rating)	API 653	—	API 510	API 570	API 598 API RP 591	API RP 576
'Support' technical documents	API 651 API 652 API RP 575	ASME RP573 ASME II, V, VI, IX	ASME B16.5 API RP 572 ASME II, V, IX	ASME II, V, IX API RP 574	API RP 574 ASME V, X	ASME PTC 25 API 527 ASME V, IX

API RP 579 'Fitness for service'
API Guides for Inspection of Refinery Equipment (IRE)

What is the status of API 510?
It is formally accepted in the USA as an ANSI standard. Some states (known loosely as *jurisdictions*) recognize API 510 as an alternative to the USA National Board Inspection Code (NBIC) and some do not. These two codes differ in some technical areas. API therefore acts as a 'legal requirement' in some states of the USA.
Outside the USA, API 510 has no special legal status at all; it has the same status as any other technical document, manual, or book.

Who uses it outside the USA?
API 510 is recognized mainly by refinery and petrochemical plant, and some offshore operators. These industries use plant which is specified to ASME/API construction codes. Some operators follow the technical requirements of API 510 directly and some use it as guidance only, preferring to follow their own inspection procedures that they feel are more relevant to them. Outside these industries, few plant operators (e.g. power, refrigeration, manufacturing plant, etc.) use it at all.

For which vessels can it be used?
- It only applies to *vessels*, not boilers.
- Strictly, it only applies to vessels that are constructed to ASME VIII. This is because the engineering methods (calculations, definitions of MAWP, design criteria, etc.) used in API 510 are based on the same engineering methods as those in the ASME code.

Fig. 11.1   API 510: Pressure vessel inspection code – a summary

methods or content of the Pressure System Safety Regulations (PSSRs) or in documents published by HSE, SAFed, or EEMUA.

In summary, API 510 works best in plants containing refinery/petrochemical equipment specified to full ASME code compliance, and in those states of the USA where it is formally recognized. Outside these limits, it is useful but probably limited in its application. As with all published in-service inspection standards, however, it provides a valuable 'back-up' resource on which the inspector can base inspection procedures and conclusions. A similar situation applies to the US National Board Inspection Code (NBIC).

The content of API 510

Figure 11.2 shows the contents list of API 510, and Fig. 11.3 shows some specific points of use to the in-service inspector. It is important to

8.2   Glossary of Terms
8.3   Inspection Programmme
8.4   Pressure Test
8.5   Safety Relief Devices
8.6   Records

APPENDIX A – ASME CODE EXEMPTIONS

APPENDIX B – AUTHORIZED PRESSURE VESSEL INSPECTOR
CERTIFICATION

APPENDIX C – SAMPLE PRESSURE VESSEL INSPECTION
RECORD

APPENDIX D – SAMPLE REPAIR, ALTERATION, OR RE-RATING
OF PRESSURE VESSEL FORM

APPENDIX E – TECHNICAL INQUIRIES

Fig. 11.2   Contents of API 510. *Pressure vessel inspection code: maintenance, inspection, rating, repair, and alteration*

remember that API 510 does not really merit the status of a 'stand-alone' document – it needs the reinforcement of the construction code ASME VIII and the support documents shown in Table 11.1 (i.e. API RP 572 and sometimes even the fitness-for-service code API 579) if it is to have real cohesion in an in-service inspection situation.

API 510 calculations

One of the cornerstones of API 510 is that it matches the engineering philosophy and methods of ASME VIII in the way that it assesses required wall thickness and stress levels. If calculations are needed as the result of an in-service inspection, the inspector has no option but to refer to the relevant paragraphs of the ASME code. These calculations can be a little confusing; Table 11.2 presents the main ones in simplified form. It is rare that an in-service inspector (working outside the USA) will need any other equations than these.

Scope
Vessels with:

volume $> 0.141\,m^3$ and $P > 250\,psig(1723.1\,kPa)$

volume $> 0.0442\,m^3$ and $P > 600\,psig(4136.9\,kPa)$

Common exclusions are:

- Pressure vessels on movable structures.
- Containers (exempt from ASME VIII-I).
- Everything else exempted by ASME VIII-I.

Inspection plan
API 510 infers that, to do a proper in-service inspection, you need to use an inspection plan (API RP 572 contains better details about what should be in the inspection plan).

Degradation mechanisms
API 510 summarizes these into only five broad categories (in paragraph 5) and suggests that they should be 'considered'. The categories are:

- Creep deformation and stress rupture.
- Creep crack growth.
- The effect of hydrogen.
- Interaction between creep and fatigue.
- Metallurgical considerations.

These categories (and others) are better explained in API 572, API 579/580/581 and API IRE

Corrosion and minimum thickness evaluation
The following principles are used:

- Corrosion is considered more important in areas of high stress and near nozzles.
- Widely scattered pits can be ignored as long as they are:

  - $\leq$ 50% wall thickness deep (excluding corrosion allowance).
  - total area $\leq 45\,cm^2(7\,in^2)$ within a 20 cm (8 in) diameter circle.
  - sum of dimensions along any straight line within the circle $\leq 5\,cm$ (2 in).

In-service inspection periodicity
Maximum period between internal or on-stream inspection is the smaller of:

- 50% of the remaining life of the vessels (based on corrosion rate). Or
- 10 years.

Remaining life

$$\text{Remaining life} = \frac{t_{actual} - t_{min}}{\text{corrosion rate}}$$

where t is the vessel wall thickness.

External inspections
These are generally required every 5 years (paragraph 6.3).

Pressure test after repair
The need for a hydrostatic test after a vessel repair is left to the discretion of the inspector (paragraph 6.5).
(See paragraph 6.4 of API 510 for possible waiving of internal inspection where corrosion rates are established as being below 0.005 in (0.125 mm) per year and other special conditions are met.

Fig. 11.3   API 510 – some useful points

API 510 reporting format

API 510 contains two sample reporting formats: a straightforward vessel inspection record in Appendix C (reproduced here as Fig. 11.4) and a record to be used for repair/re-rating activities (in Appendix D). Both concentrate on the formalized data recording requirements of ASME-coded vessels rather than any other recommended ways to record specific technical findings. They would probably not be sufficiently detailed to comply with the PSSRs or constitute a full engineering record of a complex in-service inspection. API RP 572 contains better examples.

In-service inspection of pressure vessels – engineering aspects

Despite the fact that pressure vessels are well covered by construction codes and standards, there are very few codes that address specifically

Table 11.2   Some calculations relevant to API 510 assessments

Category	Code	Section	Calculation/formula
Minimum thickness of shells (cylinders)	ASME VIII	UG-27-I.R. APP1-O.R.	$t = \frac{PR}{SE-0.6P}$ (i.r.) or $t = \frac{PR_0}{SE+0.4P}$ (o.r.)
Design pressure on shells (cylinders)	ASME VIII	UG-27-I.R. APP1-O.R.	$P = \frac{SEt}{R+0.6t}$ (i.r.) or $P = \frac{2SEt}{R_0-0.4t}$ (o.r.)
Minimum thickness of formed heads	ASME VIII	UG-32 (d), (e), (f)	Elliptical: $t = \frac{PD}{2SE-0.2P}$ Torispherical: $t = \frac{0.885PL}{SE-0.1P}$ Hemispherical: $t = \frac{PL}{2SE-0.2P}$
Nozzle reinforcement	ASME VIII	UG-37, Fig. UG-37	A_1, A_2, A_{41} must be greater than A for nozzles with no repad $A_1, A_2, A_{41}, A_{42},$ and A_5 must be greater than A for nozzles with repad
Hydro/pneumatic tests	ASME VIII	UG-99, UG-100	*Hydro:* $P = \frac{1.3\times P\times \text{stress @ test temp.}}{\text{stress @ design temp.}}$ *Pneumatic:* $P = \frac{1.1\times P\times \text{Stress @ Test Temp.}}{\text{Stress @ Design Temp.}}$

Table 11.2 (Continued).

Category	Code	Section	Calculation/formula
Weld joint efficiencies	ASME VIII	UW-11, UW-12, Table UW-12	RT-1: Full: 1.0 or 0.90 RT-2: Full on Cat A, Spot on Cat B (UW-11(a)(5)(b) 1.0 or 0.90) RT-3: Spot -1–50 foot weld − 1 for each welder 0.85 or 0.80 RT-4: Combination of above No RT: 0.70 or 0.65 Seamless: 0.85 if UW-11(a)(5)(b) is not met, 1.0 if UW-11(a)(5)(b) is met
Nozzle weld sizes	ASME VIII	UW-16, Fig. UW-16	t_c = smaller of $\frac{1}{4}''$ or $0.7 t_{min}$ t_{min} = smaller of 3.4 inches or thickness of parts t_1, t_2 = smaller of $\frac{3}{4}''$ or $0.7 t_{min}$ Leg = $1.414 \times$ throat Throat = $0.707 \times$ Leg
Corrosion rate/remaining life	API 510 and Body of Knowledge	6.4	Corrosion rate = $\dfrac{\text{metal loss}}{\text{time}}$ Corrosion allowance = actual T − min. T Remaining life = $\dfrac{\text{corrosion allowance}}{\text{corrosion rate}}$

	Form Date
	Form No
	Owner or User
	Vessel Name

Description

Name of Process	Owner or User Number
Location	Jurisdiction/National Board Number
Internal Diameter	Manufacturer
Tangent Length/Height	Manufacturer's Serial No.
Shell Material Specification	Date of Manufacture
Head Material Specification	Contractor
Internal Materials	Drawing Numbers
Nominal Shell Thickness	
Nominal Head Thickness	Construction Code
Design Temperature	Joint Efficiency
Maximum Allowable Working Pressure (MAWP)	Type Heads
	Type Joint
Maximum Tested Pressure	Flange Class
Design Pressure	Coupling Class
Relief Valve Set Pressure	Number of Manways
Contents	Weight

Special Conditions

Thickness Measurements

Sketch or Location Description	Location Number	Original Thickness	Required Minimum Thickness	Date

Comments (see note 2)

Method

Authorised Inspector

Notes:

1. Use additional sheets, as necessary

2. The location that each comment relates to must be described

Fig. 11.4   Vessel inspection record 'proforma' specified in API 510

the engineering aspects of their in-service inspections. One reason for this is the wide variety of different types and function of vessels that have to be inspected across a range of industries.

Generic areas such as corrosion and degradation, failure mechanisms, and inspection/NDT techniques are covered in other chapters of this book. These can all be used for pressure vessels, as long as they are applied in the correct context.

Peaking

One of the most common in-service defects in unfired vessels is distortion along longitudinal weld seams, known as 'peaking'. This is also one of the most common reasons for vessels retiral because of the way that it increases membrane stresses (mainly circumferential 'hoop' stress) to an unpredictable level. Peaking can occur as either *internal* peaking, where the weld peaks inwards towards the vessel centre or *external* peaking, where it projects radially outwards. Membrane stresses are increased dramatically and unpredictably, in both cases. For this reason, all pressure vessel design codes give strict limits on the amount of peaking that is acceptable. The same limits are generally used for in-service assessment – except in older vessels, or those operating in the creep range, where lower limits may be more appropriate. Peaking is measured using a bridge or chord gauge. These are used to measure profile distortion of vessel shell walls. Figure 11.5 shows the technique.

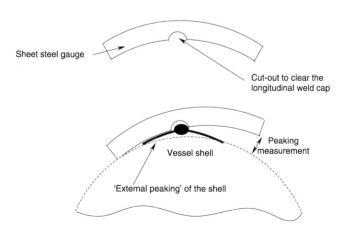

Fig. 11.5 Measuring peaking using a bridge gauge

Vessel inspection checklists

The best way to inspect vessels is using a dedicated checklist to make sure that nothing is missed. Figure 11.6 shows a typical one for a simple unlined low-pressure vessel operating at $<100\,°C$.

Name-plate (for older BS PD 5500 vessels)

- The presence of a name-plate is required, which is the operator's responsibility.
- The plate should specify the serial number.
- If 'XX' (in PD 5500 means Code Concession) is present at the end of the number, this indicates that there has been a problem at some stage, with the vessel being designated as outwith the code.
- Compare the details given on the name-plate with details on Form 'X'. They should be the same.

Pressure testing

- Check visually for distortion of the vessel. A fixed straight-edge can be used for this.
- Full visual inspection to all welded joins, looking for leaks. The vessel surfaces must be dry, to enable a proper inspection. (Can be blown dry with an air line first.)

Visual inspection
Particular attention should be paid to the following:

- Note any defects, particularly in areas of 'Discontinuity Stress', particularly the vessel-head-to-shell joints. These defects must be fully reported.
- All nozzles/attachments with $1\times$ nozzle diameter of the head-to-shell joints should be carefully inspected and reported (areas of 'Discontinuity Stress').
- The weld profile and appearance in areas of high stress, e.g. nozzle openings cut into shell, should be carefully inspected and fully reported.
- Loss of wall thickness (WT) or wastage, particularly in areas of fluid flow.
- Welded lifting lugs or hooks.

Fig. 11.6   In-service vessel inspection check list

- In outside environment vertical column-type vessels, wind loads can impose 'bending stresses' around the welds at the column base. These areas should be targeted.

Plate condition should be inspected for:

- **Corrosion.**
- **Arc strikes, spatter.**
- **Discolouration.**
- **Mechanical damage**, such as dents from external/internal. This gives 'Profile Distortion'. Check against the code for limits – to be measured over a 20° arc. Sharp edges or notches are not allowed. Accurate sketches to be made to show the damaged area, a suitable template can be made and used for this. Specify whether impacted from external or internal.
- **Bulging**. This gives rise to 'out-of-roundness'. As for Profile Distortions above, accurate sketches should be made, checking the allowable limits as in the code.
- **Peaking**. This is sometimes seen on long-seam welds. Peaking has tight acceptance limits in all vessel codes.
- **The head-to-shell welds** should be of a smooth profile, with a minimum chamfer of 1:4 on thicker plate sections.

Weld condition

- Galvanic corrosion, especially at weld toes.
- Cracks.
- Porous weld caps.
- Pitting.
- Erosion/wastage on weld roots.
- Mechanical grinding damage.
- Spatter/undercut.
- Temporary fittings/weld scars from construction or previous repairs.
- 'Unauthorized Welding'.

Comment in the report, on as-found compliance.

Internal fittings

- Check where fittings are welded to shell/head, or where they are welded to another fitting which in turn is welded to the shell/head, i.e.

Fig. 11.6   (Continued)

where there is a crack propagation path through the 'pressure-containing envelope'.

General

- Check if the vessel contains a 'Relevant Fluid', reference PSSRs 2000.
- The Safe Operating Limits (SOLs) should be clearly marked on the exterior or name-plate of the vessel.

Fig. 11.6   (Continued)

Chapter 12

Protective devices

The inspection of protective devices is a critical part of the role of the in-service inspector. 'Protective devices' is a generic term covering a variety of pieces of equipment, all of which perform the function of protecting system components such as vessels, pipework, etc. from the damaging effects of overpressure. Protective devices are safety-critical components and, in most countries, they are covered by statutory regulations requiring periodic in-service inspection. A large petrochemical/process plant may have an inventory of several thousand protective devices providing a continuous workload of inspection, testing, and overhaul.

Protective devices form a useful and tangible link with the topic of risk-based inspection (RBI) (see Chapter 7). The main reason for this is that the service conditions of protective devices vary widely, so they exist in many different areas of the 'risk spectrum'. Devices on high-pressure steam or hazardous-fluid service are considered very high risk while those in low-pressure or vacuum services are low risk. These risk classes can be addressed using RBI techniques in order to decide inspection periods and other details of the written scheme of examination used on an operating plant.

Protective device categories

The two main categories of protective device are pressure relief valves ('safety valves') and bursting (rupture) discs. The basic definition of a pressure relief valve (PRV) is a valve which is designed to open and relieve excess pressure under overpressure conditions, and then to re-close preventing further fluid flow once normal conditions have been restored. A pressure safety valve (PSV) performs a similar purpose, but is characterized by a rapid opening 'pop' action. These definitions have become mixed up in both published technical standards and manu-facturers' literature to the point where they are often used interchange-ably. Figure 12.1 shows the main types.

Pressure Relief and 'Safety' valves (PRVs)	Bursting (or rupture) discs
• Conventional pressure relief valve (enclosed bonnet).	• Conventional (domed) bursting disc.
• Balanced (bellows-type) pressure relief valve (vented bonnet).	• Scored tension-loaded bursting disc.
• Pilot-operated pressure relief valve.	• Composite bursting disc.
• Power-actuated pressure relief valve.	• Reverse-acting bursting disc.
• Soft-seat pressure relief valve.	• Graphite bursting disc.
• Vacuum vent valve (spring-loaded or weight-loaded).	
• Pressure/vacuum vent valve (spring-loaded or weight-loaded).	

*Note that the terminology of the various types of relief valves is open to interpretation. Traditionally, spring-loaded valves for boiler steam use were termed 'safety valves' and those for liquid applications termed 'relief valves'. Definitions are given in codes such as API RP 520 and API RP 576 but, in practice, the terms are used interchangeably.

Fig. 12.1   Protective device categories

PRV types

The main PRV types are as described below – note how some have specific characteristics that can affect their in-service inspection.

Conventional PRVs

The general understanding of a conventional PRV is one which has a closed barrel enclosing the spring in a pressure-tight area. This is to prevent the release of process fluid which may be toxic, flammable, or expensive. When the valve lifts, the fluid discharges to either atmosphere or closed manifold. Arrangements are required to prevent back-pressure in any closed header affecting the lift pressure of any valve.

Balanced PRVs

These are PRVs that incorporate a bellows to prevent the operation of the valve being affected by back-pressure (normally from a closed discharge collection header). The bonnets are enclosed, but contain a vent which discharges to a safe location if the bellows fail under back-pressure. They are commonly used in corrosive and flammable service in the refinery and petrochemical industries.

Pilot-operated PRVs

These use a small pilot valve to operate the main PRV. Typical applications are:

- Large low-pressure vessels [such as pressurized storage tanks (API 620)]
- Where a high back-pressure acts on the PRV so the pilot valve is designed to provide a balancing effect.
- Where large relief areas are needed. Pilot-operated PRVs allow 'full bore' discharge in which the process fluid can discharge over a wide area, unimpeded by a conventional valve seat design.

Pilot PRVs are limited to clean, low-viscosity fluid service. Most operate at fairly low temperatures as high temperatures can cause problems with the seals, diaphragms, etc. used in the pilot valve.

Power-actuated PRVs

These are an unusual type of hybrid PRV used on large power station boilers [sometimes known as ERVs (electric relief valves)]. They act as relief valves but, in reality, form part of the plant control system as they can be used to regulate steam pressure, rather than simply relieve it in the event of dangerous overpressure. Views vary (including in the design codes) as to whether these can be considered true PRVs because they need an external electric or hydraulic power source to operate them (see ASME 1 PG-674 for further details).

Soft seat PRVs

Soft-seat PRVs have resilient plastic seats (inserts or rings) instead of a conventional metal-to-metal seating surface. The objective is to achieve better pressure-tightness. The rest of the valve design is identical to other PRVs. They are finding increasing use on low-pressure 'package' steam boilers.

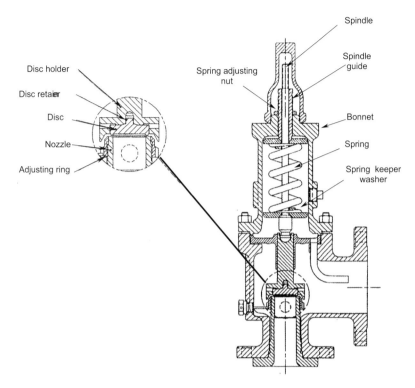

Fig. 12.2   A typical, high-pressure safety valve

Basic construction of PRVs

Safety valves are classified under Category IV in the Pressure Equipment Directive (PED), except for those that are fitted to vessels in a 'lower' category. Figures 12.2 and 12.3 show a typical high-pressure safety valve and Table 12.1 shows typical materials of construction.

PRVs – principles of operation

All conventional pressure relief valves operate on the principle of system pressure overcoming a spring load, allowing the valve to relieve at a defined capacity. The basic sequence of operation (see Fig. 12.2) is:

Fig. 12.3   A typical steam PRV

1. When the valve is closed during normal operation, fluid pressure acting against the seating surfaces is resisted by the spring force. With the system pressure below 'set pressure' by more than 1 or 2 percent, the valve will be completely leak-free.
2. As system pressure is applied to the inlet of the valve, force is exerted on the base of the disc assembly. The force produced by the compression of the spring counters this upward force.
3. When the operating system is below set pressure, the spring housing (or body) of the valve and the outlet are at atmospheric pressure (or at the superimposed back-pressure existing in the discharge header.)
4. As operating pressure begins to approach set pressure of the valve, the disc will begin to lift. This will occur within 1–2 percent of set point value and an audible sound will be produced – termed the *simmer* of the valve.

Table 12.1 Typical safety valve materials

Part	Typical material for standard service	Typical material for corrosive service?
Thread protector	Carbon steel	Carbon steel
O-ring	Rubber compound	Rubber compound
Lock nut	Alloy steel	Alloy steel
Washer	Carbon steel	Carbon steel
Thread seal	Carbon steel	Carbon steel
Adjusting screw	Carbon steel – zinc plated	316 Stainless steel
Drive screw	Stainless steel	Stainless steel
Label	Aluminium	Aluminium
Body	Carbon steel	Carbon steel
Spring keeper	304 Stainless steel	304 Stainless steel
Spring	17.7 Stainless steel	Inconel
Disc	17.4 Stainless steel	316 Stainless steel
Seat holder	303 Stainless steel	316 Stainless steel
Seat seal	Rubber/plastic compound	Rubber/plastic compound
O-ring	Rubber/plastic compound	Rubber/plastic compound
Seat guide	303 Stainless steel	316 Stainless steel
O-ring	Rubber/plastic compound	Rubber/plastic compound
O-ring	Rubber/plastic compound	Rubber/plastic compound
Seat frame	1018 Carbon steel	1018 Carbon steel

5. As the disc lifts the fluid force is transferred from the seat area to the additional area, substantially increasing the area being acted on. The result is that the amount of force being applied against the spring compression is dramatically increased. This causes the disc assembly to rapidly accelerate to the lifted position or *open* condition, resulting in a 'popping' sound.

6. The disc will not stay in the full open position and will begin to drop until an additional pressure build-up occurs. This over-pressure condition will maintain the valve in the full open position and allow it to discharge at maximum rated capacity (normally stamped on the valve body in SCFM (standard cubic feet per minute) or equivalent metric units.

7. As the system pressure begins to drop and the spring force overcomes the force created by the disc, the valve will begin to close. The reaction force becomes less as flow reduces, enabling the valve to shut clearly at its reset pressure.

8. The system pressure must drop below the set pressure before the valve will close. This process is termed the *blowdown* of the valve. Blowdown is adjustable and has a wire/seal locking arrangement to stop unauthorized interference.

PRV materials of construction vary greatly. They are chosen specifically from many hundreds of possible combinations to suit the PRVs' service conditions. Many PRVs operate in severe high-temperature, corrosive, or erosive conditions requiring specialized materials. Hence, *any* PRV components which have to be replaced during overhaul must be correctly matched to original specifications. Table 12.1 shows typical materials used on a PRV in petrochemical plant service.

PRV design codes

PRVs are classified fully as items of *pressure equipment* and so are heavily influenced by the principles of design codes such as ASME 1, ASME VIII, and similar. The PRV body (which is a cast component) forms part of the pressure-retaining envelope and so is designed to allowable stress and temperature parameters in a similar way to a pressure vessel. Codes also cover aspects of the sizing of a PRV, mainly parameters relating to its discharge capacity, i.e. its ability to relieve an overpressure condition quickly enough to prevent danger.

In-service inspection of PRVs

Most PRVs are subject to extensive in-service inspection. In the UK, they are classed as protective devices under the Pressure System Safety Regulations (PSSRs): SI 128 and so have to be formally incorporated in the written scheme of examination (WSE) covering an operating plant. The critical inspection activities are much the same for all PRV designs – with minor practical differences, depending on the type and service. These in-service inspection activities divide neatly into several well-defined types. Figure 12.4 shows the situation. Note that the cumulative reasons for all these inspection activities are to minimize the possibilities of:

- Failure of the pressure boundary.
- Failure to open correctly.
- Leakage across the seat.
- Failure to reseat correctly after lifting.

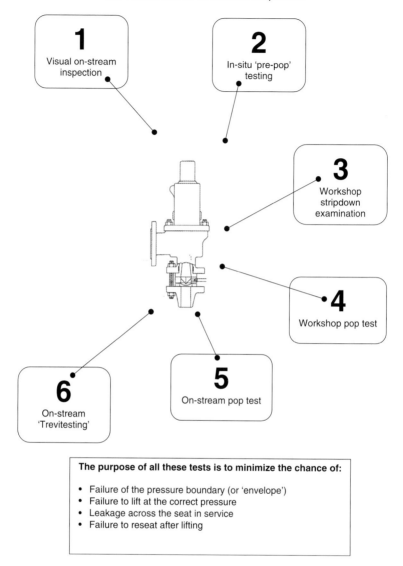

1
Visual on-stream
inspection

2
In-situ 'pre-pop'
testing

3
Workshop
stripdown
examination

4
Workshop pop test

5
On-stream pop test

6
On-stream
'Trevitesting'

The purpose of all these tests is to minimize the chance of:

- Failure of the pressure boundary (or 'envelope')
- Failure to lift at the correct pressure
- Leakage across the seat in service
- Failure to reseat after lifting

Fig. 12.4   In-service inspection activities for PRVs

In-service inspection plays a key part in making sure that all PRV inspection and test activities are done correctly. Serious accidents can result if PRVs are neglected, or testing is done in an incorrect or haphazard way.

On an operating plant, each PRV should:

- Have an identification plate containing its tag number and set pressure.
- Have the PRV manufacturer's identification plate containing a string of digits representing model number, design and coding information. A typical example could be (but varies between manufacturers):

8CX3HINFU

where

8: a 'series' number (from the manufacturer's catalogue data);
C: size of the orifice, e.g. C = 0.125 in^2;
X: material of construction, e.g. all stainless steel;
3: pressure class, e.g. up to 1500 lb/in^2 set pressure;
H: spring material, e.g. Inconel X-750;
I: inlet size, e.g. I = 25 mm;
N: codes representing details of inlet/outlet connections;
F: accessory code, e.g. F = packed lever;
U: code requirement, e.g. ASME 'u' standards.

- Be stamped with the cold differential test pressure (CDTP).
- Be referenced in the WSE and have a comprehensive file containing its datasheet, previous inspection reports, and overhaul/test results

Fig. 12.5   PRV identification and documentation – what to look for

Before you start – PRV identification and documentation

In most large plants (particularly refinery and petrochemical sites), each PRV should be separately identified, with individual marking and labelling. Figure 12.5 shows the things to look for. The first check before any inspection activity is to review the WSE and documentation file for each PRV in question to make sure that all the necessary information is present.

PRVs without identification
On poorly operated sites it is not unusual for some PRVs to have no identification at all, and (very often) no documentation records or previous inspection reports either. PRVs that do not have adequate identification (a site tag number or WSE references is absolutely essential) *are a safety risk.*

In the event of a failure incident or accident the lack of identification will probably be classed as a direct contravention of the PSSRs. Whether the accident was a direct result of the lack of identification or due to something totally different, the user/operator will be in a serious position. Any inspector who has allowed the situation to exist without opposition can also expect some embarrassing questions and could face accusations of negligence. Most competent plant owners/operators (and in-service inspectors) will withdraw PRVs found without identification instantly from service for either replacement or full inspection, retesting, and re-identification. If faced with opposition from a plant owner/ operator, some in-service inspectors would go so far as to class this a 'condition of imminent danger' and act accordingly under the PSSRs or other relevant legislation in force.

PRV visual on-stream inspections

On-stream visual inspections of PRVs are normally carried out before plant shutdowns and at regular periods between PRV testing and overhaul. The frequency is specified in the WSE. Being visual only, such an inspection does not provide a full examination, and is perhaps best seen as a control measure against which only limited conclusions can be drawn. Note from Figs 12.6 and 12.7 how the visual on-stream inspection covers several different areas – it is *not* just a check for leakage. The individual points identified in these two figures form the basis of a good WSE and should be itemized in any inspection report proforma used for recording the results of on-stream inspections.

PRV lift pressure ('pop') testing

The testing of the pressure at which a PRV lifts is generally referred to as 'pop' testing, i.e. the pressure at which the PRV 'pops open'. Pop testing is an important witness and reporting point for the in-service inspector. On most large plants pop testing is carried out at several different stages (see Fig. 12.8). The precise activities performed at each stage depends on individual plant operators' procedures and preferences. General guidance is provided in API RP 576. The PSSRs do not give specific recommendations – relying on their usual goal-seeking approach that PRVs should be tested 'as required'.

The basic parameter – cold differential test pressure (CDTP)
Cold differential test pressure (CDTP) is the pressure at which a PRV is adjusted to open under cold test conditions, i.e. ambient temperature on the test stand. It incorporates correction factors to allow for the

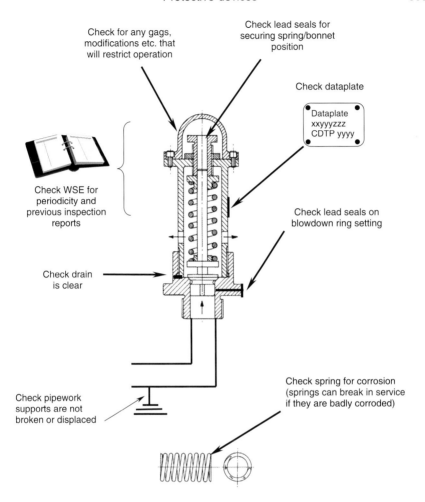

Check for any gags,
modifications etc. that
will restrict operation

Check lead seals for
securing spring/bonnet
position

Check dataplate

Dataplate
xxyyyzzz
CDTP yyyy

Check WSE for
periodicity and
previous inspection
reports

Check lead seals on
blowdown ring setting

Check drain
is clear

Check pipework
supports are not
broken or displaced

Check spring for corrosion
(springs can break in service
if they are badly corroded)

Fig. 12.6   PRV visual on-stream inspection – areas to check

differences between ambient temperature and the service temperature but it is *not* a test under service temperature conditions. The main use of CDTP is to specify a test pressure that can be used as a reference point for an individual PRV – hence the CDTP is normally stamped on the nameplate, or sometimes on the valve body casting itself.

The 'pre-pop test' itself

The pre-pop test is carried out either hot while still on the plant or, more commonly, cold, on a test stand in the workshop, after the PRV has been removed from the plant but *before* any cleaning etc. This is a key

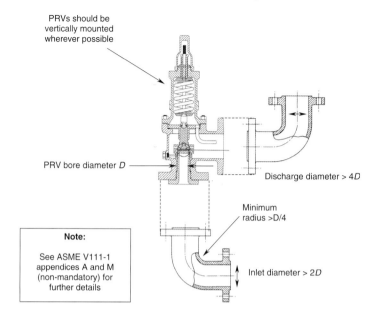

Fig. 12.7   PRV inlet/discharge connections – points to check

point; a shop pre-pop test is only properly valid if the PRV is in 'as-removed' condition. The purpose of the pre-pop test is to:

• Check if the PRV lifts at the correct CDTP, or is sticking due to:

  – 'blocking' by contaminants in the process fluid;
  – mechanical damage to the PRV component;
  – general corrosion.

• Provide data to justify the inspection interval in the WSE.

The pre-prop test therefore acts as a basic condition assessment for a PRV, providing a guide as to how *reliable* it is likely to be in service.

The number of tests at each stage depends on operator procedures and preferences. General guidance is available in API 576

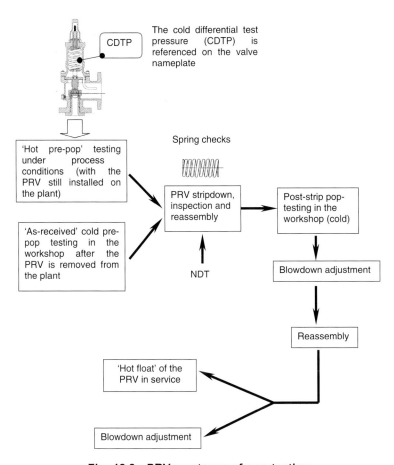

Fig. 12.8   PRVs – stages of pop testing

On most sites, the pre-pop test is given a pass/fail result. Again, the details vary depending on the operators' experience, with only general guidance available in published documents such as API RP 576 and the ASME code (see later in this chapter). Figure 12.9 shows a typical pre-pop procedure used in a petrochemical plant. Figure 12.9 shows a corroded nozzle ring on a PRV that failed its pre-pop test.

1. Pressurize progressively to 100% CDTP: if the PRV operates correctly at < 100% CDTP the result is recorded as 'satisfactory'.
2. If the initial pop is at a pressure > 100% CDTP, the PRV is tested a second time. If during this test it pops at or near 100% CDTP the original problem is concluded as being due to minor process deposits.
3. If the initial pop is at a pressure < 100% CDTP it is concluded that the spring is weakened or the PRV was incorrectly set (check previous records).
4. If the initial pop is at > 110% CDTP, this is considered a default, indicating significant corrosion or process deposits. The WSE inspection internal may need to be reviewed, depending on findings during the PRV stripdown.
5. There is a limit on maximum pre-pop pressure that can be applied to a PRV that is not opening (see below).

PRV set pressure	Maximum allowable pre-pop pressure
Set ≤16 barg	115% CDTP
Set > 16 barg	110% CDTP
Under special agreement with inspector present	130% CDTP

A PRV that does not pop at 130% CDTP is considered 'stuck shut' and requires further diagnosis during stripdown.

Fig. 12.9   A typical PRV pre-pop procedure used on a petrochemical plant

Workshop stripdown and inspection

Under most good WSEs, all PRVs are subject to workshop stripdown and inspection, irrespective of the pre-pop test result. The stripdown inspection consists of a comprehensive inspection of all the major components, including measurement and NDT where necessary. From the inspector's viewpoint, this should be a documented inspection with all findings carefully itemized and recorded in the relevant inspection reports. Figure 12.11 shows the major points to check during the stripdown inspection. In practice there are very few subcomponents of a PRV that can be safely re-used if defects or corrosion are found – parts are nearly always replaced with new ones. Figure 12.12 shows PRV components ready for inspection during stripdown.

Fig. 12.10   A PRV corroded nozzle ring

Workshop reassembly and testing

Following stripdown and inspection the next stage (a common witness point for the in-service inspector) is workshop testing. The extent of this depends on an individual valve's history. If, for any reason, the PRV body casting has been repaired or its pressure rating changed, the body will probably require a hydrostatic test (at 150 percent MAWP or similar code requirement). This is known as a 'body test' – the same as that done during new manufacture of the PRV.

A further interim test that may be carried out during workshop rebuild is the 'seat leakage test', also termed a bubble test. This uses low-pressure air on one side of the seat and water on the other side. Leakage across the seat appears in the form of bubbles. In practice, this test is more common on isolation and control valves than PRVs but some sites do use it. Published standards API 598, BS 6759 Part 3 and FCI 70-2 give technical details.

Setting of PRV lift pressure

Following rebuild a PRV is installed on the test stand, pressurized using air or nitrogen, and adjusted so that it relieves at the desired CDTP. It is then pop tested again (twice in some cases) to ensure that it lifts correctly. More than one pop test is generally considered desirable as it helps align the internal components of the PRV, ensuring smooth operation. Figure 12.13 shows typical tolerances. It is normal practice to

Preliminaries

Activity	Points to check
Identification check	• Parts from each PRV should be kept separate. • Obtain PRV documentation file with drawings and previous inspection/test reports.
Cleaning	Chemical and mechanical cleaning to remove corrosion deposits etc.
Inspection of main spring	• Check for external corrosion of spring. • Check uncoiled length against manufacturer's tolerances (as a rule, a spring should not show a permanent set of $> 0.5\%$ unloaded length). • Check flatness of spring ends. • Springs should only be replaced with properly identified 'original' items. • New springs may need to be 'scragged' ie. compressing several times in a test rig before use to eliminate any relaxation that has occurred in storage.
Inspection of PRV cast body	• Ultrasonic testing (UT) check of wall thickness. • Visual examination of internal and external surfaces for significant corrosion. • Visual examination of flange faces for pitting, cracking or erosion of seating surfaces. • Check for visible cracks (or use DP/MPI) around small radii. • Replication test to detect creep [high-temperature PRVs only ($> 400\,^{\circ}\text{C}$)].
Inspection of bellows (for balanced-type PRVs)	• Check for leaks.

Fig. 12.11   PRV workshop stripdown – main inspection points

	• Visual examination for stress corrosion cracking (SCC) on tight radii of bellows made from stainless or high nickel alloy grade materials. • Check for deformation (confirm measurements against manufacturer's catalogue data).
Inspection of PRV seating surfaces (seat and disc/lid)	• Check seat dimensions with 'seat gauge' to check they do not exceed allowable dimensions (usually specified as a maximum). • Check fit of seat in body (interference fit or locking ring, depending on the pressure class of the PRV). • Check seat flatness using lap ring gauge. • Visual examination of seat surfaces for 'wire-drawing' (sharp grooving of the seat faces caused by high-velocity fluid leakage paths). • Visual examination of soft-seat materials for indentation, cracking, and perishing.
Inspection of guide and disc	• Check fit for correct clearance. • Visual examination for scoring of clearance fit surfaces (galling can be caused by valve chatter or vibrating as well as corrosion).
Inspection of inlet and discharge nozzles	• Visual/DP/MPI examination for cracks or corrosion (inlet nozzles are normally in the worst condition). • Visual check for erosion of wall thickness around sharp radii. • MPI check for wall thickness reduction over critical areas.
Inspection of PRV spindle	• Visual check for corrosion (and SCC on stainless steel shafts). • Measurement check for shaft straightness (or by rolling on a surface plate). • Visual DP/MPI test for cracks and galling.

Fig. 12.11   (Continued)

General notes

In the vast majority of cases, the main degradation mechanism found during stripdown inspection of PRVs is corrosion-related. Attack by H_2S, Fe_2S, and general chloride contaminants is common for PRVs in chemical plants. High-temperature steam PRVs ($> 400\,°C$) can suffer from creep.

Fig. 12.11   (Continued)

hold the pressure at 90 percent CDTP for a while, to make sure that there is no seat leakage. Figure 12.14 shows a PRV test rig.

Storage and handling of PRVs before reinstallation

Most petrochemical plant operators place a limit on the time that a rebuilt PRV may be held in storage before it is reinstalled on the plant. This varies from 3 to 6 months. If the PRV is not used within this time it is treated as if it had been in service, and must be subject to a further 'pre-pop' test before deciding if it is suitable for reinstallation. All stored PRVs must retain their correct identification and documentation records, in order to maintain their position in the WSE.

The handling of rebuilt PRVs is an important area. A surprisingly large number of problems experienced with PRVs shortly after reinstallation are the result of poor handling after rebuild. General guidelines are:

- PRVs should be transported vertically (i.e. in the upright position).
- Flange faces should be protected by wooden blanks or similar.
- The correct rigging arrangement (slings etc.) should be used when manoeuvring the PRV into position.
- If a rebuilt PRV is bumped or dropped, it must be treated as if it has just been removed from the plant, i.e. returned to the prescribed procedure for pre-pop test, stripdown/inspection, and resetting before it can be legitimately reinstalled.

Trevitesting

'Trevitesting' is a proprietary name referring to a method of testing of PRVs on-line, i.e. in the 'hot' condition. It can be used, with adaptation, on conventional spring-type and pilot-operated PRVs. The objective of the technique is to assess the set pressure of the PRV and so identify (in

Fig. 12.12   Stripdown inspection of a bellows-type PRV

a similar way to the traditional pre-pop test) whether a PRV is in reliable condition.

Figures 12.15 and 12.16 show the arrangement. A hydraulic jack and force transducer are attached to the main valve spindle and a tensile force applied. This overcomes the compression force of the spring keeping the valve shut until it reaches a point where the force is sufficient to cause the valve to lift. From data collected, the equipment produces a documentary record of the set pressure, spring adjustment, and characteristic of the PRV. It is sometimes claimed that information can be gained on the reseating pressure (i.e. allowing determination of the valve *blowdown*) but in practice this is difficult.

The acceptability of this type of testing as a replacement for traditional pre-pop and stripdown inspection is not accepted by all plant operators. It is included in many WSEs as part of an overall risk-based approach to PRVs but mainly as an interim 'control measure' rather than as a full replacement for periodic stripdown and rebuilds.

PRV technical standards

The situation regarding published technical standards relating to PRVs is not as simple as for some other types of pressure equipment. The scope of existing published standards means that the role of in-service inspection requires involvement both with construction codes and those relating to in-service inspection. The situation is further complicated by

GUIDELINES

From ASME 1 UG 134(d)(1)
Maximum deviation of pop pressure from the set pressure is:

for set pressure $\leq 70\,\text{lb/in}^2$ deviation $\leq \pm 2\,\text{lb/in}^2$
for set pressure $> 70\,\text{lb/in}^2$ deviation $\leq \pm 3\%$

From ASME VIII-1 UG 125 (c) (3)
Maximum deviation is $+10$ to 0%

Some plant owners apply much wider deviation limits than these, e.g. $\pm 10\%$ on low-pressure PRVs (less than about 17 barg) and perhaps $\pm 5\%$ on those designed for set pressures greater than 17 barg.

In all cases, any adjustments to CDTP required to compensate for in-service high temperatures or back-pressure have to use the information presented on the PRV manufacturers' datasheet, to minimize the probability of errors.

Fig. 12.13   Setting of PRVs – tolerance guidelines

Fig. 12.14   A PRV test rig

the fact that the scope of both of these families of standards is incomplete – PRV standards show the characteristics of having developed organically over time, rather than to a well-defined logical plan. Figure 12.17 shows some general code-related guidance.

PRV construction standards

In the American family of design codes, the ASME code recognizes PRVs as essential items of pressure equipment and provides some guidance towards achieving operational integrity. Unfortunately the

'Trevitesting' is a method of determining PRV **lift pressure only** on-site (under operational conditions). A hydraulic tensioner pulls the spindle, compressing the spring until the valve lifts.

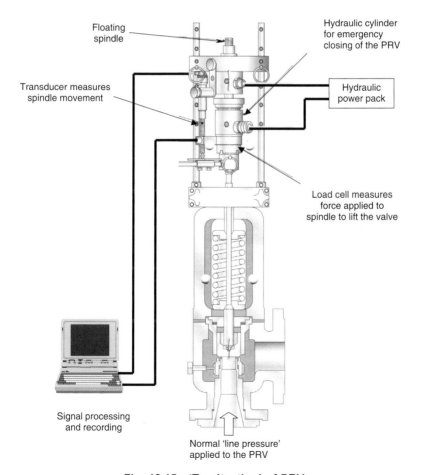

Floating spindle

Hydraulic cylinder for emergency closing of the PRV

Transducer measures spindle movement

Hydraulic power pack

Load cell measures force applied to spindle to lift the valve

Signal processing and recording

Normal 'line pressure' applied to the PRV

Fig. 12.15   'Trevitesting' of PRVs

Fig. 12.16   Trevitesting

main sections of the code do not provide comprehensive design rules for valve wall thicknesses so, in practice, PRV design draws heavily from two other standards:

- ASME/ANSI B16.5 *Pipe flanges and flanged fittings.*
- ASME/ANSI B16.34 *Valves; flanged threaded and welding end.*

API RP 520 Sizing, selection, and installation of pressure-relieving devices in refineries

This is divided into two parts: Part 1, *Sizing and selection*, and Part 2, *Installation*. Neither contains much information essential to the in-service inspection. Of the two, API RP 520 Part 1 contains the most useful general 'background' information. Figure 12.18 shows the table of contents.

API RP 576: 2000 Inspection of pressure-relieving devices

API 576 is a recommended practice (RP) published document rather than a formal API code. In common with other API RP documents, it provides wide-ranging technical coverage of PRVs without going into too much prescriptive detail. Useful parts of direct relevance to the in-service inspector are:

- Chapter 5: *Causes of improper performance* (includes details on corrosion, sticking, spring failures, etc.).

PRV relieving capability

ASME 1 (power boilers) specifies, with a few exceptions, that, when a PRV on a boiler lifts, the boiler pressure must remain below 106% of the PRV set pressure, and in all cases below 106% maximum allowable working pressure (MAWP).

ASME VIII-1 (unfired vessels) specifies that maximum pressure after PRV lifting must remain below 110% MAWP (minimum of 3 lb/in^2). If multiple PRVs are fitted the allowance is 116% maximum allowable working pressure (MAWP).

PRV set pressure

If a single PRV is fitted, it is normally adjusted to a set pressure at or below 100% MAWP. If more than one PRV is fitted, the other can be set to a higher lift pressure.

Fig. 12.17   PRVs – some useful code content points

- Chapter 6: *Inspection and testing* (covers the reasons for, and extent of, PRV visual inspection and testing).
- Chapter 7: *Records and reports.* This provides a good indication of the types of proforma reporting forms that are used for recording PRV in-service inspection details and results.

API RP 521: Guide for pressure-relieving and depressurizing systems

This RP is more biased towards the *design* of pressure-relieving systems, rather than inspection of individual components. The main content deals with the subject of relieving rates (i.e. how PRVs are sized) and sample calculations. There is little in the document that is essential reading for the in-service inspector.

API 526: Flanged steel pressure relief valves

Again, this is mainly to do with the manufacture of new PRVs, covering designs, materials, shop tests, and component identification. The details of shop tests (body hydrostatic test, seat leakage test, and set-pressure test) are very similar to those done during PRV site rebuild, so are of relevance. The concepts are not that different from those explained in API RP 576.

General
Section 1 – Introduction
1.1 Scope
1.2 Definition of Terms
    1.2.1 Pressure Relief Devices
    1.2.2 Dimensional Characteristics of Pressure Relief Devices
1.3 Reference Publications

Section 2 – Pressure relief devices
2.1 General
2.2 Spring-Loaded Pressure Relief Valves
    2.2.1 Safety Valves
    2.2.2 Relief Valves
    2.2.3 Safety Relief Valves
    2.2.4 Pressure Relief Valves
2.3 Pilot-Operated Pressure Relief Valves
2.4 Rupture Disc Devices
2.5 Rupture Discs
    2.5.1 General
    2.5.2 Types of Rupture Discs
    2.5.3 Application of Rupture Discs
    2.5.4 Terminology of Rupture Disc Devices
2.6 Rupture Disc Devices in Combination with Pressure Relief Valves
    2.6.1 General
    2.6.2 Rupture Disk Devices at the Inlet of a Pressure Relief Valve
    2.6.3 Rupture Disk Devices at the Outlet of a Pressure Relief Valve
2.7 Other Types of Devices

Section 3 – Causes of overpressure
3.1 General
3.2 Process Causes (Excluding Fire)
3.3 Fire
    3.3.1 General
    3.3.2 Effect of Fire on the Wetted Surface of a Vessel
    3.3.3 Effect of Fire on the Unwetted Surface of a Vessel

Section 4 – Procedures for sizing
4.1 Determination of Relief Requirements
4.2 Retrieving Pressure
    4.2.1 General
    4.2.2 Operating Contingencies

Fig. 12.18   API RP 520 part 1 – table of contents

Table 12.2 ASME I and IV operating requirements for PRVs

Application	Allowable vessel overpressure (above MAWP or vessel design pressure)	Specified pressure settings	Set pressure tolerance with respect to set pressure	Required blowdown
Section I Boilers	6% (PG-67.2)*	One valve $\leq$ MAWP Others up to 3% above MAWP (PG-67.3)	$\pm 2\,\text{lb/in}^2$ up to and including $70\,\text{lb/in}^2$ $\pm 3\%$ for pressures above $70\,\text{lb/in}^2$ up to and including $300\,\text{lb/in}^2$ $\pm 10\,\text{lb/in}^2$ for pressures above $300\,\text{lb/in}^2$ up to and including $1000\,\text{lb/in}^2$ $\pm 1\%$ for pressures above $1000\,\text{lb/in}^2$ (PG-72.2)	Minimum: 2% of set pressure or $2\,\text{lb/in}^2$ whichever is greater Maximum: 4% of set pressure or $4\,\text{lb/in}^2$ whichever is greater, with some exceptions (see PG-72)
Forced-flow steam generators	20% (PG-67.4.2)	May be set above MAWP, valves must meet overpressure requirements (PG-67.4.2)	Same as above (PG-72.2)	Maximum: 10% of set pressure (PG-72)

Table 12.2 *(Continued)*

Application	Allowable vessel overpressure (above MAWP or vessel design pressure)	Specified pressure settings	Set pressure tolerance with respect to set pressure	Required blowdown
Section IV				
Steam boilers	$5\,\text{lb/in}^2$ (HG-400.1e)	$\leq 15\,\text{lb/in}^2$ (HG-401.1a)	$\pm 2\,\text{lb/in}^2$ (HG-401.1k)	$2\text{--}4\,\text{lb/in}^2$ (HG-401.1e)
Hot water boilers	10% for a single valve 10% above highest set valve for multiple valves	One valve at or below MAWP Additional valves up to $6\,\text{lb/in}^2$ above MAWP for pressures to and including $60\,\text{lb/in}^2$ and up to 5% for pressures exceeding $60\,\text{lb/in}^2$ (HG-400.2)	$\pm 3\,\text{lb/in}^2$ up to and including $60\,\text{lb/in}^2$ $\pm 5\%$ for pressures above $60\,\text{lb/in}^2$ (HG-401.1k)	None specified

*Note: References in parentheses are Code paragraphs. $1\,\text{lb/in}^2 = 6.89\,\text{kPa}$.

Table 12.3 ASME VIII operating requirements for PRVs

Application	Allowable vessel overpressure (above MAWP or vessel design pressure)	Specified pressure settings	Set pressure tolerance with respect to set pressure	Required blowdown
Section VIII *Divisions 1 and 2* All vessels unless an exception specified Exceptions: when multiple devices are used	10% or $3\,\text{lb/in}^2$ whichever is greater (UG-125c) (AR-150a) 16% or $4\,\text{lb/in}^2$ – whichever is greater (UG-125c1) (AR-150b)	$\leq$ MAWP of vessel (UG-134a) (AR-141) One valve $\leq$ MAWP, additional valves up to 105% of MAWP (UG-134a) (AR-142a)	$\pm 2\,\text{lb/in}^2$ up to and including $70\,\text{lb/in}^2$ $\pm 3\%$ over $70\,\text{lb/in}^2$ (UG-134d1) (AR-120d)	None specified Note: Pressure relief valves for compressible fluids having an adjustable blowdown construction must be adjusted prior to initial capacity certification testing so that blowdown does not exceed 5% of set pressure or $3\,\text{lb/in}^2$, whichever is greater (UG-131c3a) (AR-512)

Table 12.3 (Continued)

Application	Allowable vessel overpressure (above MAWP or vessel design pressure)	Specified pressure settings	Set pressure tolerance with respect to set pressure	Required blowdown
Supplemental device to protect against hazard due to fire	21% (UG-125c2) (AR-150c)	Up to 110% MAWP (UG-134b) (AR-142b)		
Device to protect liquefied compressed gas storage vessel in fire (Div. 1 only)	20% (UC-125c3a)	$\leq$MAWP (UG-125c3b)	$-0\% + 10\%$ (UG-134d2)	
Bursting disc	Same as above	Stamped burst pressure to meet requirements noted above (UG-134 – Note 56) (AR-140 – Note 15)	$\pm 2\,lb/in^2$ up to and including $40\,lb/in^2$ $\pm 5\%$ of stamped burst pressure at specified coincident disc temperature (UG-127a1a) (AR-131.1)	

$1\,lb/in^2 = 6.89\,kPa.$

PRVs – glossary of terms

Figure 12.19 shows the definitions of common terms used in PRV codes and technical documents.

- **Back pressure**. Static pressure existing at the outlet of a PRV device due to pressure in the discharge line.
- **Blowdown**. The difference between actual popping pressure of a PRV and actual reseating pressure (expressed as a percentage of set pressure).
- **Bore area**. Minimum cross-sectional area of the nozzle.
- **Bore diameter**. Minimum diameter of the nozzle.
- **Chatter**. Rapid reciprocating motion of the moveable parts of a PRV, in which the disc contacts the seat.
- **Closing pressure**. The value of decreasing inlet static pressure at which the valve disc re-establishes contact with the seat, or at which lift becomes zero.
- **Disc**. The pressure-containing moveable member of a PRV which effects closure.
- **Inlet size**. The nominal pipe size of the inlet of a PRV, unless otherwise designated.
- **Leak test pressure**. The specified inlet static pressure at which a quantitative seat leakage test is performed in accordance with a standard procedure.
- **Lift**. The actual travel of the disc away from closed position when a valve is relieving.
- **Lifting device**. The device for manually opening a PRV, by the application of external force to lessen the spring loading which holds the valve closed.
- **Nozzle/seat bushing**. The pressure-containing element which constitutes the inlet flow passage and includes the fixed portion of the seat closure.
- **Outlet size**. The nominal pipe size of the outlet passage of a PRV unless otherwise designated.
- **Overpressure**. The pressure increase over the set pressure of a PRV, usually expressed as a percentage of set pressure.
- **Popping pressure**. The value of increasing inlet static pressure at which the disc moves in the opening direction at a faster rate as compared with corresponding movement at higher or lower pressures. It applies only to PRVs on compressible fluid service.

- **Pressure-containing member**. A pressure-containing member of a PRV is a part which is in actual contact with the pressure media in the protected vessel.
- **Pressure retaining member**. A pressure retaining member of a PRV is a part which is stressed due to its function in holding one or more pressure-containing members in position.
- **Rated lift**. The design lift at which a PRV attains its rated relieving capacity.
- **Safety valve**. A PRV actuated by inlet static pressure and characterized by rapid opening or pop action.
- **Set pressure**. The value of increasing inlet static pressure at which a PRV displays the operational characteristics as defined under 'popping pressure'. It is often stamped on the PRV's nameplate or body.
- **Seat**. The pressure-containing contact between the fixed and moving portions of the pressure-containing elements of a valve.
- **Seat diameter**. The smallest diameter of contact between the fixed and moving members of the pressure-containing elements of a valve.
- **Seat tightness pressure**. The specific inlet static pressure at which a quantitative seat leakage test is performed in accordance with a standard procedure.
- **Simmer**. The audible or visible escape of fluid between the seat and disc at an inlet static pressure below the popping pressure and at no measurable capacity. It applies to PRVs on compressible fluid service.

For further details see ASME PTC 25.3.

Fig. 12.19   Terminology – pressure relief valves (PRVs)

ASME I and ASME VIII

Although not directly related to the construction or testing of PRVs, ASME sections I, IV, and VIII contain extensive requirements for their functions and settings. Tables 12.2 and 12.3 show the details.

Chapter 13

Pipework and pipelines

Pipework is one of the main problem areas of in-service inspection. Technical aspects apart, it is a problem because:

- There is a *lot of it* – large plants contain several hundred miles of piping of various diameters.
- It is absolutely *impossible* to inspect it all.

The problem is not just logistical, pipework systems suffer from a wide and multi-faceted variety of corrosion and erosion problems, many of which are composite mechanisms that are difficult to both predict and detect with any degree of certainty. Various studies sponsored by HSE, plant operators, etc. over the past few years have confirmed what plant engineers have long realized intuitively – that pipework is responsible for a very large percentage of plant forced-soutage incidents. Colloquial reports from various studies suggest that:

- Pipework systems are responsible for more than 60 percent of unplanned fluid releases from operating plant.
- At least 20 percent of leaks are from gasketed joints on pipework systems.

Note how these are *leakage* incidents – catastrophic structural failure of pipework is (as for pressure vessels) fairly rare. All pipework constructed to recognized design codes (EN 13480, ANSI B31, ANSI B16.5 (flanges), etc.) incorporates high factors of safety. Where the diameter is below about 75 mm design factors of safety increase well above those found in more complex types of pressure equipment owing to the 'as-manufactured' schedule wall thickness being much thicker than required by code design calculations. The result is that true principal stress failure caused by design errors are very rare in pipework – problems are more likely to be caused by corrosion and other mechanisms.

Pipework types and classifications

Pipework systems are difficult to sort into simple classifications as they include many different categories of pressure, temperature, process fluid, and construction materials. Figure 13.1 shows the basic jurisdiction of what is generally known as pipework and Fig. 13.2 gives an idea of the three main subdivisions that can be identified. These subdivisions are realistic but arbitrary; there is no single classification that is recognized across all industry sectors. Pipework systems are also commonly classified in two other ways:

- *The PSSR definition.* The PSSRs subdivide systems into minor, intermediate, and major categories. This is based on a consideration of process fluid, and pressure volume product. These categories include 'normal' pipework included in the system but the situation is less clear regarding small-bore instrumentation pipework and similar.
- *By construction code.* This is a popular method originating from the classification of new pipework. Such subdefinitions contain common aspects; ANSI B16.5, for example, is used for forged flanges.

Note how Fig. 13.1 simplifies both of the above methods into three generic categories: high integrity, general process/utility, and small-bore systems. These categories are useful because they reflect the engineering realities of in-service inspection aspects formed in each. The categorization is *inspection friendly*, and can form the basis of the structure of a written scheme of examination (WSE) that deals properly with pipework, rather than treating it as subordinate to the inspection of vessels, heat exchangers, and suchlike. It also fits well with the concept of risk-based inspection (RBI).

Pipework systems and RBI

Pipework systems are the areas of process plant that are best suited to the application of RBI. The methodology of RBI techniques works best when the following conditions prevail:

- The plant items exhibit a spread of different risk levels.
- There are lots of separate systems, and bits of systems, to analyse rather than a few large discrete items (e.g. vessels).
- The plant items are of varying ages, contain different process fluids (giving different degradation mechanisms), and are in various conditions of (dis)repair.

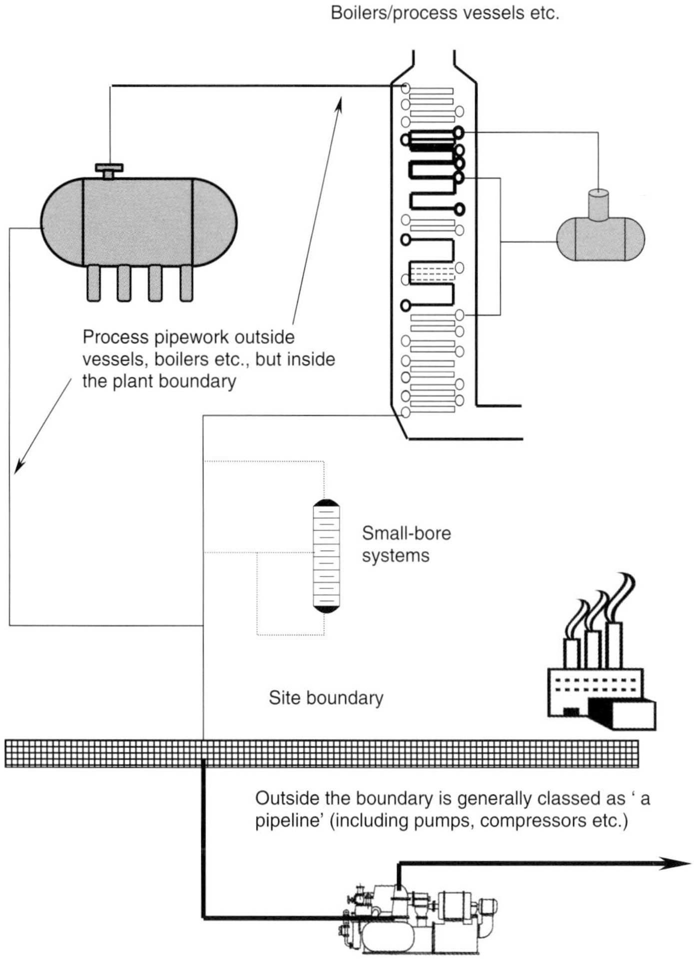

Boilers/process vessels etc.

Process pipework outside
vessels, boilers etc., but inside
the plant boundary

Small-bore
systems

Site boundary

Outside the boundary is generally classed as ' a
pipeline' (including pumps, compressors etc.)

Fig. 13.1   Pipework scope – the general understanding

Paradoxically, many large process plant organizations start their RBI
implementation on vessels, rather than pipework, perhaps because
vessels are more easily identifiable and easy to deal with. This is one
reason why RBI programmes often become ineffective and end up being
quietly sidelined. It is a much better idea to start with the pipework
systems, which can produce a better-organized inspection regime.

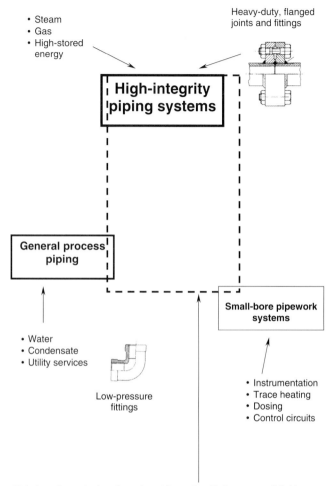

- Steam
- Gas
- High-stored energy

Heavy-duty, flanged joints and fittings

High-integrity piping systems

General process piping

Small-bore pipework systems

- Water
- Condensate
- Utility services

Low-pressure fittings

- Instrumentation
- Trace heating
- Dosing
- Control circuits

Note how these designations do not fit neatly with the system definitions in The PSSRs. e.g. this boundary shows the pipework types encompassed in a typical 'major system' under The PSSRs

Fig. 13.2   Pipework system classification – a way of looking at it

Pipework in-service inspection

Pipework systems suffer from the common problem of being difficult to inspect. The main problems are shown in Fig. 13.3 – note how most of them apply equally to both high-integrity and process/utility systems.

The biggest practical restriction on pipework inspection is the cost of scaffolding and lagging removal. Both are done by specialist contractors and are highly labour intensive. For plants that rise high above the ground (power station, chemical/refinery plant, etc.) it is not uncommon

Access: full access for inspection on large installations needs extensive scaffolding and staging.

Lagging: thermal lagging and metal cladding has to be removed before any visual inspection or NDT can be carried out

Identification and information: it can be difficult to fully identify and discriminate between pipework systems, particularly complex process/ utility systems. P&IDs and isometric drawings are not always up-to-date on older plants.

Redundant items: live pipework systems are often interspersed with redundant parts of old disused pipe systems. This causes confusion.

The PSSRs: although the scope of the PSSRs does include pipework systems, they are only applicable to those containing relevant fluid. Plants contain many pipework systems that are not a relevant fluid.

Shutdown: any kind of intrusive inspection of pipework requires plant shutdown.

The end result?
Pipework systems are very rarely subject to 100% inspection and many are rarely inspected properly *at all*.

Fig. 13.3 Pipework inspection – the difficulties

for the cost of scaffolding and lagging removal/replacement during a shutdown to exceed the cost of the NDT and mechanical inspection work that follows.

High-integrity pipework has its own, mainly technical, problems. Degradation mechanisms such as fatigue and high-temperature creep are not easy to detect by visual inspection or basic NDT. It is possible (and common) for a plant operator to commit to extensive shutdown inspection and NDT and *not find* defects that cause failure a short time after the plant is recommissioned. Both the form of the degradation mechanisms and the practical restrictions on NDT of pipework system components contribute to this. These difficulties of pipework inspection are, however, not insurmountable – they can all be overcome by adopting a robust approach to the process, rather than considering pipework as an 'add-on' to the vessels and larger components of a plant. In many plants, this means designing an inspection programme that exceeds the statutory requirements of the PSSRs, and then applying the principles of RBI (and engineering common sense) to best effect.

Codes and standards

As for other types of engineering plant, there are fewer published standards covering in-service inspection of pipework than there are dealing with its design and new construction. Pipework construction codes are published by most of the main international standards bodies (EN, ISO, BS, ASME, API, AWWA, etc.) but these deal purely with design and manufacturing issues rather than in-service degradation. At the other end of the technical scale, detailed fitness-for-purpose (FFP) analysis of pipework is covered by dedicated FFP assessment standards such as BS 7910 and API 579 (see Chapter 9).

For in-service inspection of pipework, the main standards that are available are exclusively American; published by the American Petroleum Institute (API). These have developed from experience in the refinery industry. They are:

- API 570 *Piping inspection code – inspection repair, alteration, and re-rating of in-service piping systems.*
- API RP 574 *Inspection practices for piping system components.*

While these are perhaps not 'cutting-edge technology' standards, they do have practical applications, and can provide some useful technical background to in-service inspection of industrial pipework of all types.

API 570 Piping inspection code

API 570 is the most commonly used technical standard to provide guidance on pipework inspections. It is a sister standard to API 510, which gives similar coverage for pressure vessels. Figure 13.4 shows the scope of the content of API 570.

Note that not all pipework systems fall within the scope of API 570. It concentrates on higher-risk systems rather than small-bore or 'utility' pipework. The main technical content of API 570 covers the 11 separate risk areas shown in Fig. 13.5. It explains the scope and nature of recommended inspections in each of these areas. These 11 areas are identified in API 570 with the *inference* that they should be covered by the written scheme of examination. They are chosen based on the background of degradation mechanisms experienced in petrochemical/refinery plant. The scope of these degradation mechanisms is, however, only a fraction of those included in for example API 580/581 (see Chapter 7). Note also that some of the terminology used, e.g. 'environmental cracking' is not the same as that used in other, non-API standards.

API 570 – what's in it?

1. Inspection and testing practices.
2. Frequency and extent of inspections.
3. Thickness calculations.
4. Repair, alteration, and re-rating.
5. Inspection of buried pipelines.

The scope of API 570

Included: almost any metallic piping system	Excluded
• Raw and finished oil products. • Raw and finished chemical products. • Hydrogen/fuel/flare gas systems • Sour/waste systems etc.	• Water, steam, and condensate. • Piping related to mechanical equipment (pumps/ compressors etc.). • Pressure vessels. • Domestic sewers etc. • Everything $< \frac{1}{2}$ in diameter.

Fig. 13.4   The scope of API 570

One of the more important inspection areas specified in API 570 is injection points. These are a common source of corrosion. API 570 defines the extent of the injection point circuit (IPC) that surrounds the injection point and should be inspected on a regular basis. Figure 13.6 shows the IPC, as used by most process and refinery plants.

API 570 pipework classes

API 570 defines three levels of pipework *class*. Although these are generalizations, they are generally applicable to many process plant types. Some plant users alter them slightly, while retaining general compliance with the spirit of API 570. Table 13.1 and Fig. 13.7 summarize the situation. These risk classes are a feature of API 570 and are not necessarily recognized as being an accurate representation of the degree of risk of pipework systems in all operating plants. In practice, they rarely correspond to the site-specific results of an RBI analysis. The main reason is individual differences in process fluids and pipework systems rather than any inherent weakness in the API 570 approach.

Checks are needed for specific types of corrosion and cracking:
1. Injection points.
2. Dead legs.
3. Corrosion under insulation (CUI).
4. Soil/air (S/A) interfaces.
5. Erosion and erosion/corrosion.
6. Environmental cracking.
7. Corrosion under linings.
8. Fatigue cracking.
9. Creep cracking.
10. Brittle fracture.
11. Freeze damage.

Fig. 13.5   The eleven 'risk areas' identified by API 570

API 570 retains its place as a document that can be used for general guidance rather than a set of prescriptive step-by-step instructions. Figure 13.8 shows, the index of API contents.

API 570 calculations

As with API 510 *In-service inspection of pressure vessels* (see Chapter 11), API 570 is closely linked to the design philosophy of the corresponding pipework construction codes used in the American Petroleum Industry. The calculation philosophy of API 570 is linked to that of ANSI B31 and the ASME code, with the same emphasis placed on remaining wall thickness as an indicator of fitness for purpose and remnant life. Although acceptable in some industries, this is seen as an oversimplified approach in others and one which needs to be qualified by more advanced FFP analyses such as API 579 and BS 7910 (see Chapter 9).

API 570 has no direct statutory significance in countries outside the USA. It is not mentioned in the UK PSSRs but is not expressly prohibited either so it can be used, with common-sense limitations, as a general guide for in-service inspection, when the occasion demands.

API RP 574 Piping system components

This is one of the API RP (recommended practice) series of documents that act as technical support to the formal published API standards. This recommended practice covers the inspection of piping, tubing, valves (other than control valves), and fittings used in petroleum

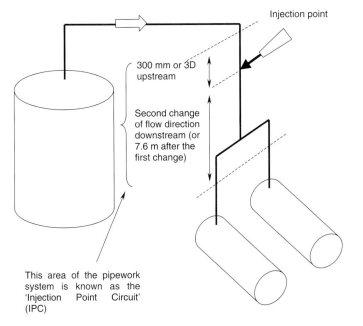

Injection point

300 mm or 3D
upstream

Second change
of flow direction
downstream (or
7.6 m after the
first change)

This area of the pipework
system is known as the
'Injection   Point   Circuit'
(IPC)

Fig. 13.6   The API 'injection point circuit'

refineries and chemical plants. Although not specifically intended to
cover speciality items, many of the inspection methods described are
applicable to items such as control valves, level gauges, instrument
control columns, etc. The document covers:

- A description of various piping components (piping, tubing, valves,
  fittings, joints).

Table 13.1   The basis of the API 570 risk classes

| Type of circuit | Maximum API inspection intervals (by class) | |
	Thickness measurements	*External visual*
Class 1	5 years	5 years
Class 2	10 years	5 years
Class 3	10 years	10 years
Injection points	3 years	By class
S/A interfaces	–	By class

The maximum permitted inspection interval depends on pipe *class*.
Class 1: highest risk/consequence.
Class 2: medium risk.
Class 3: low risk.

Class 1

- Flammable services/flash-off.
- Explosive vapours after flash-off.
- H_2S gas.
- Hydrofluoric acid.
- Piping over public roads.

Class 2

- Most other process piping.
- On-site hydrocarbons.
- H_2, fuel gas, and natural gas.
- On-site acid and caustics.

Class 3

- Fluids that will not flash-off (even if they are flammable).
- Distillate/product storage/loading lines.
- Off-site acids and caustics.

Fig. 13.7   API 570 pipe class definitions

- Reasons for inspection (safety, reliability, efficiency, and regulation).
- Inspection for deterioration in piping, including corrosion monitoring and various specific types of corrosion and cracking.
- Inspection tools and safety precautions.
- Procedures for various methods of inspection.
- Determination of retirement thickness for piping, valves, and flanged fittings.
- Necessity of keeping and reviewing complete records of inspections.

It acts as a useful support to API 570 but is probably not detailed enough to act as a stand-alone standard. It follows closely USA inspection practice and is not widely used in Europe and other countries outside the petroleum/refinery industry.

High-integrity pipework

The in-service inspection of *high-integrity* pipework is not well covered by API 570, which is more suited to process/utility systems. Included in this category are systems containing:

GENERAL
SECTION 1 – GENERAL

- 1   Scope
- 1.1 General Application
- 1.2 Specific Applications

SECTION 2 – REFERENCES

SECTION 3 – DEFINITIONS

SECTION 4 – OWNER–USER INSPECTION ORGANIZATION

- 4.1   General
- 4.2   API Authorized Piping Inspector Qualification and Certification
- 4.3   Responsibilities

SECTION 5 – INSPECTION AND TESTING PRACTICES

- 5.1   Risk-Based Inspection
- 5.2   Preparation
- 5.3   Inspection for Specific Types of Corrosion and Cracking
- 5.4   Types of Inspection and Surveillance
- 5.5   Thickness Measurement Locations
- 5.6   Thickness Measurement Methods
- 5.7   Pressure Testing of Piping Systems
- 5.8   Material Verification and Traceability
- 5.9   Inspection of Valves
- 5.10 Inspection of Welds In-Service
- 5.11 Inspection of Flanged Joints

SECTION 6 – FREQUENCY AND EXTENT OF INSPECTION

- 6.1   General
- 6.2   Piping Service Classes
- 6.3   Inspection Intervals
- 6.4   Extent of Visual External and CUI Inspections
- 6.5   Extent of Thickness Measurement Inspection
- 6.6   Extent of Small-Bore Auxiliary Piping, and Threaded-Connections Inspections.

SECTION 7 – INSPECTION DATA EVALUATION, ANALYSIS, AND RECORDING

- 7.1   Corrosion Rate Determination

Fig. 13.8   API standards 570, Piping inspection code – contents

Fig. 13.8   (Continued)

- High-pressure steam (ANSI B31.1) (see Fig. 13.9).
- High-pressure gas (ANSI B31.3).
- Other high-pressure fluids containing a gaseous phase, i.e. that contain high levels of stored energy.
- Fluids at high temperature, above about 400 °C, and so require piping materials to operate in the creep range.

High-integrity systems require the use of sensitive NDT techniques to detect small defects. The most onerous requirements are on systems that operate at high temperatures as they are subjected to severe thermal fatigue and creep conditions, in addition to the normal pressure stresses exerted by the high-pressure process fluid. In all systems of this type the high pressure and temperature exaggerate the intensity of degradation mechanisms that act on the material, so increasing the importance of using the correct inspection and NDT techniques.

High-integrity NDT techniques

NDT of high-integrity pipework requires a very stringent approach to the use of NDT techniques and the acceptance criteria used. The result of the UK PANI study – see Chapter 6) showed that site NDT is *not* a particularly reliable process even when well-qualified technicians perform the work. This is an important point to keep in mind when specifying site NDT on high-integrity pipework as part of an in-service inspection programme. Figures 13.10 and 13.11 show some key points.

Advanced UT techniques such as time of flight diffraction (TOFD) and corrosion mapping can be usefully applied to pipework inspection, in preference to more traditional techniques. They are more expensive and take longer to set up, but are much more effective in finding the types of defects that can threaten FFP in high-integrity applications.

Written schemes of examination (WSEs) for pipework

WSEs for pipework systems are more likely to be too broad and general rather than too detailed. Even in plants that have had a full RBI analysis the essential link between the output of the RBI analysis and the detailed contents of the WSE is easily lost. In too many cases the WSE actually *regresses* over time, losing its technical detail until it contains nothing but neutral statements of generality. This is especially the case with pipework where proper inspection is never popular, because of the cost and inconvenience. One of the greatest dangers is the inclusion of high-integrity pipework in the same WSE 'category' as

Fig. 13.9   A high-temperature boiler power piping elbow with lagging removed for inspection

process/utility pipework. This leads to a detailed inspection scheme that has little chance of discovering complex defects until they have progressed to an advanced state. Figure 13.12 gives some guidance points for the pipework section of WSEs.

Figure 13.12 indicates the importance of a comprehensive WSE. Because statutory regulations such as the PSSRs do not specify WSE content in detail, plant operators and their Competent Person organization often treat the WSE as a formality, making it so general as to lose its cutting edge. This leads to poor, lacklustre in-service inspections and is a major contributor to the large number of failures of pipework components. Most of these failures are *avoidable*, with a bit of planning.

Pipeline inspection

Pipeline inspections are governed heavily by the question of whether they are above-ground or buried. In many developing countries, particularly desert or jungle regions, pipelines are laid on the surface – many extend for several hundreds of kilometres. In more urbanized countries, long-distance pipelines are nearly always buried underground. Types of pipeline vary widely from low-pressure water and unrefined petroleum products to high-pressure gas, chemical and distillate fractions such as aircraft fuel.

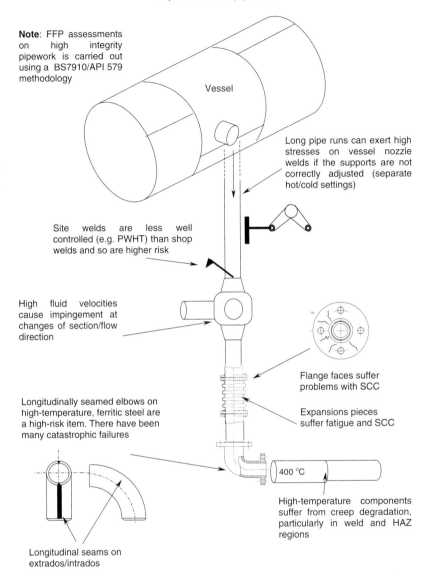

Note: FFP assessments on high integrity pipework is carried out using a BS7910/API 579 methodology

Vessel

Long pipe runs can exert high stresses on vessel nozzle welds if the supports are not correctly adjusted (separate hot/cold settings)

Site welds are less well controlled (e.g. PWHT) than shop welds and so are higher risk

High fluid velocities cause impingement at changes of section/flow direction

Flange faces suffer problems with SCC

Longitudinally seamed elbows on high-temperature, ferritic steel are a high-risk item. There have been many catastrophic failures

Expansions pieces suffer fatigue and SCC

400 °C

High-temperature components suffer from creep degradation, particularly in weld and HAZ regions

Longitudinal seams on extrados/intrados

Fig. 13.10   High-integrity pipework – risk areas

Pipelines and the PSSRs

Pipeline process fluids group fairly easily into those that are relevant fluids, i.e. contain stored energy, and those that are not. In practice, these differences are academic, as low-pressure pipelines are excluded from the PSSRs. Notwithstanding this, many pipelines *are* included in

- **Surface non-destructive techniques** dye penetrant (DP)/magnetic particle inspection (MPI) are in reality little more than enhanced visual techniques. They will not reliably find creep or fatigue cracks in their early stages.
- **DP/MPI** techniques do not give good results on corroded surfaces. High-temperature components need a lot of mechanical cleaning (wire brushing etc.) to remove the scale in order to get good results.
- **Sample ultrasonic testing (UT) wall thickness measurements** are not very effective at finding localized internal corrosion. A corrosion mapping technique, guaranteeing 100% coverage, (see Chapter 6) is much better.
- **Weld cap grinding** is necessary in order to obtain a full *fitness-for-purpose* UT scan of pipework welds. Partial scans 'around' a weld cap can easily miss defects.
- **Specialized UT techniques** are required for known high-risk items such as longitudinally seamed high-chromium (SA335-P22 and P91) creep-resistant alloys. The techniques require increased sensitivity and defect sizing methods.
- **Radiographic testing (RT) and UT in combination,** is the best way to find any defects. This is expensive, but gives much greater level of confidence in the integrity of the system.
- **Site welds** *generally* develop more in-service defects than shop welds.

Fig. 13.11   NDT of high-integrity pipework – some key points

the written scheme of examination for the plant that manages the installation (either the one at the inlet or the outlet end). The main reason for this is good practice and completeness, rather than any specific legislative interpretation.

Pipeline inspection techniques face two main difficulties:

- *Size.* Compared with pipework within the boundary of plants, pipelines are long, containing substantial volumes of material and large numbers of welds.
- *Accessibility.* Buried pipelines run several metres underground, with only a few access chambers along their length. Hence most of the pipeline is inaccessible from the outside for its normal working life.

To balance the situation, pipelines have the advantage that they are straightforward, and of regular construction, with few novel design features or variants. They are also well covered by design codes and

- **Separate pipework categories** are essential. The WSE should not include all pipework in a single category with equal periodicities and the same inspection/NDT techniques.
- **PSSR system classifications** (minor, intermediate, major systems) are not sufficiently detailed to use as the basis of a WSE. Subdivisions are needed so that different risk categories are properly addressed.
- **High-integrity** and process/utility pipework systems should be differentiated.
- **NDT techniques** must be properly defined, to make sure that the most suitable technique is used.
- **Thickness measurement locations** (TMLs) should be stated, to make sure that important ones are not overlooked. Guidance is given in API 570 *In-service inspection of pipework*.
- **Degradation mechanisms** should be mentioned in the WSE. This will inform inspectors about what they are looking for. This means that WSEs need some technical *forethought*. API 580/581 contains detailed guidance.
- **API 570**. The 11 degradation mechanisms and locations outlined in API 570 should be addressed (see Fig. 13.5). Outside this, API 570 is useful more as guidance rather than a set of step-by-step instructions.
- **Defect acceptance criteria.** A WSE that does not reference defect acceptance criteria or corrosion severity standards is weak. It is better to include them in the WSE rather than cross-reference some obscure company acceptance criteria that nobody even reads (at least not during inspections).
- **FFP assessments**. Reference should be made to assessment levels and techniques of FFP standards such as BS 7910 or API 579 so these can be used if required.

Fig. 13.12   Pipework WSEs – some guidance points

their supporting technical standards. Inspection techniques vary from simple internal examination of selected areas using a borescope, to advanced long-range NDT methods and the use of intelligent pigs which can provide a detailed corrosion and defect survey of an entire long-distance pipeline. Pipeline NDT is a good example of a part of in-service inspection that has advanced rapidly in recent years, with several new techniques becoming commercially available.

Pipeline pigging

Pigging is a method of testing in-service pipelines for corrosion and defects. It is in common use in the offshore and onshore petrochemical industry and has developed to a high level of technical sophistication. A predominant feature of pipeline pigging is that, because it deals with long lengths of pipeline (and hence large linear areas of material), it is often forced to deal with results on a *statistical basis*. This means that the results are often presented in terms of 'probabilities', 'average wall thickness', and suchlike. Despite these limitations, pipeline pigging is fairly well established as one of the best long-range NDT processes.

Principles of operation

The process involves sending a travelling tool known as a 'pig' (or sometimes just 'the tool') through the entire length of the pipeline. The pig carries on-board batteries and is 'intelligent', i.e. it can perform various types of NDT techniques, and record and store the data. To be suitable for pigging, a pipeline must be of special design, with particular geometrical features and be fitted with a pig launcher and catcher at each end of its run. Figure 13.13 shows the situation. The launcher and catchers are themselves constructed as coded pressure vessels (ASME-VIII-I or similar) with hinged doors. Figure 13.14 shows an example.

The pig is pushed through the pipeline solely by the pressure of the process fluid itself. It is therefore an 'on-line' operation. For decommissioned (short) pipelines, alternative systems are available in which the pig is pulled by cable or crawls through under its own power. Figure 13.15 shows some general 'rule-of-thumb' points about the pigging procedure.

The pig design

Figure 13.16 shows a typical design of a 'live recording' (or intelligent) pig. It consists of several sections: a drive section that provides a pressure seal against the pipe wall, an NDT section (magnetizing or UT-based, depending on type), and data recording section. The cleaning and dummy pigs are much simpler. The overall length and geometry of the pig is chosen to match the type of pipeline to be surveyed – minimum radii and changes of section are important considerations.

Defect detection
Methods

There are two major NDT principles used for pipeline pigging: ultrasonic detection and magnetic flux leakage detection. Of these, magnetic flux leakage is the most common, used on most modern

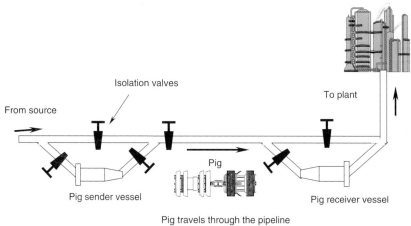

From source

Isolation valves

To plant

Pig

Pig sender vessel

Pig receiver vessel

Pig travels through the pipeline
under system fluid pressure

Fig. 13.13   Pipeline pigging – the principles

designs of intelligent pig. Figure 13.16 shows the principle. A unit on the
pig saturates the pipe material with magnetic flux and continually
monitors the flux level emitted from the pipe wall. If no defects exist, all
the lines of magnetic flux remain in the body of the material, i.e. none
are 'lost'. If defects do exist, they redistribute the lines of flux, causing
some to be lost into the process fluid. This will be sensed by the pig's

Fig. 13.14   Visual inspection of a pipeline pig catcher

Pipeline diameter. Pigging is normally restricted to pipelines between 100 and 500 mm internal diameter.

Changes of section. Pigs are available that can handle changes in pipeline sections, but they are more complex (expensive). Bend radii > 1.5 diameter are required for most pig designs.

Valves and internal fitting. These need to be of specific design in order to be compatible with pigging operations.

Pigging procedure. The pigging procedure consists of several 'runs' using different pigs. The usual sequence is:

1. Cleaning runs – a special pig to clean debris and check for obstruction.
2. Gauging runs – to check the magnetic 'locators' that identify the pig's location in the pipeline.
3. Dummy runs – an identical pig to the 'live' one but without electronics and data recording.
4. 'Live recording' pig runs – the main tool that records full survey data.

Fig. 13.15   Pipeline pigging – some key points

electronic circuits in the form of an electrical signal proportional to some function of the size and shape of the defect. In a similar way, loss of pipe wall thickness will also result in flux leakage, owing to the reduction in cross-sectional area.

Detection effectiveness

Magnetic flux leakage techniques have the capability to detect the main types of in-service defects expected in pipelines, i.e. cracks (and similar linear defects), corrosion voids, and wall thinning. It can also detect various common manufacturing (mill) defects that have been present since the pipe was made. Invariably, owing to the large amounts of material used in a long pipeline, it is necessary to collect such defect data on some kind of *averaging* or statistical basis. The technique also has the capability to differentiate between internal and external corrosion of the pipe wall, particularly when the corrosion loss is extensive (known as 'clustering loss').

From an inspection viewpoint it is not unusual for the defect, acceptance threshold levels, and the nature of the averaging procedure itself to vary from job-to-job, depending on the type, design code, and age of pipeline, and any known damage mechanisms. Figure 13.17 shows a typical way that defects are categorized using a magnetic flux leakage technique, along with some typical defect threshold values and

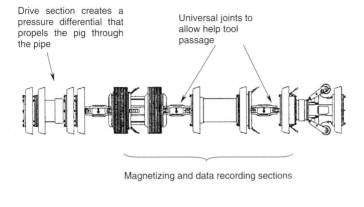

Drive section creates a pressure differential that propels the pig through the pipe

Universal joints to allow help tool passage

Magnetizing and data recording sections

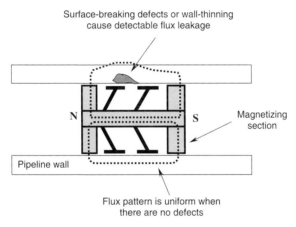

Surface-breaking defects or wall-thinning cause detectable flux leakage

N   S   Magnetizing section

Pipeline wall

Flux pattern is uniform when there are no defects

Fig. 13.16   A magnetic flux leakage 'intelligent pig'

averaging method for wall thinning. Note also how the detection of defect methodology is qualified by a 'probability of detection' (POD). This is basically a confidence limit (90, 95, or 99 percent depending on the technique) that is placed on the ability of the equipment to actually find the size and orientation of defects it is programmed to detect. These limits can be changed by adjusting the 'filtering' capability of the sensing equipment itself in the software used to process its output data.

Presentation of results
Pigging results are presented in two stages: preliminary results from new data collected immediately after the line pigging runs and refined results

presented later, after processing of the data in the laboratory. They are normally presented in three forms:

- A *flaw list*, showing a complete list of indications greater than a pre-set size threshold.
- A *summary sheet* showing the full length of the pipeline to scale (1:50 or 1:100), summarizing the extent of defects and wall thinning along the length.
- *Various histograms* with coloured areas representing priority areas for external examination (requiring excavation for buried pipelines), to confirm the extent of wall-thinning corrosion.

Remember that all result sets have to be heavily qualified by reference to the limitations of the defect threshold levels and statistical techniques that underlie them. Changes in such boundary condition assumptions result in vastly different graphical results. Figure 13.18 shows a typical form of presentation used by several pigging organizations.

Interpretation and sentencing
Despite the ability of pigging software to produce highly analytical statistical results, the real decision about the need for repairs, re-rating, etc. of the pipeline rests with the inspection engineers involved. In many cases, the existence of wall thinning or outside-code defects is not necessarily a reason for excavating a buried pipeline and carrying out major repairs. The first step is normally to carry out a risk assessment qualifying, as far as is possible, the likelihood of failure and the associated consequences (see Chapter 7 on RBI). This leads to an overall assessment of the *criticality* of various areas of the pipeline – and the various benefits in pipeline integrity, overall cost saving, etc. for each of the possible engineering solutions.

Ultimately, pipelines are probably easier to assess accurately than many other types of engineering plant because, using pigs, they can be subject to 100 percent in-service NDT, rather than a sample. The criticality of defects and wall thinning is frequently considered to be reduced if a pipeline is buried, or runs across remote desert regions, as long as there is no immediate risk to integrity.

Typical acceptance criteria
Over the past 10 years, a lot of work has been done on the acceptability of pipeline corrosion. This has taken the form of industry-sponsored studies involving leak and burst tests on corroded sections of pipeline to try to validate the actual failure performance against a set of assessment

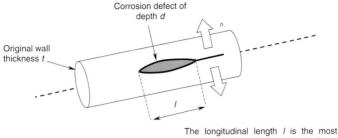

The longitudinal length *l* is the most critical as it is resisting the dominant hoop (circumferential) stress σ_h

Assessment chart using the principles of DNV RP-F101

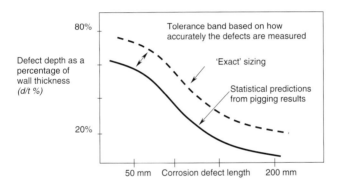

Typical acceptance curve for a defined pipe diameter, wall thickness and operating pressure

Fig. 13.17   Pipeline corrosion assessment

criteria that can be used in the field. Such acceptance criteria are heavily dependent on the accuracy with which defects can be sized so they inevitably contain various 'factors' to try to allow for any uncertainty that exists. A typical level of assumption used is:

- Defects sized to an accuracy of ± 10 percent wall thickness.
- A general 80 percent confidence limit of all measurements.
- Defects exist as normal distributions.

One of the best recently published documents that provides corrosion assessment criteria is DNV RP-F101 *Corroded pipelines*: 1999. This is a recommended practice document from the Norwegian marine classifica-

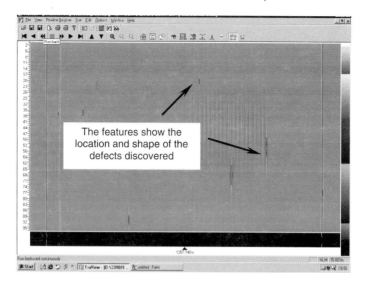

The spreadsheet shows the location and 'clock
position' of each defect

	Event Name	Event Dist (m)	Joint Len (m)	Depth	A. Len (mm)	A. Width (mm)	Dist. To Weld (m)	Clock	Speed (kph)	Comments
1	W - 208	1250.419	0.648						8.5	Weld A
2	Anomaly	1250.980		28%	10	23	0.561	6.20	8.4	External
3	W - 209	1251.067	12.672						8.2	Weld B
4	Anomaly	1252.197		43%	13	28	1.130	6.00	8.0	External
5	Anomaly	1252.865		33%	13	41	1.798	3.50	8.2	External
6	Anomaly	1254.430		41%	18	38	3.363	12.50	8.2	External
7	Anomaly	1254.973		27%	13	23	3.907	7.40	8.2	External
8	Anomaly	1257.737		23%	8	20	6.670	3.10	7.6	External
9	Anomaly	1257.798		28%	8	38	6.721	3.10	7.6	External
10	Anomaly	1259.352		60%	15	76	8.285	9.00	7.4	External
11	Anomaly	1261.029		60%	15	76	9.962	7.10	8.2	External
12	W - 210	1263.739	12.456						9.5	Weld C

Fig. 13.18   A typical pigging results display

tion society Det Norske Veritas and specifically addresses the issue of
external corrosion on pipelines. It can be used with equal validity on
BS 8010 or API-coded pipelines. Figures 13.19 and 13.20 show the
details.

Is there a link with construction code acceptance criteria?
The link between acceptance criteria suggested by DNV RP-F101 and
those quoted in pipeline construction codes (BS 8010, API 1104, etc.) is

a controversial one. In all cases the limits suggested by RP-F101 are more relaxed than those in the construction codes. There are two feasible reasons for this:

- Construction code criteria have their own origins in the control of the manufacturing process. *Whereas*
- Documents such as DNV RP-F101 are structured around a fitness-for-purpose assessment approach (i.e. a similar philosophy to the API 579 and BS 7910 methods described in Chapter 9). *Or*
- Construction code criteria do not claim to attempt a consideration of in-service corrosion defects, preferring to wrap their entire approach with large all-encompassing factors of safety. This will help them compensate for any big (and largely unpredictable) risks that arise from inaccuracies, poor inspection, and in-service conditions.

On balance, there is little disagreement with the general concept that an FFP-based assessment produces *more accurate* defect acceptance criteria than construction codes. Pipelines provide a good example of this working in practice. Figures 13.19 and 13.20, although they are only 'typical examples' shows that significant corrosion defects *can exist* in a pipeline without compromising its fitness for purpose.

Glass re-inforced plastic (GRP) re-lining

In the offshore, marine, and chemical industries, where heavy corrosion is commonplace, large-bore pipework, ductwork vessels, and fabricated equipment are frequently lined with GRP following the discovery of corrosion during a shutdown inspection. The monitoring of this lining is a key inspection task to make sure that it is done properly. GRP lining is the common term used to describe the family of glass-reinforced plastic coatings suitable for providing a protective coating on steel and some non-ferrous metals. GRP consists of a plastic material 'body' reinforced by a matrix of glass or ceramic fibres or flakes. There are a large variety of types of GRP each with its own particular properties. GRP materials are used to line equipment items that are prone to a mixture of corrosion and erosion, such as seawater pump casings, condenser waterboxes, flue gas ductwork, and chemical reaction vessels.

Fitness for purpose (FFP) criteria

The FFP criteria for GRP linings are similar to those for rubber and other linings, although there are some small differences in the technical detail. The key criterion is *integrity* of the lining – GRP performs its

Assessment level	Defect interaction equation: level of assessment
Level 1 (screening)	Circumferentially supported corrosion defects do not interact if the angular spacing, ϕ, between them is $$\phi > \frac{360 \times 3}{\pi}\sqrt{\frac{t}{D}} \quad \text{degrees}$$ where D is the outer diameter (OD) of the pipe (mm) and t is the nominal wall thickness of the pipe (mm). Longitudinally supported corrosion defects do not interact if the axial distance, s, between them is $$s > 2\sqrt{Dt} \quad mm$$ Adjacent corroded area in which the maximum depth $< 20\%$ original wall thickness can be considered as not interacting at all, i.e. can be treated as separate isolated defects.
P_o = predicted failure pressure for plain pipe (N/mm^2)	To calculate failure pressure P_f. $$P_f = P_o \times R_s$$ $$P_o = \frac{2\sigma_u}{\left(\frac{D}{t}-1\right)} \quad \text{for} \quad \frac{D}{t} > 18$$
R_s = remaining strength factor (dimensionless) D = OD of pipe (mm) L = axial length of corrosion defect (mm) A = maximum depth of corrosion defect (mm) t = original wall thickness of pipe (mm) σv = ultimate tensile strength of pipe material (N/mm^2)	$$R_s = \frac{1-\left(d/t\right)}{1-\left(d/t\right)\frac{1}{1+0.31\left(L/\sqrt{Dt}\right)^2}} \quad \text{for } \frac{d}{t} \leq 0.85$$
	Note the use of non-dimensional parameters: D/t, d/t, and $L/\sqrt{Dt}$.

Fig. 13.19   Pipeline corrosion assessment defect interaction equations

Allowable pressure for a pipeline with isolated corrosion defects is given by:

$$\text{Allowable pressure} = \frac{\gamma_m 2tR\,[1 - \gamma_d(d/t)]}{(D - t)\left[1 - \frac{\gamma_d(d/t)}{Q}\right]}$$

where:

$$Q = \sqrt{1 + 0.31\left[\frac{1}{\sqrt{Dt}}\right]^2}$$

D = outer diameter
d = depth of corrosion defect
t = nominal wall thickness
l = length of corrosion defect (axial direction)

γ_m = safety factor
γ_d = safety factor $\Big\}$ See DNV RP – F101

R = specified minimum tensile strength

Some typical strength values for API/BS pipeline materials are given in BS 8010. They vary from 241 up to about 480 MN/m^2 for the higher strength 'X-grade' alloys.

Fig. 13.20 Assessing pipeline corrosion to DNV RP F101 'corroded pipelines'

protective function by isolating a component's surface from the corrosive and erosive effects of its environment and even a small breach will allow corrosion to start. The quality of surface *preparation* is vitally important.

Basic technical information

Most GRP lining 'systems' consist of three coats. The first coat is typically a vinyl-ester-based resin containing fine glass flakes. It is applied by brush or roller and provides a good 'key' to the prepared surface. The intermediate coat is typically a similar vinyl ester resin but containing a ceramic filler to provide durability. The choice of top coat depends on the particular environmental conditions but, because of the frequently erosive conditions (one of the reasons for choosing GRP lining), it often contains a filler with lubricating properties. This gives the so-called 'low-friction' lining surface useful on rotating fluid machinery such as pumps. Total lining thickness is normally 1–1.5 mm but, unlike paint, there are few disadvantages in having a much thicker lining. Each GRP coat needs a drying time (1–3 days) before it can be overcoated, so the full lining process can take up to 7–8 days. Figure 13.21 shows the details.

Acceptance guarantees

It is difficult to specify meaningful acceptance guarantees for a GRP lining. The main objective of the purchaser is the lining's longevity but it is rarely practical to use this as a contractual requirement, because of uncertainties in the environmental conditions. You will find that in most contracts, the choice of lining material is left to the equipment contractor or manufacturer (perhaps after a brief review by the purchaser) with the only overt requirement of the specification guarantees clause being that the integrity of the lining be proven by a high-voltage spark test. This means that as an inspector, it is not productive for you to spend a lot of time worrying about and verifying the *choice* of lining materials. First, you may not have a solid specification clause to work from, and second, even if there is a mistake in system choice or compatibility you will have difficulty in finding it – it is likely that you will draw the wrong conclusions. Do not waste your time. Concentrate on the more productive issues.

Special design features

The design features of equipment to be GRP lined do need some attention. The main areas are:

- *Remove sharp corners.* All sharp corners and edges must be radiussed or blended in with the contours of the surrounding metal – GRP will erode quickly if it presents a sharp edge to an abrasive fluid.
- *Fit surface pits and recesses.* Whereas rubber can be glued down over surface craters, lips, or recesses in the parent metal, GRP adheres better if it is presented with a flat surface. Check that surface marks more than 1 mm deep (in castings for instance) are filled with a suitable epoxy compound before the GRP lining is applied. Likewise, weld caps should be ground to a smooth convex or flat profile.

Specifications and standards

The main technical standard relevant to GRP linings is BS 2782. This is a wide-ranging standard which comes in many different parts, most of which contain detailed technical information on the analysis and chemistry of various plastics, but nothing about inspection or practical

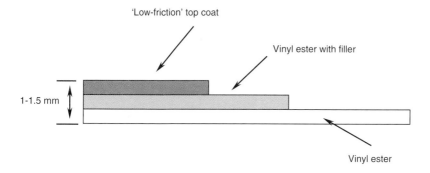

GRP – SURFACE CONTAMINATION TEST

Performed before lining or repairs

1.   Shotblast to Sa2½
2.   Wash with distilled water then litmus test
3.   Leaching period (3–4 hours)
4.   Final wash and litmus test
5.   Flash shotblast to Sa3
6.   Dry in warm air

Fig. 13.21   GRP lining of metallic components

test methods. It is unlikely that you will need to use this standard in relation to a works inspection. The best technical information on GRP materials is to be found on manufacturers' datasheets, they give clear information on the suitability of linings for various environmental conditions – some of it hard and objective, other parts optimistic. Treat it as useful background information – but no more.

Test procedures and techniques

There are three main test techniques for the inspection of GRP linings: surface contamination tests, the visual inspection, and the spark test.

Surface contamination tests

The surfaces of metal to be lined with GRP must be completely free of contamination by salts. Salt contamination is common on equipment which has already been in use with seawater but also on new castings or steel plate that has been exposed to a marine environment during sea transport or storage. It is important to check that any salt has been removed by washing. The procedure must be done in the correct order, as follows:

- An initial shotblast of the surface to $Sa2\frac{1}{2}$.

- Washing the surface with distilled water and testing for salts with pH (litmus) indicator paper (salts may also be visible as dark smudges on the surface).
- Further washing until all traces of salts have gone.
- A waiting period of 3–4 hours during which the surface is left wet to encourage any remaining salts to leach out.
- Final wash and pH paper test.
- Light 'flash' shotblast to Sa3 to finally prepare the surface.
- Drying in warm air (10–20 °C) with low humidity.

The time taken between the final flash shotblast and the GRP base coat should be as short as possible. The oxidation process can start within 1 or 2 hours so it is bad practice for any surface which has been left standing longer than this to be lined without a further light shotblast. Try to witness these key preparation stages – they have an important effect on the longevity and therefore the fitness for purpose of the lining.

Visual inspection
The visual inspection is an important part of the FFP assessment of any GRP lining system; there is little difference in technical 'risk' between different types of GRP material so all need a close visual inspection. The best time to do this is after the final coat has been allowed to cure fully. Interim inspections are sometimes possible but it can be uneconomical to attend multiple inspections on GRP-lined equipment, many of which will be small items. As with rubber linings, the watchword with GRP is *close* visual inspection – you should use a methodical approach, working slowly over all areas of the lining under a good light. You are looking for five main things.

- *Adhesion to design features.* Check that the lining has adhered well to design features such as lips, radii, and edges. There should be no evidence of the lining being loose or 'bulged', particularly over tight radii edges, and no big voids or air bubbles. Such defects are rare if design features have been properly prepared and radiussed, but common if they have not.
- *Surface finish.* Nearly all GRP top-coats are designed to be shiny – special low-friction top-coats are designed to have a particularly high gloss finish. Watch for dull, uneven areas that can indicate contamination by dampness, incorrect mixing, or incomplete curing – these are often symptomatic of a problem with the application of the lining and can have a significant effect on its fitness for purpose. Large components such as gas ductwork and tanks are more likely to

have such defects than small equipment items. If the top-coat surface is absolutely matt make sure you review the manufacturer's datasheet to confirm whether this is intended (there are a few lining systems like this – although they are not in common use).

- *Staining.* Be wary of staining. On clear or light-coloured linings, dark discoloration (black, blue, or sometimes brown) can be indicative of leached contaminants from the surface, particularly in castings. Treat this as a serious FFP issue – if it is contamination then the lining will probably flake off within a short time. Be careful not to confuse this with dust that has been picked up by the surface when wet, which is not so serious. Feel the surface – if it is slightly rough, it may be caused only by dust.
- *Thickness.* Unlike rubber, GRP linings are actually quite forgiving to variations in application thickness. A standard system thickness of 1–1.5 mm can vary up to 4 or 5 mm with no negative effects as long as it is properly mixed, applied, and cured. The only situations in which overthickness can cause problems is if it either causes interference with locating parts or machined surfaces or is so thick that it protrudes off the surface, sagging under its own weight. It is important, however, that the lining is not below its *minimum* thickness, because this can reduce its longevity under erosive conditions. Thickness is measured using a magnetic or eddy current instrument. Make sure it is calibrated on a test piece before use and that the lining is completely cured and hard – it is difficult to obtain accurate readings from a lining which is still soft.
- *Mechanical damage.* This has nothing to do with the application of the lining, it is simply caused by physical bumps or mishandling during storage, but *is* one of the most common causes of failure in service of GRP linings. Mechanical damage, however small, in which the lining is breached will cause rapid failure as the underlying unprotected metal is exposed to the aggressive process conditions.

The spark test

The spark test checks for continuity of the lining and can detect small pin-holes not easily visible to the naked eye. The important point to check is that the correct voltage is used (normally between 4 and 10 kV, depending on the coating) and that the test is carried out in a dry atmosphere for accurate results. Concentrate the spark test on areas of lining around sharp edges where damage is more likely and around holes, fittings, lugs, and bolted connections particularly in difficult to access areas. Repaired areas should be given special attention – check carefully around the periphery of each repair to check for leakage paths. It is not uncommon to find 'new' pin-holes caused by overgrinding or mechanical damage during the repair procedure itself. The objective of using GRP lining is to achieve a coating which isolates the base material totally from its aggressive process environment. This infers that the lining must be *spark free* if it is to properly meet its FFP criteria (see Fig. 13.22).

Small-bore tubing systems

Statistically, small-bore tubing systems (ranging from 4 to about 50 mm diameter) are the largest single contributor to accidents involving the loss of pressure containment in hazardous industrial plants. Various UK Health and Safety Executive (HSE) reports have highlighted the need for increased inspection of these systems in order to reduce the

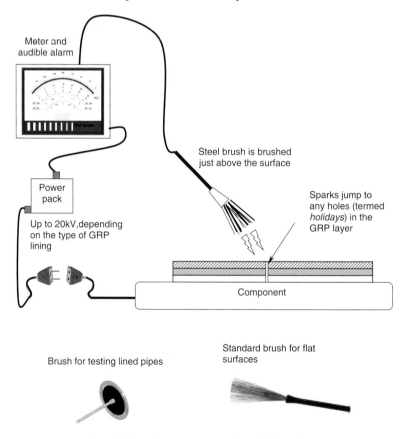

Fig. 13.22   The spark test for GRP linings

number of accidents. Figure 13.23 shows the main uses of these small-bore systems. Figure 13.24 shows a typical application.

Applications	Types of plant
• Control systems (pressure-sensing lines etc.).	• Offshore hydrocarbon.
• Instrumentation tappings for remote readings.	• Onshore hydrocarbon.
	• General process plant.
• Chemical injection lines.	• Manufacturing assembly lines.
• Hydraulic fluid lines.	• Power plant.
• Sampling lines.	• Ships and other marine applications.
• Pneumatic control lines.	• Aircraft.
	• Food processing and brewing.

Most small-bore tubing systems are made from stainless steel. They use cold-formed bends (often done on-site 'to fit') and cold compression fittings. Some severe-service systems use titanium for its strength and corrosion resistance.

Fig. 13.23   Application of small-bore tubing systems

Fig. 13.24   Typical application of small-bore pipework

Small-bore tubing – the problem

There are several main issues at the core of the problems of small-bore tubing. They all tend to be most serious in coastal and offshore environments. We can look at these in turn.

Inclusion in written schemes

Many plants, particularly offshore installations, contain large and complex small-bore tubing systems. Many are in low stored energy systems, i.e. those which do not reach the 250 bar litre threshold of the Pressure System Safety Regulations (PSSRs), and so do not have to be included in the formal written scheme of examination (WSE). Even if they are included, the systems are often so extensive that only a small sample can realistically be inspected. The end result is that despite inspectors' best intentions, small-bore systems can become neglected.

The corrosive environment

The combination of a chloride-rich (salt) environment and the presence of moisture traps under lagging produces an environment which encourages stress corrosion cracking (SCC) of stainless steel material, of which most small-bore tubes are constructed. SCC is a composite corrosion mechanism that starts with thin branched (known as 'bifurcated') cracks resulting finally in unexpected failure. Simple visual inspection and dye penetrant (DP) testing can find SCC in its early stages, before it becomes an integrity risk.

Mechanical damage

Because of their size (most are <15 mm diameter) small-bore tubing systems are very susceptible to mechanical damage. Most of the damage comes from personnel and includes:

- *Standing* on the tubes when maintaining other components.
- *Bending* tubes 'out of the way' when maintaining gauges, pressure sensors, etc.
- *Overtightening* of compression fittings causing kinking and over-stressing of the tubes.

The use of incorrect components

The use of incorrect non-standard components is a big problem, particularly with older systems. Although a small-bore tubing system may be designed to have all standard components (compatible tubes, single/double ferrule compression fittings, tube clamps, etc.) the reality can be somewhat different, with many non-standard parts being used during site assembly and ongoing maintenance of the system. This happens in all industries, but particularly those such as offshore and petrochemical plants which traditionally use a mixture of imperial and metric fittings. Figure 13.25 shows some common fitting types and the types of problems that occur.

A typical small-bore, stainless steel tubing system application

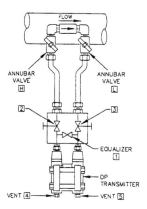

The correct type of high-pressure ferrule fitting

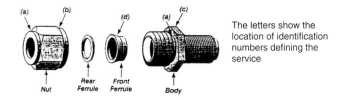

The letters show the location of identification numbers defining the service

Older types of small bore fittings

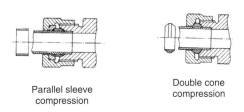

Parallel sleeve compression

Double cone compression

Fig. 13.25   Small-bore tubing systems – what to expect

Inspection procedures

The best way to ensure effective in-service inspection of small-bore tubing systems is:

- First, make sure that they are properly included in the WSE *in detail* rather than just as a general mention. In large plants, small-bore systems should have their own section of the WSE.
- Second, inspect a *significant sample* of the systems at each inspection so that the whole installation is covered comprehensively, over time.
- Third, inspect (and report) to *detailed checklists* to make sure that nothing is missed.

Figure 13.26 shows a sample checklist. The only way to use this properly is to have one per system, item of plant, or whatever, so that nothing is missed.

Points to check	Method
Tubes	
Check for correct material grade.	Tube etching and certification.
Check correct tube diameter, (*d*).	Vernier calliper.
Check for tube ovality and kinking or bends (<10% *d*)	Vernier calliper.
Check tube runs are correctly secured/clamped to minimize vibration.	Visual examination.
Remove sample lagging for SCC/ corrosion check.	Visual examination and DP of selected areas.
Compression fittings (single and double ferrule)	
Check the ferrules are correctly 'seated' on the tube wall and not deformed.	Visual check of alignment of ferrule with the tube axis.
Check for ferrule size and compatibility.	Metric and imperial fittings (e.g. NPT/BSP metric threads) can look very similar. Non-compatible components can be forced together, giving a poor joint that will loosen in service.

Valves

Check that valves are of the   Visual.
correct type for the application,
and have the correct compression
fitting.

Fig. 13.26   Small-bore tubing fittings – inspection checklist

WSE administration

For plant that has a history of small-bore tubing failures, the WSE should contain a separate section devoted to these systems. While this can mean that a large number of systems have to be inspected, most of the inspections are visual only, and so quick and easy to do. In many cases, even though small-bore tubing systems can be adjudged to be outside PSSR jurisdiction, it is still good practice to include them in the WSE. Some plants publish documents called 'PSSR implementation precedents' (see Chapter 5) to make sure that these systems are properly registered as site pressure systems, and therefore subject to periodic inspection under a WSE. Failure to do this will result in an installation that lacks integrity and which could fail in the short or long term with indeterminate consequences. Most failures will be in the form of plant shutdowns (owing to fail-safe trips) rather than serious pressure-release incidents but the possibility of serious incidents cannot be ignored.

Chapter 14

Storage tanks

The most common type of storage tank you will encounter as an inspection engineer is the above-ground type of cylindrical steel construction. There are many other types, for example, underground tanks constructed of steel, concrete, or a combination of both, but the above-ground steel type is the most common. There is nothing particularly complicated about the design and construction of storage tanks. Tank inspections, however, are not so predictable, because the nature of the contents of storage tanks is very varied, leading to many different possible corrosion regimes and in-service degradation problems.

Above-ground steel storage tanks divide broadly into two groups: *atmospheric tanks* for petroleum and chemical products, and *refrigerated tanks* designed to hold liquefied gases at low temperatures. The construction and range of in-service problems are different, leading to different techniques and procedures for their in-service inspection.

Codes and standards

Codes and standards for storage tanks subdivide neatly into those concerned with design and construction of the tank, and those that cover in-service inspection. Inspection-related standards are produced by both 'national' standards bodies and 'industry group' bodies in several countries and, technically, are very similar to each other. This means that there is not the problem of competing inspection philosophies that can be found with some other types of engineering plant. In general, inspection standards for storage tanks are comprehensive, containing clear and straightforward technical detail. They have been developed iteratively over the past 50 years or so and there is not much in them that can be considered new. They provide good guidance for in-service inspectors.

EEMUA 159 Users guide to the maintenance and inspection of vertical cylindrical steel storage tanks

This is published by the Engineering Equipment and Material Users Association (EEMUA) which is a UK-based organization comprising mainly power and petrochemical utilities. EEMUA 159 has the status of a published document rather than a formal code or standard but is nevertheless well established in these industries as a document that sets a high technical standard for tank inspections. Figure 14.1 summarizes the contents of EEMUA 159; note how it addresses both atmospheric and refrigerated tank types and important inspection topics such as corrosion, hydrotesting, and inspection checklists.

Being predominately a UK-compiled document, EEMUA 159 displays a tendency to refer to tanks built to the British Standard tank construction code BS 2654. In content, however, it draws heavily on the technical and procedural details of API 653. The method of calculating areas of corrosion is effectively the same as that used in API 653.

API 653 Tank inspections, alteration, repair, and reconstruction

API 653 is the complementary code to API 650 *Welded steel tanks for oil storage* which specifies design and manufacturing aspects of atmospheric storage tanks. This is typical of API practice in, for example, pressure vessels and pipework systems, in which design/manufacture and in-service inspection codes are used almost as a 'matched set'. In common with the general philosophy of API codes, API 653 provides both general technical information and a certain level of acceptance criteria that can be used as a sound technical basis for in-service inspections.

API 650 Welded steel tanks for oil storage

This is the main design/manufacturing code for atmospheric storage tanks. It has widespread use in the petrochemical industry worldwide. From an inspection viewpoint, the code gives information on the way that tanks are constructed, material thickness, and similar.

BS 2654 Manufacture of vertical steel storage tanks with butt-welded shells, for the petroleum industry

This is a well-established technical code commonly used for older (and new) atmospheric tanks. It follows similar technical lines to API 650 and links with many of the inspection requirements of EEMUA 159.

EEMUA 159: *Users guide to the maintenance and inspection of above-ground vertical cylindrical steel storage tanks*

Section	Contents	Comment
2	**Atmospheric storage tanks**	
2.1	Codes and history	References BS/API/DIN standards for tank construction.
2.2	Tank foundations	Describes problems of foundation settlement causing tank tilting and leakage.
2.3	Tank corrosion	Identifies corrosion locations on tank shells, roofs and bottom plates.
2.4	Tank bottoms	Describes settlement shapes and tolerances.
2.5	Tank shells	Gives calculations for minimum acceptable shell thickness (based on BS 2654).
2.6	Fixed roofs	Details of rejection limits, roof pressure tests.
2.7	Floating roofs	Describes types of floating roofs/drains/seals and acceptable tolerances.
2.8	Ladders, platforms etc.	No detailed technical information provided.
2.9	Tank instrumentation	Outlines typical tank instrumentation. No detailed technical information provided.
2.10	Determination of effective plate thickness	Shows how to calculate 'corrosion areas' and decide acceptability. The content and calculation method is very similar to that in API 653.
2.11	Paint coatings	References NACE and API publications and gives guidelines for internal and external coatings.
2.12	Insulation	Reference BS 2654 guidelines on tank insulation practices.

Fig. 14.1 Using EEMUA 159

2.13	Inspection checklist	Internal and external inspection checklist and suggested inspection periodicities.
3	**Refrigerated storage tanks**	
3.1–2	Introduction and scope	References EEMUA 147 and BS 7777.
3.3	Tank data	Suggests operating history data that are needed prior to inspections.
3.4	External inspection	Checklist for tank shells, foundations and fittings.
3.5	Internal inspection	Checklist of inner tank and suspended deck.
4	**Hydrotesting**	
4.1	General requirements	Itemizes the objectives of the hydrotest. References BS 2654.
4.2	Hydrotest after repairs	Explains when a hydrotest after repair is (or is not) required.
4.3	Filling rates	General statements only. No actual filling rates quoted.
4.4	Test temperature	Specifies 4 °C minimum temperature to avoid brittle fracture.
4.5	Holding time	Specifies 24 h holding time at maximum fill height.
4.6	Emptying	Emphasizes checks on vacuum vents to avoid shell buckling.

Fig. 14.1   (Continued)

European national storage tank codes

The following two codes have been in use as national standards:

- Germany: DIN 4119 parts 1 and 2: *Above ground cylindrical flat-bottomed storage tanks of metallic materials.*
- France: *Code Francais de construction des reservoirs cylindrique verticaux en acier VCSIP et SNCT.*

Construction features

The design and construction features of atmospheric storage tanks are well proven and do not change much between individual design codes.

The nature of some of the common design features, however, has direct relevance to the activities of in-service inspection, and to the practicalities of the diagnosis of in-service problems and their repair.

Foundations

Foundations are a common source of problems, particularly on older tanks. Most tanks built before about 1970 used simple rubble foundations, without piling. The rubble layers and underlying subsoil are compressible, leading to settling of the tank over time. The nature of the settlement can take several forms causing various types of in-service defects. Figure 14.2 shows the arrangement of a typical rubble foundation and Fig. 14.6 the in-service defects that can result from various types of settlement.

Many modern tanks are now built with concrete foundations. This has almost eliminated the problem with settlement (particularly non-uniform settlement, which is the worst type). A layer of 'bitsand' (a sand/bitumen mix) is set between the tank and the concrete foundation block.

Floor plates

One unfortunate feature of all tank foundation types is that the tank steelwork sits directly on top of the foundation, i.e. the underside of the tank floor is normally not accessible for in-service inspection. Special designs of refrigerated tank can be mounted on concrete plinths or stilts, which solves this problem. Tank floors consist of an arrangement of rectangular 'infill' floor plates set within a ring of *annulus plates*. Figures 14.3 and 14.4 show typical details. The annulus plates are 25–30 percent thicker than the infill plates and are butt welded. The infill plates (which are themselves lap-welded to each other) are lapped on top of the annulus ring. The tank floor is rarely flat, the infill plate laps being arranged so that the floor slopes away from the centre, towards the annulus. Floors are nearly always distorted owing to the heat input from welding and general rippling caused during manufacture.

Shell plates

Figure 14.3 shows the general arrangement of the shell plate construction. Shell plate thickness is greatest for the bottom 'strake' of the tank to resist the stresses caused by the hydrostatic head of the tank contents. As a rule of thumb, the lower strake is about twice the thickness of the annulus floor plate and is welded to it using simple fillet

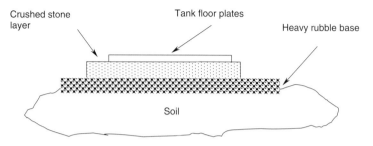

Fig. 14.2   A typical rubble foundation

welds on the inside and outside of the tank. The plates then diminish in thickness as they progress up the shell. Shell plates are butt-welded (not lapped) using double-vee preparations for the thicker, lower strakes and single-vee preparations (nearly always welded from the tank outside) for the thinner, upper strakes. Non-destructive testing (NDT) requirements depend on the design code but are generally 100 percent visual inspection plus radiographic testing (RT) or ultrasonic testing (UT) of sample seams.

Wind girders

Most large tanks have one, or sometimes more, wind girders fitted around the upper part of the shell (Fig. 14.3 shows details). These prevent the common failure mechanism of buckling of the top part of the shell – caused by wind gusts on floating-roof tanks when the fluid level is low. Additional girders may be either to guard against buckling due to corrosion later in the life of the tank, or to help strengthen an already bulged tank. Wind girders are normally attached to the shell plates by a continuous fillet weld around the top plate surface and an intermittent 'stitch' fillet weld underneath.

Shell manholes and nozzles

Shell manholes are typically 600 mm in diameter and located in the lower region of the shell. The manhole nozzle itself is normally 'set-through' the shell but welded directly only to a compensation plate (i.e. rather than to the shell and compensation plate as in pressure-vessel practice). Welds are, again, fillet welds rather than full penetration.

Fixed roofs

Atmospheric storage tanks containing non-volatile substances are normally designed with a simple fixed roof. The roof structure is

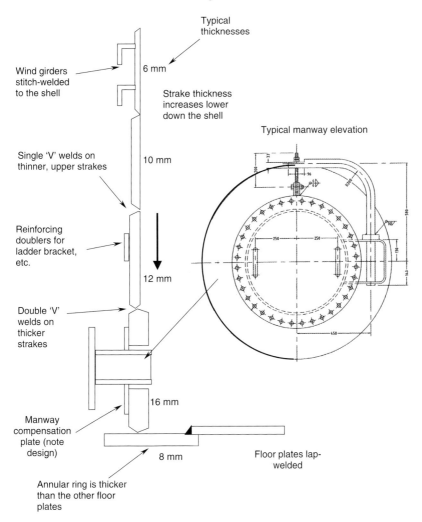

Fig. 14.3   Storage tank shell and floor construction

made of lap-welded plates (the thinnest material used anywhere in the tank) and the assembly connected to the top curb angle by a single seal weld (see Fig. 14.5) rather than the shell itself. This weld is weak, so enabling the roof/curb fixing to act as a frangible joint. In the event of serious overpressure or internal explosion, the roof joint will fracture, rather than the shell itself rupturing. Separate roof vents and pressure/vacuum valves are also fitted.

Fig. 14.4   Storage tank floor construction

Floating roofs

Floating roof tanks are used mainly for volatile petroleum products.
The roof structure, rather than being fixed to the top of the shell, floats
on top of the tank contents and moves up and down as the level varies.
This eliminates the void space above the product surface in which an
accumulation of flammable gases could cause an explosion risk. Figure
14.5 shows the two main types of floating roof design.

The pontoon type

This is the most common type. The buoyancy of the roof assembly is
supported by a ring of pontoons spaced around the annulus. The
pontoon compartments are welded airtight 'boxes' which occupy 20–25
percent of the roof area. The remainder of the roof is made of thin lap-
welded plates. These sag slightly below the surface level of fluid in the
tank.

The double-deck type

In this type the entire roof structure is of fabricated double-deck
construction. This gives a much more rigid roof than the pontoon type;
it is therefore suitable for large tanks > 40 m diameter, or for much
smaller tanks < 15 m diameter, in which pontoon roofs would be less
cost-effective to manufacture. In contrast to the pontoon type, all the

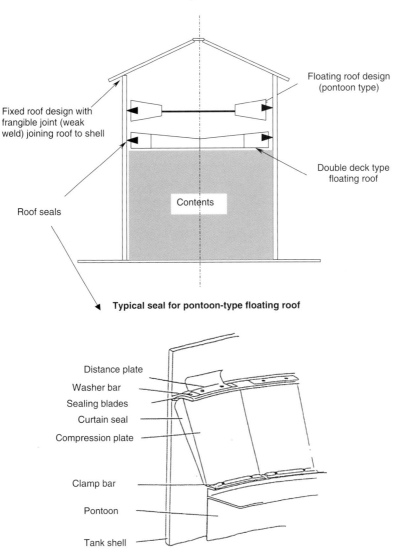

Fig. 14.5   Storage tank roof construction

upper surface of a double-deck roof is above the surface level of the tank contents. This makes roof draining easier.

Roof drains

Both types of tank designs have to be fitted with a method of draining off rainwater from the roof. Rainwater collects in sumps in the roof and discharges via either flexible hoses or a series of articulated steel pipes connected by 'swing joints'. The drain pipes pass through the tank contents and discharge via an isolator valve to ground level. In double-deck roofs an emergency drain is fitted in case the primary drains become blocked. This allows excessive rainwater to discharge directly into the tank where it settles at the bottom and has to be drained off manually. With pontoon-type floating roofs, because the roof surface at the centre of the tank is slightly below the surface level of the contents, emergency tank drains cannot be fitted. Such roofs are designed, therefore, to accommodate about 250 mm depth of rainwater over the entire roof surface, without causing damage. This gives sufficient time to allow blocked roof drains to be cleared.

Floating roof seals

All floating tank roofs need a seal to:

- Allow the roof to move up and down.
- Centre the roof structure within the shell.
- Prevent evaporation losses.
- Accommodate any out-of-roundness (OOR) of the tank shell.

There are several types of seal; most use some type of flexible shoe or wiper arrangement, or a liquid/foam seal. Earthing straps (known as 'shunts') are fitted between the roof and shell at several points to discharge any build-up of static electricity that could cause sparks. For tanks containing highly volatile products, double seals are used. The primary (lower) seal is in direct contact with the fluid while the secondary (upper) seal provides a back-up, preventing minor evaporation losses from any fluid that has passed the primary seal.

Tank inspections

In-service problems with storage tanks are not new. Failure and corrosion mechanisms are well defined and covered in both API 653 and EEMUA 159. In-service inspections of tanks are therefore not difficult, but do benefit from being carried out in a structured way.

Settlement of foundations

Settlement of storage tank foundations is a major problem, particularly on older tanks that use rubble foundations. There are two possible types of settlement:

- *Uniform settlement.* The foundations and subsoil settle evenly under the weight of the tank and its contents. Uniform settlement is only dangerous by degree, i.e. when it reaches an excessive amount.
- *Non-uniform settlement.* Some parts of the foundations and subsoil sink more that others. This is more dangerous and causes the tank either to tilt, or suffer from various types of distortion, which causes excessive stresses.

Figure 14.6 summarizes the types of settlement, and resulting failure mechanisms that can occur.

How is settlement measured?
Settlement measurement is a specialist task, commonly carried out by land survey companies. It is therefore a civil engineering-related task, i.e. cannot be performed by the mechanical inspection engineer during an in-service examination. Optical alignment 'dumpy' gauges, or laser techniques are used to measure tilt and deflections (mainly at the tank bottom) at regular intervals (5–10 years) or after settlement-related problems become evident in service. For new tanks, measurements are taken as a benchmark and to identify any settlement during the hydrostatic test. Note that it is next to impossible to check tank settlement by casual visual observation alone; the *effects* of settlement, i.e. cracking, leaks, etc. will show themselves long before the settlement can be reliably detected by eye.

Recessed foundations
This is a very common defect on older rubble-based tanks. The edge of the tank bottom sinks into the foundation creating a trough in the tank pad shoulder around the shell. Rainwater lies in the trough, causing localized corrosion at the bottom of the lower shell strake. A similar defect can also be caused on concrete-based tanks by distortion of the lower tank lip. The corrosion rate in these areas can exceed 1 mm/year once the paint layer is penetrated and many old tanks have visible thinning (sometimes more than 50 percent plate thickness) in this area. Figures 14.6 and 14.7 show a typical situation. Note the additional corrosion of the tank floor annulus plates – once the water penetrates under the floor annulus plates, the plates will start to thin. Although the annulus plates are thicker than other floor plates (see Fig. 14.3) the rate of corrosion can be rapid, causing the plates to lose their strength and deform under the weight of the tank contents. This type of degradation mechanism has the additional disadvantage that, under API 653, floor annulus plates cannot be 'patch welded' to restore their strength. The

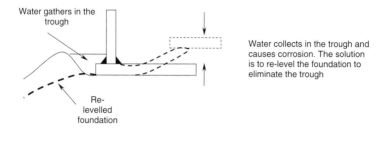

Settlement into the foundation

Water gathers in the trough

Water collects in the trough and causes corrosion. The solution is to re-level the foundation to eliminate the trough

Re-levelled foundation

Edge settlement

Settlement causes deformation of the annulus plate near the joint with the shell. As a guide, edge settlement becomes excessive when A/B> 0.16 .If the settlement occurs over a very short length of tank circumference, this is reduced to A/B >0.1 to reduce the risk of weld cracking

Annulus plate

Floor plate

A

B

Fig. 14.6   Storage tank settlement and bottom defects

only solution is to conduct a repair procedure to cut out and replace the floor annulus, and usually the lower shell strake plates also. This involves raising the tank shell on jacks – a well-established but expensive process.

For tanks in which lower strake/annulus floor plate corrosion is discovered before it has become too serious, a repair procedure is possible. The foundation recess is re-levelled by local excavation and the corroded plates shot blasted and repainted.

Uneven settlement

There are several types of in-service defect that can be caused by uneven settlement of the tank foundation. Unfortunately, not all of these can be detected by visual examination from outside of the tank. For a proper survey, it is necessary to check the tank bottom from the inside – this

Bottom plate rippling

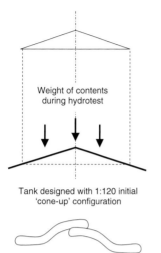

Ripples form in the centre region of the tank floor (mainly parallel to the longitudinal line of the lap welds). Maximum allowable ripple height is 75 mm over a 500 mm span, or when they form a sharp-edged crease

Weight of contents during hydrotest

Tank designed with 1:120 initial 'cone-up' configuration

Bottom plates ripple as they are pushed flat

General settlement bulges

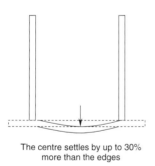

Smooth 'bulges' that follow the general profile of 'cone down' centre settlement are not a cause for concern. Sharper bulges can be an indication of voids or 'washout' of the foundations. This can lead to tank instability

The centre settles by up to 30% more than the edges

Fig. 14.6   (Continued)

means that the tank has to be emptied and cleaned to a gas-free condition.

Figure 14.8 shows the range of in-service defects that can be caused by uneven settlement. The acceptance limits shown are based on those given in API 653 and EEMUA 159. Remember, however, that these acceptance limits are guidelines only and should not be considered as absolute criteria for rejection. Many tank owners continue to use tanks that are outside these limits, but under increased periodicity of

Fig. 14.7   Local thinning of storage tank base

inspection (every 12 months in some cases) so that the settlement is monitored over time.

One occurrence that does need immediate action is *foundations washout*. This is where the tank contents leak into the foundations and wash them away. It is a dangerous condition as the resulting settlement can be rapid and uneven (the worst sort) and lead to catastrophic failure at the floor/shell junction.

Tank bottom defects

Apart from wind-bulging in the upper regions of a tank, most storage tank in-service defects occur around the tank bottom. The annulus plate and the 'infill' bottom plates suffer from a variety of problems that can lead to cracking, minor leaks or, in extreme cases, wholesale spillage of the tank contents.

Edge settlement

Edge settlement, as well as causing a longer-term corrosion risk due to recessed foundations (see Figs 14.6 and 14.8), also has shorter-term implications. The bottom annulus plate deforms downwards, causing cracking in the shell-to-annulus plate fillet weld. This normally occurs in the outside weld as it is in tension. The limit of acceptability of this type of deformation depends on whether the settlement occurs over a small or large length; deformation over a short length (perhaps 1000 mm or

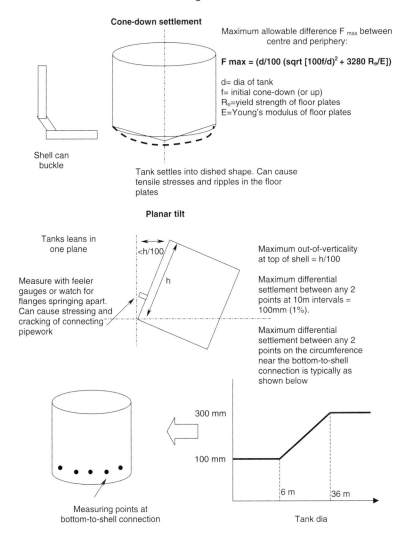

Cone-down settlement

Maximum allowable difference F_{max} between centre and periphery:

$$F_{max} = (d/100 \text{ (sqrt } [100f/d]^2 + 3280 \, R_e/E))$$

d= dia of tank
f= initial cone-down (or up)
R_e=yield strength of floor plates
E=Young's modulus of floor plates

Shell can buckle

Tank settles into dished shape. Can cause tensile stresses and ripples in the floor plates

Planar tilt

Tanks leans in one plane

<h/100

h

Maximum out-of-verticality at top of shell = h/100

Measure with feeler gauges or watch for flanges springing apart. Can cause stressing and cracking of connecting pipework

Maximum differential settlement between any 2 points at 10m intervals = 100mm (1%).

Maximum differential settlement between any 2 points on the circumference near the bottom-to-shell connection is typically as shown below

300 mm

100 mm

6 m   36 m

Measuring points at bottom-to-shell connection

Tank dia

Fig. 14.8   Storage tank settlement and tilt limits (EEMUA 159)

so) will result in higher stress and an increased risk of cracking. For more gradual settlement a distortion ratio of ≤ 0.16 (see Fig. 14.6) is unlikely to result in cracking or leakage problems. If significant edge settlement is discovered during an in-service inspection from the outside of the tank, it is advisable to check the lap welds between the annulus plate and the first floor plates on the inside of the tank. These often crack as a secondary effect of edge settlement.

Shell distortions

Distortion of the shell of storage tanks can occur in new tanks (because of poor manufacture or problems during hydrotesting) or in-service as a result of corrosion, wind loads, foundation settlement, overpressure, or excessive vacuum. Storage tanks are designed using simple membrane stress theory that assumes the shell is circular – hence any distortions need to be carefully assessed. Shell distortion acceptance criteria are given in tank construction codes, i.e. BS 2654 or API 650, rather than in inspection-related documents such as EEMUA 159. While these documents provide useful reference information for in-service inspectors, it often happens that their tolerances cannot be achieved during in-service inspections (after major repairs to tank shell steelwork, for example). Tank repair is an imperfect science and it can be difficult to meet construction tolerances on shell out-of-roundness. In such cases, the role of the in-service inspector becomes one of interpretation.

Wind bulging
Buckling caused by wind gusts is one of the most common forms of tank shell buckling. It mainly occurs on the windward side of the tank in the thinner, upper courses of the tank when the tank content level is low, i.e. there is no liquid hydrostatic pressure acting in the top areas of the shell. To prevent buckling, floating roof tanks are fitted with a wind girder positioned near the top of the shell. If a tank does experience wind buckling (immediately below the wind girder is the most common location) then additional girders can be retrofitted to prevent the situation getting worse. Excessive out-of-roundness of the tank shell increases both the changes of wind buckling, and its severity when it occurs.

Elephant's foot buckling
This is a unique form of buckling that occurs near the shell-to-bottom connection, rather than near the top. It has two main causes:

- *Seismic movement.* Axial movement during earthquake activity causes a sideways 'sloshing' movement of the tank contents. This causes a compressive stress in one side of the tank (exaggerated by any existing out-of-roundness) and a buckling 'bulge' results.
- *Tank overpressure.* If a fixed-roof tank suffers overpressure or an explosion in the vapour space, the result is often a non-uniform compressive load in the shell. This type of elephant's foot buckling can be dramatic with large unstable bulges opening at several points around the lower periphery of the tank.

Vacuum buckling

Fixed-roof tanks have vacuum relief valves installed in the roof. If these fail and a partial vacuum is induced in the vapour space (either by the contents being lowered or because of ambient temperature changes) compression stress can be induced in the shell. Roof and shell are pulled downwards and inwards, respectively, producing the conditions for classical external pressure buckling. The clue that vacuum buckling has occurred lies in the number and location of the resulting deformations – the plates show distortions into multiple waves spaced more or less uniformly around the tank shell.

Floating roof distortion

Distortions in tank roofs are rarely seen during in-service inspections. One exception however is deformation of the centre decks of large tanks fitted with pontoon-type floating roofs. Prolonged strong winds can cause plastic deformation of the single plate thickness centre deck (i.e. the 'infill' roof plates inboard of the pontoon annulus). You will see this first as permanent rippling or waviness of the deck surface (see Fig. 14.9).

Storage tank corrosion

In-service inspections of older storage tanks nearly always reveal some corrosion. In the longer term, corrosion is the main cause for retiral of steel tanks – once the coating has broken down or suffered mechanical damage, corrosion rates of up to 0.5 mm/year are not uncommon on tanks in coastal or estuary environments. The locations of tank corrosion are almost totally predictable. Figure 14.10 shows a summary of the most common features.

Corrosion assessment

Corrosion assessments are mainly concerned with corrosion of the shell plates. Owing to the large size of storage tanks it is impractical during routine in-service inspections to perform thickness measurements on 100 percent of the shell plates. Because of this, both API 653 and EEMUA 159 standards specify a system of averaging thickness readings over selected planes and treating the results on an averaging basis. In addition, the general principle is applied that *isolated areas of pitting* can be ignored – the theory being that this rarely contributes to wholesale strength reduction of the shell.

Tank corrosion assessment is based around two thicknesses: the minimum thickness, t_{min}, of a corroded area (excluding isolated pits)

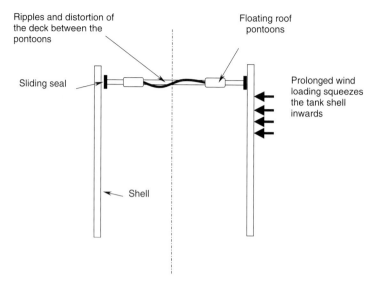

Fig. 14.9   Storage tank floating roof deck distortion

and the average thickness, t_{av}, as measured over an assessment length, L, which is itself calculated using a consideration of t_{min} identified in the corroded area. The concept is based on the assumption that hoop stress over a corroded area will average out over the assessment length. Once L has been determined then t_{av} is calculated over the length and compared with acceptance criteria. This sounds complicated but is straightforward in practice. Figures 14.11 and 14.12 shows the procedure.

In practice this type of corrosion assessment only gives part of the picture of tank condition. It is mainly about deciding an acceptable level of material thinning with respect to the tank's resistance to bulging and wind buckling. Isolated areas of corrosion (particularly under welded attachments etc.) will cause leakage long before the overall strength of the tank is affected. Thorough inspection of localized corrosion is therefore also an important part of an in-service inspection programme. It is sometimes not sufficient to rely on the code acceptance criteria alone.

Hydrostatic testing

API and BS design codes require that new and substantially repaired tanks are hydrotested by being completely filled with water (see Fig. 14.13). The test imposes stresses greater than those when the tank is in

Location	Useful points
Floor plate underside: severe pitting corrosion.	• Visual inspection impossible. Thickness loss can be detected by eddy current 'floor scanner' from the tank inside. • Most corrosion occurs within 1 m in from the shell. • Corrosion rates are higher at higher temperatures. • Broken or brittle mastic seals that allow rainwater to creep under the tank bottom are a common cause of underfloor pitting.
Lower shell strake pitting corrosion.	• Caused by water sitting in the trough resulting from foundation settlement or shell lip distortion. Expect corrosion rates of 0.5 mm/year once the paint layer has broken down.
Internal corrosion of tank bottom: preferential weld attack.	• Corrosion starts around the lap welds where water inevitably gathers if the tank is left empty. Corrosion rates can be unusually high, > 1 mm/year in extreme cases.
Water draw-off sump pits: pitting or galvanic corrosion.	• These are often cathodically protected. Expect medium-high corrosion rates if this breaks down. Failure is rare as sump pits are normally constructed of thicker steel than the floor plates.
Internal shell corrosion in the vapour space: general corrosion and chemical attack.	• Look for this in tanks that are filled and emptied regularly, or contain aggressive fluids (particularly acid salts). • Check in floating-roof tanks around areas scraped by roof seals/shoes.

Fig. 14.10   Storage tank corrosion – where to look

	• Not normally a problem in crude oil storage tanks except when water contaminates the contents and sits in a layer just above the tank bottom.
Corrosion on tank roof.	• Caused by pools of water getting caught in traps between stiffeners etc. (Note API 653 gives an absolute minimum acceptable roof plate thickness of 2.5 mm.) • Vapours from tank contents increase external roof corrosion rates.
Internal corrosion of roof.	• Look for this in cone-roof tanks storing some crude products. (high sulphur content = high corrosion rate).
Corrosion under insulation (CUI).	• Endemic on the horizontal surfaces of insulated wind girders and stiffening rings (and the tank roof, if it is insulated). If the whole tank is insulated, look for CUI near the bottom where it is *always* worse.

Fig. 14.10   (Continued)

its design use as water is heavier than the petroleum/chemicals etc. normally stored.

The purpose of a tank hydrotest

- *Foundation integrity*. This really applies only to refurbished tanks rebuilt into re-levelled rubble foundations. If the foundation is insecure the tank will experience initial settlement during hydrotest. Optical alignment checks are carried out before and after the test (see Fig. 14.8 for typical settlement limits). Concrete foundations rarely suffer from settlement unless there has been a major problem with stability of the subsoil.
- *Shakedown of critical welds*. Similar to the situation with pressure vessels, hydrotesting of a tank has the secondary objective of introducing *shakedown*. Plastic yielding occurs in the most highly stressed welds introducing a local compressive stress. This helps guard against future tensile weld cracking and slightly increases the effective factor of safety. It also improves the circularity of the tank. This

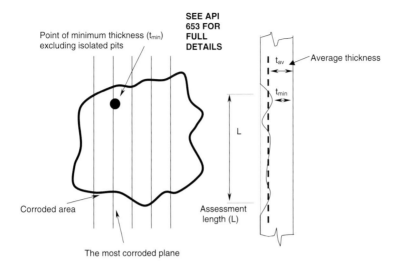

Assessment steps (based on API 653 and EEMUA 159)

1. Find t_{min} anywhere in the corroded area
2. Calculate the assessment length L=3.7 sqrt(D t_{min}) ;whereD=Tank diameter in feet and t_{min} is in inches
3. Measure/estimate the average remaining thickness (t_{av}) over the assessment length using at least 5 equally spaced locations
4. Compare with the minimum t_{av} in the design code

Minimum acceptable t_{av} = D/20SE [98W(H-0.3)+P] or 25 mm,whichever is the lower

where

S> 80% yield R_e (use R_e =215 MPa if it is not known)
t=Thickness in mm
S= Max allowable stress (MPa)
E=Original joint efficiency of the tank design (use E=0.85 for pre-1965 or E=1 for post-1965 designs)
D=Tank nominal diameter (m)
H=Height from assessment point to maximum fill level
W=Specific gravity of tank contents
P=Design vapour pressure (mBar)

Fig. 14.11 Tank shell corrosion assessment

yielding is not visible so a close check is necessary for cracking at weld toes around the shell-to-bottom joint.

- *To check for leaks.*

Not all tanks require hydrotesting. Figure 14.14 shows the exemptions limits proposed in API 653.

Hydrotest of tank repairs

Minor repairs to tanks do not always require the tank to be re-hydrotested. It *is* normally required, however, when:

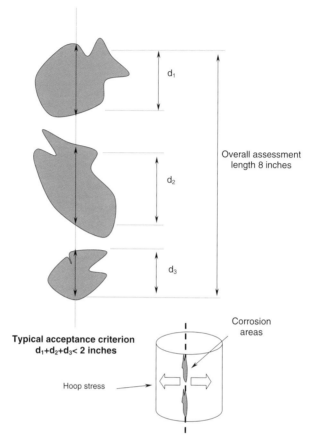

The general approach is to assess the aggregate length of corrosion over a measurement length as shown above. The most critical plane for corrosion in a storage tank is that which weakens the **vertical plane** i.e. because this the area resisting the dominant hoop (circumferential) stress in the tank wall.

Fig. 14.12   Tank shell corrosion assessment

- The tank has been jacked-up for foundation re-levelling or bottom plate replacement.
- Nozzles or manholes (and their associated compensation pads) have been replaced or substantially repaired (say > 25 percent of the welding).
- The annulus plates have been replaced. Note that under the API codes, patch welding of annulus plates is not permitted. If annulus plates are cracked or thinned they must be cut out and replaced.
- Shell plates have been replaced (or extensively patch-welded) near existing seams.

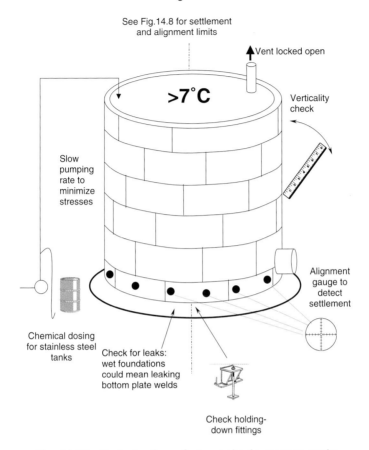

Fig. 14.13   Hydrotesting of atmospheric storage tanks

These are general guidelines only – smaller tanks with maximum plate thickness < 10 mm are often not re-hydrotested to full hydrotest pressure on the basis that the advantage to be gained from shakedown is less for thinner welds. They are subject to a 'partially full' test instead, to check for leaks.

Hydrotest conditions

API 653 gives the clearest guidance on hydrotest temperature constraints. The test should not be performed when the ambient fluid temperature is less than 7 °C, to avoid the risk of brittle fracture. For stresses of less than about 7 ksi however, the standard concedes that brittle fracture is rare.

EEMUA 159 is more conservative, specifying a minimum temperature of 5°C. The holding time for the hydrotest is normally 24 hours.

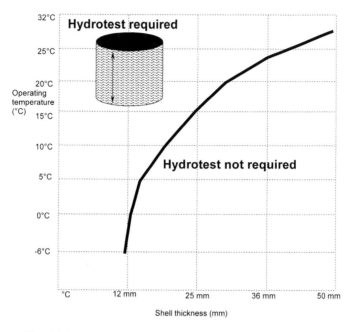

Tanks which meet the criteria shown below do not necessarily need to be hydrotested. The rationale is that the combination of high operating temperature and low material thickness means that the risk of brittle fracture is very small, hence the hydrotest is not essential. Many are hydro tested anyway, to check for leaks.

Fig. 14.14   Storage tank hydrotests – exemption limits

Refrigerated storage tanks

Refrigerated tanks are used for bulk storage of liquefied gases and some chemical products. Their construction is more complex than atmospheric tanks, and the amount of in-service inspection that can be done on a routine basis is limited by the existence of lagging and cladding around the shell and roof. There are several generic types of refrigerated storage tank classified broadly by the method of containment that is used. Single containment types have a single insulated shell while 'double containment' and 'full containment' types have multiple shells. Figures 14.15 and 14.16 show typical in-service checklists for both types of storage tank.

Figures 14.17 to 14.21, and Tables 14.1 and 14.2, show useful data relating to in-service inspection of storage tanks.

Inspection item	External	Internal
A FOUNDATION		
1. Even soil settlement under tank shell	x	
2. Uneven soil settlement under tank shell	x	
3. Tilting of tank	x	
4. Out-of-roundness of tank shell	x	
5. Pipe connections and bellows	x	
6. Drainage of rainwater from tank pad shoulder	x	
7. Condition of tank pad shoulder	x	
8. Leakage of tank bottom	x	x
9. Ripples of tank bottom due to soil settlement		x
10. Necessity of re-levelling	x	x
11. Foundation bolts (where appropriate)	x	
12. Cathodic protection system	x	
13. Earthing strip	x	
B TANK BOTTOMS		
1. External corrosion, outer edge of bottom, underside corrosion	x	x
2. Internal corrosion	x	x
3. Shape of tank bottom		x
4. Ripples in tank bottom (see 9 above)		x
5. Heating coils (if installed)		
C TANK SHELLS		
1. General thickness of shell plates	x	x
2. Buckling of shell plates	x	
3. Local corrosion of shell plates/welds	x	x
4. Corrosion of wind stiffeners	x	x
5. Pipe connections	x	x
6. Manholes and clean-out doors	x	x
7. Side entry mixers		
8. Shell-to-bottom connection around the periphery	x	x
9. Out-of-roundness of shell rings	x	

Fig. 14.15 In-service checklist for atmospheric storage tanks

10.	Insulation of tank shells, including assessment of corrosion under insulation	x	

	D	**FIXED ROOFS**		
1.	Corrosion of roof plates	x	x
2.	Corrosion of roof supporting structures		x
3.	Seal weld of roof-to-shell connection	x	x
4.	Pipe connections	x	x
5.	Manholes	x	x
6.	Roof vents and pressure/vacuum valves	x	
7.	Firefighting lines	x	
8.	Railings	x	
9.	Insulation of roof including assessment of corrosion under insulation	x	
10.	Measuring devices	x	x
11.	Flame arrestors	x	x
12.	Internal floating decks (if installed)	x	x

	E	**FLOATING ROOFS**		
1.	Corrosion of centre deck and pontoon compartments	x	x
2.	Inspection for leakage inside pontoon compartments	x	x
3.	Inspection for correct position of floating roof (high or low)	x	
4.	Drainage of floating roof blockage of centre drain	x	
5.	Movement of roof seal, rim gap check for long wide gaps and possible jamming of roof.	x	
6.	Broken shunts	x	
7.	Rolling ladder, including wheels, railtrack self-levelling treads	x	
8.	Roof supports and automatic including wheels, railtrack self-levelling treads	x	
9.	Level measuring device	x	x
10.	Fatigue cracks in large centre decks, pontoon, and double decks	x	
11.	Articulated pipe drains, including checks for leaking swing joints	x	x

Fig. 14.15   (Continued)

12. Hose drains, including check on drainage for fatigue at shell connection	x	x
13. Valve of roof drain in tank shell. Check on correct position. Check tight shut-off	x	
14. Syphon drains, priming	x	
15. Foam drains	x	
16. Standing water/product on roof		

F	PAINTING		
	Paint condition on tank shell, roof, bottom, and attachments	x	x

G	FIRE FIGHTING SYSTEM		
	Condition of all items	x	x

H	LEVEL MEASURING SERVICES		
	Condition of all items	x	x

I	PIPE CONNECTIONS		
	Condition, level, and settlement of all connecting piping to the tank shell	x	x

Reference EEMUA 159

Fig. 14.15   (Continued)

Inspection plan

A   **CONDITION OF TANK FOUNDATION**

1. Carry out level survey of outer shell to bottom junction or the foundation. This should be compared with original level survey when tank was constructed. Examine for any signs of buckling at the base.
2. Check condition of ring wall or pile cap. (cracked, spalled, exposed, etc.)
3. Check drainage away from ring wall.
4. Check erosion settlement and frost heave of ground around tank foundation.
5. Check condition of grout around the outer shell.
6. For elevated piled tanks check the underside of the pile cap for ice in winter. The junction of the pile to cap should be examined for spalling, cracking, and traces of corrosive product from the reinforcing bar.

B   **FOUNDATION HEATING SYSTEM**

1. Examine randomly selected conduit junction boxes for damage etc.
2. Check condition of fuses.
3. Check foundation heating controls.
4. Compare ammeter readings for each phase with original design values.
5. Record voltage readings for each phase.
6. Measure each thermocouple or remote indicating device reading used to control foundation heating. Compare with temperature indications in control room.
7. Check switch gear operation.
8. Check records of power consumption of heating system.
9. Check foundation heating records.

C   **CONDITION OF OUTER TANK**

1. Check outer tank paint surface condition.
2. Check paint condition on shell stiffeners, stairways, structural members, and nozzles.
3. Check for signs of distortion or damage to the shell or roof.
4. Check for corrosion or mechanical damage at the bottom plate extension beyond the shell and at the anchor bolts and chairs.
5. Check visually for frost or ice spots on the outer shell or roof.
6. Carry out a thermographic survey on the outer tank to verify the general condition of the insulation.
7. Check condition and integrity of insulation cladding.

D   NOZZLES AND PIPING
1. Check for frost or ice build up on thermal distance pieces.
2. Check the condition of any expansion bellows for corrosion, unusual distortions, ice build up or other damage.
3. Check for rotation of nozzles. Movement would indicate frost heave on the inner tank or deterioration of the foundation.
4. Check pipe support connections to roof or shell.
5. Check piping insulation.
6. Check condition and settings of spring supports and compare with data sheet settings.
7. Check trace-heated nozzles and adjacent piping for localized corrosion.

E   STRUCTURES
1. Check condition of roof platforms, walkways, and handrails.

F   PRESSURE AND VACUUM RELIEF VALVES AND INSTRUMENTATION
1. Check pressure and vacuum relief valves for icing, mechanical damage, corrosion, and leakage.
2. Check and test that steam lances are functional (where fitted).
3. Check sources of possible liquid escape, which could impinge on the tank and check condition of any protection devices provided, e.g. mats, catchtrays, etc.

Fig. 14.16   In-service checklist for refrigerated storage tanks

Table 14.1   EEMUA 159 inspection frequencies for storage tanks

		External		Internal	
Classification	Service conditions	Regular visual (years)	Ultra-sonic thickness (years)	Sample or single tank (years)	Maxi-mum for group (years)
Group 1	Slops, corrosive or aggressive chemicals, raw water, brine; no coating	3	1	3	10
Group 1A	With proven internal coating		5	7	12
Group 2	Refrigeration storage for butane/propane, natural gas, ammonia, nitrogen	3		Not required	
Group 3	Crude oil storage, intermediate pro-ducts, i.e. white oil storage, treated water	3	5	8	46
Group 4	Fuel oil storage, gas oil and lub, grease storage, inert or non-aggressive chemicals	3	8	16	20
Group 5	JET A1 (fully coated)	3	10	15	30
Group 6	Light products, kerosene, gasoline cracked distillates, and JET A1 (uncoated)	2	5	10	20

Note that these are EEMUA 159 maximum recommended intervals between in-service inspections. These periods are not part of any statutory requirement in the UK.

API 653: Tank inspection, repair, alteration, and reconstruction

Section	Contents	Comments
1	Introduction	References other relevant API and ASME standards.
2	Suitability for service	The main section: it deals with the evaluation of the tank roof, shell, bottom, and foundations. Gives the acceptance criteria for: • Minimum thickness of shell and roof. • Maximum thickness of annulus plates. • Maximum thickness of tank bottom plates.
3	Brittle fracture consideration	Flowchart for assessing the risk of brittle fracture during hydrotest. Confirms that the risk of brittle fracture is minimal if: • Maximum shell stress <7 ksi, 48.26 MPa). • Shell thickness <12.5 mm. • Minimum shell temperature $> 17\,°C$ ($62\,°F$). Shows the exemption curve for tank hydrotesting (see Fig. 14.11).
4	Inspection	A short section (3 pages) covering inspection techniques and recommended intervals. Cross-references to appendices of inspection, checklists, etc.
5	Materials	One page only. Cross-references the mechanical tests (e.g. ASTM A6 and ASTM A370) specified in the storage tank construction standard API 650.
6 and 7	Design considerations for reconstructed tanks' and Tank repair and alteration	Specifies a maximum shell stress of 2/3 yield or 3/8 UTS for repaired shells. Prohibits patch-weld repair to the tank bottom annular ring. Gives reinforcing plate diameters for 'hot tapping' of existing tanks. The remainder of the technical details are cross-references to the requirements of API 650.

Fig. 14.17 API 653 contents

8	Dismantling and recon- struction	Section 8.5 specifies dimensional tolerances for reconstructed tanks (see Fig. 14.21).
9	Welding	Cross-references ASME IX to be used for repair welding.
10	Examination and testing	Specifies the scope and type of NDT to be used on tank shells, penetrations, and bottoms. This section specifies the definitions of major and minor repair and includes details of hydrotest procedures and settlement limits (see Fig. 14.8).
Appendix B	Evaluation of tank bottom settlement	Defines types of elevation measurement, edge settlement, bottom settlement, and bulging, along with acceptance levels (see Fig. 14.6).

Fig. 14.17   (Continued)

Plumbness: maximum of 1/100th of the total tank height, to a maximum of 127 mm (5 in).

Roundness: Tank diameter (m)	Radius tolerance (measured at 300 mm above the shell-to-bottom weld)
<12.2	±12.7 mm
12.2–45.7	±19 mm
45.7–76.2	±25.4 mm
> 76.2	±31.75 mm

Peaking: maximum of 12.7 mm over a horizontal sweep of 914 mm.

Banding: maximum of 25.4 mm over a horizontal sweep of 914 mm.

Fig. 14.18   Dimensional tolerances to be used on reconstructed tanks (as per API 653)

A full hydrostatic test (held for 24 hours) is required when:

1. The tank has been subject to reconstruction (e.g. such as jacking and replacement of the lower shell strake).
2. The tank has been subject to *major repairs* – defined as any of the activities below:
   - Replacement of the annular floor plate ring on the entire tank bottom.
   - Replacement of the shell-to-floor plate weld.
   - Replacement of any shell plate with a length dimension > 300 mm.
   - Cutting/replacement of any vertical shell plate weld > 300 mm in length.
   - Installation or replacement of any shell penetration > 300 mm diameter, below the design liquid level.

API 653 provides a series of conditions in which tanks are exempt from hydrotesting after major repairs. Figure 14.20 shows these conditions.

Fig. 14.19   Hydrostatic test on reconstructed tanks (as per API 653)

The tank is exempt if:

1. The material impact strength (toughness) complies with API 650 and conditions a, b, c, and d apply (see below).
2. The material is of unknown toughness but falls within the 'exempt' region of Fig. 14.20 **and** conditions a, b and c apply.
3. The repairs have been limited to the bottom plates or annular plate ring (excluding the shell-to-bottom weld) and conditions a and b apply.

Condition a. The repair has been reviewed and approved as compliant with API 650.
Condition b. The repair materials themselves are compliant with API 650.
Condition c. Shell welds are full penetration and are subject to 100% RT and the root run checked using DP or MPI.
Condition d. Shell penetration welds are full penetration and subject to 100% UT, and the root runs checked using DP or MPI.

Fig. 14.20   Storage tanks – exemption from hydrotesting after 'major repair'

Table 14.2 **BS 2654: Minimum shell thickness (including corrosion allowances) for tank construction**

Nominal tank diameter, D (m)	Minimum shell thickness, t (mm)
$D < 15\,w$	5
$15 \leq D < 30$	6
$30 \leq D < 60$	8
$60 \leq D < 75$	10
$75 \leq D < 100$	12
$D \geq 100$	14

Maximum allowable thickness of shell plates is 40 mm.

- **Out-of-verticality:** 1 in 200.

- **Local 'form defects' e.g. bulges and depressions:**

Plate thickness, t	Deviation over a length of 1 m
$t \leq 12.5\,mm$	10 mm
$12.5 \leq t < 25\,mm$	8 mm
$> 25\,mm$	6 mm

- **Allowable misalignment, a_{max}, of butt welds (after welding):**

  – Vertical shell welds.
  The larger of:
    10% of plate thickness, t, or
    for $t \leq 19\,mm$, $a_{max} = 1.5\,mm$
    for $t > 19\,mm$, $a_{max} = 3\,mm$
  – Horizontal shell welds.

  The smaller of:
    20% of the upper plate thickness, t, or
    for $t \leq 8\,mm$, $a_{max} = 1.5\,mm$
    for $t > 8\,mm$, $a_{max} = 3\,mm$

- **Floating roof tolerance:** maximum variation in gap between roof and shell around the tank circumference = 13 mm.

Fig. 14.21   BS 2654 specified shell tolerances

Chapter 15

Heat recovery steam generators (HRSGs)

The worldwide trend in the power generation industry over the past 15 years has been towards the use of combined cycle gas turbine (CCGT) power stations fired on either natural gas or light distillate (gas oil). These plants use so-called heat recovery steam generators (HRSGs) or waste heat recovery boilers (WHRBs) to generate steam using the waste heat from the gas turbine exhaust gases. There are currently many hundreds of these units in operation worldwide ranging in size up to approximately 260 MW per unit. In most countries HRSGs are subject to strict pressure systems legislation and are subject to periodic on-stream and cold shutdown in-service inspection.

In comparison to older designs of coal- and oil-fired power boilers, HRSGs are simple in design and quick to manufacture and erect on site. They are of modular construction so the same sizes of tubes, headers, and pipework can be used across a range of HRSG designs and sizes. Most HRSGs are designed and manufactured to the same design and construction codes used for traditional power boilers – ASME I is the most commonly used worldwide.

HRSG construction

There are two basic types of HRSG design, each subdivided into several variants. Each variant is marketed as a proprietary design by its licensor, the three main groups of licensors being based in Europe, USA, and Japan. The manufacture of individual parts of the HRSG tends to be extensively subcontracted to boiler manufacturing works, often in the country in which the plant will be used.

The two design types are based on the orientation of the gas flow path: either horizontal or vertical. In the horizontal type (Fig. 15.1) the gas flows from the GT exhaust, passing horizontally over the tube banks termed 'harps' or 'modules' which hang vertically between headers located above and below. In most horizontal designs the headers are

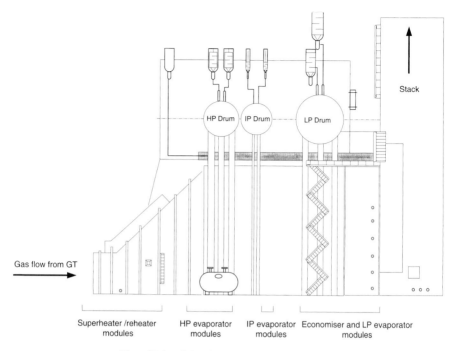

Fig. 15.1   A horizontal gas flow HRSG

located outside the gas path in a dead space and so are accessible for inspection during cold shutdown. Figure 15.2 shows the alternative type, in which the gas flow passes vertically (upwards) over tube banks which are orientated horizontally. The headers are therefore stacked vertically up opposing sides of the HRSG. Some headers (normally the higher temperature ones) towards the bottom of the HRSG are located in dead spaces outside the gas path while others may be inside the gas path, near the side wall.

Drum layouts

HRSGs (both horizontal and vertical types) normally have either two (HP/IP) or three, high, intermediate, and low (HP/IP/LP) pressure stages in order to supply the appropriate cylinder of the steam turbine. This requires separate steam drums (Fig. 15.3).

HRSG header design

Most HRSGs have similar types of header construction. The headers comprise seamless tube containing multiple rows of drilled hole *penetrations* into which are welded the tube stub ends of the harp

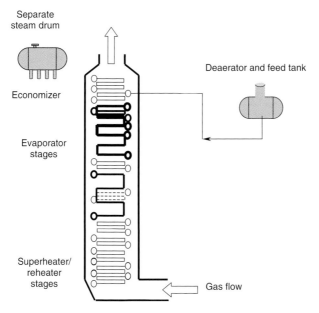

Separate
steam drum

Economizer

Evaporator
stages

Superheater/
reheater
stages

Deaerator and feed tank

Gas flow

Fig. 15.2   A vertical gas flow HRSG

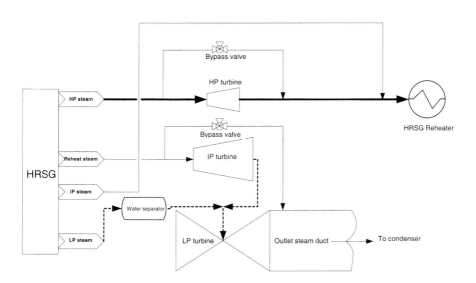

Bypass valve

HP turbine

HP steam

HRSG Reheater

Bypass valve

IP turbine

Reheat steam

HRSG

IP steam

Water separator

LP steam

LP turbine

Outlet steam duct

To condenser

Fig. 15.3   HRSG steam turbine circuits

assemblies. Most headers are not of sufficiently large diameter to be accessed from the inside – the tube stubs being a clearance fit in the header holes, then seal welded from the outside only. Headers may be 15 m or more in length, consisting of two or three separate spools connected by circumferential welds. In most designs the header ends are flat, secured into the header by a circumferentially welded spigot joint. Figure 15.4 shows a typical arrangement.

HRSG harp design

HRSGs use harp assemblies made of finned tubes in order to maximize heat transfer from the GT exhaust gas. The fins are resistance-welded into a shallow helical groove extending round the tube periphery. The fins are often split or castellated to increase gas turbulence and so increase heat transfer. The main tube length is circumferentially welded to stub pieces at each end, using an automated welding process. The stubs penetrate into the header wall as described previously. In modern designs tube spacing is very close (pitch ~ 1–1.25 tube diameter) with the result that minimal access is available between adjacent tube banks.

HRSG operating conditions

Table 15.1 shows typical operating conditions in a triple-pressure HRSG.

HRSG corrosion

Unlike many components in petrochemical and refinery plants, HRSGs operate in a low-corrosion environment. Both natural gas and light distillate fuel used for GTs have a low sulphur content which minimizes acid dewpoint corrosion. They are also virtually free of ash and other constituents that cause problems with fireside slagging and corrosion in traditional oil- and coal-fired power boilers.

Water-side corrosion is also low in HRSGs. Many HRSG installations have access to very high purity feed water from modern ion-exchange or reverse osmosis (RO) plant, installed mainly to meet the high purity requirements of NO_x-reduction water injected into the GT combustion chambers. This high purity feed water, coupled with close purity monitoring via a distributed control system (DCS), results in a clean water-side operating environment with few scaling problems.

These low corrosion environments on both fire and water-sides of the HRSG system means that there is little chance of serious primary

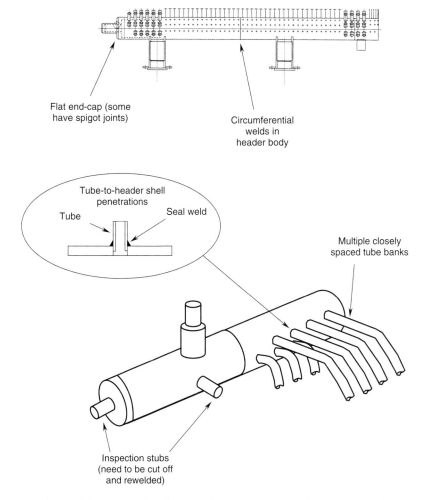

Fig. 15.4 HRSG header construction – general arrangement

corrosion that leads to wall thinning of pressure envelope components. Hence traditional methods of calculating component lifetime using an average annual corrosion rate (i.e. wall thinning) rarely need to be applied to HRSGs. Even the HRSG variants with a high evaporation rate (and associated high unit heat transfer rates) remain virtually corrosion-free, if operated correctly.

Temperature considerations

HRSG temperature conditions range from about 150–180 °C at the cold end to 600–620 °C at the entry to the HRSG from the GT exhaust.

Table 15.1   Typical operating temperatures and pressures – triple-stage power station HRSG

Parameter	Operating condition
LP drum	9 barg, 180 °C
IP drum	35 barg, 253 °C
HP drum	133 barg, 343 °C
Reheater inlet	26 barg, 343 °C
Reheater outlet	25 barg, 545 °C
Gas inlet to HRSG	620 °C
Gas outlet from HRSG (to stack)	84 °C
Circulation ratio	4–5
Full load HP steam flow	80 kg/s @ 103 barg
Full load IP steam flow	10 kg/s @ 25 barg
Full load LP steam flow	10 kg/s @ 3.5 barg

These are high temperatures comparable with those attained in other power boiler designs, requiring special creep-resistant alloys. External components related to the HRSGs such as the feed heaters [once-through coolers (OTCs) using bled GT compression air as the heating medium] also experience similar high temperatures. The mechanism of *creep* is therefore an important consideration in the design and inspection of HRSGs (see damage mechanisms later in this chapter).

A further problem in HRSGs is that of temperature *variations*. These exist in two forms: *distributions* and *transients*. Temperature distributions exist:

- Mainly along the headers – these can be up to 15 m long in some vertical gas flow variants. Temperature difference of perhaps 100 °C can exist along the length, depending on prevailing gas and steam flow conditions. They can result in local imposed stresses and strains which are not always predictable.
- In header walls. HRSG headers are thick walled in comparison to their diameter and so exhibit a complex temperature distribution between their inside and outside (diameter) surfaces.

Temperature transients result from two features of HRSG operation; operational cycles (*two-shifting*) and GT air-purging. Both cause fast heating and cooling transients and associated thermal stresses and strains. Again, the high-temperature thick-walled headers (final stage superheat and reheat) are the worst affected.

Fatigue conditions

Most high-cycle fatigue in HRSGs originates from gas-flow-induced vibration of the tube module 'harp' assemblies. Although the gas flow path is reasonably straight and uniform, the flow velocity is high which, coupled with long unsupported tube lengths, leads to vibration. Pressure fatigue (mainly low-cycle) is compounded by thermal fatigue conditions existing in the headers. This causes header ligament cracking (see later) over time.

Most HRSG problems are related to various combinations of the operating environments outlined above. Together, they result in a selection of damage mechanisms that affect the materials of construction. It is the purpose of in-service inspection to find and identify these mechanisms, before they progress far enough to constitute a risk of failure.

HRSG materials of construction

HRSG materials can be subdivided into the following three main groups, based predominantly on temperature capability:

- Low-carbon steel: low temperatures.
- Low-alloy steels (1–3 percent Cr): medium temperatures.
- High-alloy steels (8–9 percent Cr): high temperatures.

The choice of these materials does not vary *too* much between modern HRSG variants. Preferences have changed over time, as experience has been gained with HRSG operation. The choice of individual materials is based on the concept of *utility*, rather than engineering perfection. Although material temperature resistance, strength, and lifetime are important factors, these are tempered by the practical issues of ease of fabrication and cost. Figure 15.5 shows the situation.

One of the most significant but less tangible issues in material choice is that of *predictability*. As a general rule, materials that have properties such as high creep or impact resistance have a tendency to be *less predictable* than those that have more simple properties. This can apply to the material's mechanical properties, chemical resistance, weldability, ability to resist the formation of weld defects, or any combination of the four. This is by no means a perfect engineering rule, but it does fit reasonably well with the experience of material performance in HRSGs. The main effects become apparent at the high-temperature end of the HRSG where the materials are chosen mainly on the basis of their resistance to creep. Hence, the *predictability* of the material performance

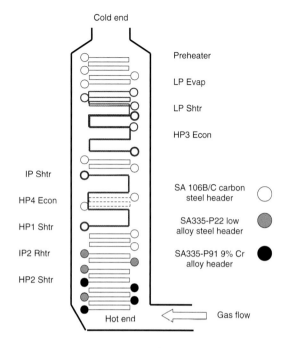

Factors influencing the choice of HRSG materials

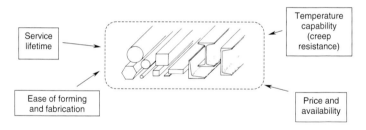

Fig. 15.5   Typical material usage in HRSGs

of the high-chromium alloys is much lower than at the low-temperature end where low-alloy and low-carbon steels are used.

The net result of the unpredictability is a reinforcement of the need for effective in-service inspection of HRSG components, particularly those at the high-temperature end of the gas path (superheater and reheater headers and harps). It also means that the results of inspections cannot always be accurately predicted. Experience shows that the incidence of damage mechanisms will vary significantly across a range of HRSGs that may be of nominally the same design and with similar operating conditions.

Typical HRSG materials

Figure 15.5 and Table 15.2 show a sample material breakdowns for vertical and horizontal gas-flow HRSGs. Figure 15.5 shows a common combination of SA 106B/SA 335-P22/SA 335-P91 alloys for the headers, with similar materials for the horizontal tube-bank 'modules'. Table 15.2 shows typical specified properties. Of these, the most challenging material is the SA 335-P91, 9% Cr alloy used in the high-temperature regions. In this design it is used for four of the headers (HP1 Superheater outlet, IP2 Reheater outlet and HP2 Superheater inlet and outlet). SA 335-P91 is a high-alloy martensitic hardenable alloy used for creep resistance. It has onerous 760 °C post-weld heat treatment (PWHT) requirements and experience in the UK shows that the ASME-specified PWHT procedure during manufacture does not always give the best results. In the 'as-welded' condition it exhibits hardness values > 400 HV which reduces to <250 HV by correct heat soaking at 760 °C. Hardness values greater than 270 HV have been found to cause problems with *reheat cracking*. This is a mechanism by which the alloy becomes 'creep brittle', leading to a situation where the material can eventually fail by advanced (type IV) creep cracking. The situation is normally most serious in the heat-affected zone (HAZ) of header welds and is complicated by the existence of 'back-fill solidification cracking', which also occurs in P91 material (see later discussions on damage mechanisms). For these reasons, P91 is considered a high-risk material and needs to be subject to thorough inspection during HRSG cold shutdowns.

As a general 'rule of thumb', material choice is based on a material's oxidation limit, i.e. the temperature above which it will start to oxidize sufficiently to produce a risk of initiation of corrosion and fatigue.

HRSG damage mechanisms

In-service inspection is the search for damage mechanisms (DMs). In the world of metallurgy they are often amalgamated for convenience into damage mechanism groupings (DMGs). Figure 15.6 summarizes some points about damage mechanisms. It is a fact that the damage mechanisms that occur in HRSGs are much more well-defined and predictable than those that occur in, for example, refinery and petrochemical plant. The predominance of fatigue and creep-related mechanisms means that standard metallurgical theory used to quantify these mechanisms fits HRSGs well.

Table 15.2 Typical HRSG material properties

Application	Standard/grade	Material form	Mechanical properties				Main chemical constituents
			Yield, R_e (MPa)	UTS, R_m (MPa)	Elongation, E (%)	Impact	
Low-temperature headers	ASTM A530/SA 106 (B)	Seamless carbon steel pipe (Grade B)	240	415	12–30	Not specified	0.3 C, 0.29–1.06 Mn, 0.1 Si
LP drums	ASTM SA 285 (B)	Carbon steel pressure vessel plates – low and intermediate tensile strength	185	345–485	25–28	Not specified	0.22 C, 0.9 Mn
LP/IP drums	ASTM SA 299	Carbon steel, manganese–silicon pressure vessel plates	275	515–655	16–19	Not specified	0.3 C, 0.9–1.5 Mn, 0.15–0.40 Si
IP/MP drums	ASTM SA 302 (C)	Mn/Mo/Ni alloy steel for pressure vessel plates	345	550–690	17–20	Not specified	0.25 C, 1.15–1.5 Mn, 0.5–10.4 Si, 0.45–0.6 Mo, 0.7–1.0 Ni
Low-temperature headers	SA 335-P11 (UNS K11597)	Seamless ferritic alloy steel pipe for high-temperature service	205	415	14–30	Not specified	0.05–0.15 C, 0.3–0.6 Mn, 0.5–1.0 Si, 1.0–1.5 Cr, 0.44–0.65 Mo

Table 15.2 (Continued)

Application	Standard/ grade	Material form	Mechanical properties				Main chemical constituents
			Yield, R_e (MPa)	UTS, R_m (MPa)	Elongation, E (%)	Impact	
Medium-temperature headers	SA 335-P22 (UNS K21590)	As above	205	415	14–30	Not specified	0.05–0.15 C, 0.3–0.6 Mn, 0.05–0.5 Si, 1.9–2.8 Cr, 0.87–1.13 Mo
High-temperature headers	SA 335-P91 (no UNS)	As above, tempered at 730 °C minimum	415	585	12–20	Not specified	0.06–0.12 C, 0.3–0.6 Mn, 0.2–0.5 Si, 8.0–9.5 Cr, 0.85–1.05 Mo

Note: P11/P22/P91 material for HRSG tubes is given in SA 213. The chemical and mechanical properties are essentially the same as for the header material specified in SA 335.

- DMs are a feature of a specific material's natural behaviour. **However**, they are always influenced by:
  - design and construction features;
  - manufacturing techniques.
- Some DMs are repairable while others are not.
- HRSG DMs are more often found by inspectors during in-service inspection rather than by metallurgists or corrosion engineers in the laboratory.

Fig. 15.6   HRSG damage mechanisms – some points to remember

From an in-service inspection viewpoint, the techniques used to find DMs are also well established. Non-destructive testing (NDT) techniques used are conventional and well supported by EN and ASME published standards. There are a few limitations – one of the main ones being the design features of the HRSG itself, which can make some types of inspection and NDT difficult. In summary, the search for damage mechanisms in HRSGs is an activity which is *predictable*. Hence it is easier to plan and prioritize in-service inspection activities of HRSGs with a degree of confidence. This is best done in the form of a written scheme of examination (WSE). This summarizes all the necessary inspection-related information and is used to plan the activities of a shutdown inspection. WSEs are required for statutory items covered under the UK Pressure System Safety Regulations (PSSRs) and comparable in-service legislation used in other countries. Table 15.3 shows a typical example for a 5-year-old vertical gas flow HRSG installation.

High-temperature HRSG headers

The high-temperature headers (superheater outlet and reheater outlet) suffer from the most complex damage mechanisms (DMs) to affect HRSG boilers. They have been the centre of much technical attention over the past 20 years and have been subject to several design trends and changes of both configuration and materials of construction. There have also been several notable failures in service, some of them with fatal consequences.

Two lowest common denominator DMs affect these headers: creep and cracking. These act together in several different ways and combine

Table 15.3   Typical WSE/NDT requirements for HRSG inspection

HRSG component	Damage mechanism	Inspection technique used	Typical extent	Comments
Low-temperature headers (low-carbon steel, e.g. SA 106B)	Internal scaling and metal loss	Visual	100%	Mainly on older inspection
	Cyclic damage (economizer harps)	DP/MPI of tube-header welds.	10%	Economizer inlet header is subject to damage from cyclic effects/thermal shock
		Borescope examination	5%	
	Flow-accelerated corrosion	UT wall thickness check	5%	Mainly happens in economizer headers.
High-temperature headers (low- and high-alloy steel, e.g P11, P22, P91)	High-temperature creep (general)	Dimensional measurement for bulging (difficult). Replication tests	Minimum 4 points per header	Bulging indicates serious advanced creep damage. Replication tests provide early warning.
	Localized high-temperature creep (at discontinuities)	Replication tests	Locations across tube/nozzle-to-header welds	Tests must extend across parent materials, weld and the HAZ

Table 15.3 (Continued)

HRSG component	Damage mechanism	Inspection technique used	Typical extent	Comments
High-temperature headers (low- and high-alloy steel, e.g P11, P22, P91) (continued)	Thermal/mechanical fatigue (stub welds)	DP/MPI of tube-header welds and UT of longitudinal seam welds	25–100% depending upon severity	More severe in welds and support attachments owing to the stress-concentrating effect
	Oxide notching	Visual (borescope) to look for spalling	10%	Difficult to identify until it has progressed far enough to initiate ligament cracking
	Ligament cracking (longitudinal cracks in bore holes)	Borescope or specialist DP technique	10%	Often a result of temperature distributions along the header
Steam drums	Internal metal loss (wall thinning)	UT thickness measurement	20 measurement points per drum	Rarely a life-limiting mechanism on HRSGs if feed water quality has been well maintained
	Thermal stress cracking of shell/head welds	DP/MPI from inside of drum	Circumferential and longitudinal seam welds	Fairly uncommon under normal HRSG operating conditions at < 80 000 h life

Table 15.3   (Continued)

HRSG component	Damage mechanism	Inspection technique used	Typical extent	Comments
	Feedwater nozzle weld cracking	DP/MPI from inside and outside of drum	100%	Cracking possible owing to the high temperature differential. Rarely a problem in early HRSG life
	Steam outlet nozzle weld cracking	DP/MPI and UT angle probe	100%	Often exhibits erosion rather than cracking
	Rolled joint leakage	Visual DP/MPI for weepage. UT for ligament cracking	100% if problems evident	Once weepage starts, the high residual stress in the tubes (from rolling) can lead to caustic embrittlement and ligament cracking
Boiler tubing (superheater)	Oxide growth on inside (water) surfaces of tube banks	Visual. Specialist UT oxide thickness measurement techniques	10%, with emphasis around tube welds	Most common in superheater banks. Causes significant reduction in service lifetime. Can lead to hydrogen damage

Table 15.3 *(Continued)*

HRSG component	Damage mechanism	Inspection technique used	Typical extent	Comments
Boiler tubing (superheater) (continued)	Creep	Visual (sagging or bulging) then replication tests	<1%	Limited to superheater/reheater tube banks >400 °C. Can lead to catastrophic 'fish-mouth' tube failures
Boiler tubing (low temperature)	Internal diameter cracking caused by corrosion fatigue	Difficult to detect by standard NDT methods. Borescope examination can be used	5%	Caused by operational cycling
Attemporator (superheat and reheat temperature control) containing high alloy steel e.g. P91	Liner weld or shell damage (fatigue caused by thermal shock). Spray head	Disassembly of the attemporator to allow visual/DP/MPI of the spray nozzle and liner shell. Borescope examination if full internal access of the liner shell is not possible	100%	Main damage occurs in the welds near the water spray nozzle
High-temperature external steam piping (seamed high-alloy steel, e.g. P91)	Creep	Dimensional check (difficult) and replication testing	10%	Longitudinally seamed P91 pipework is increased risk

Table 15.3   (Continued)

HRSG component	Damage mechanism	Inspection technique used	Typical extent	Comments
	Cracking of seam welds	Longitudinal and transverse UT scans (angle probe)	10%	Longitudinal UT scan requires grinding of weld-cap
	Corrosion under insulation	Visual and UT wall thickness check	10%	More common in older HRSGs. Rarely results in life limitation owing to large wall thickness of B31.1 steam pipework.
Internal link pipework	General corrosion fatigue	UT wall thickness measurements and UT of welds (angle probe)	5%	The main damage mechanisms are flow-accelerated corrosion and fatigue-accelerated corrosion caused by vibration and pressure/temperature cycling

with thermal stress, pressure stress, and fatigue effects to form a raft of possible DMs, each of which, acting alone or in combination, has the potential to cause failure. The final failure mode of high-temperature headers is invariably *cracking* but the mechanisms that precede this final stage can vary. This is a complex, three-dimensional picture – one that provides a real challenge to the practice of in-service inspection.

What are the damage mechanisms?

Figure 15.7 shows the four main DMs. The important point is that while each is shown as a separate mechanism they exist, in reality, in combination with the other three. The true situation is best represented by the conceptual diagram at the bottom of Fig. 15.7 – although in reality even this is oversimplified – the true representation should perhaps be three-dimensional. This complex combination of DMs means that degradation of high-temperature headers is one of *degree*. There is always a combination of generic DMs waiting to happen – and it is left to the prevailing combination of plant design and operating characteristics to decide which ones prevail on an individual header at a certain stage of its lifetime.

Where do these header damage mechanisms occur?

Hot header DMs are mainly *later life* mechanisms. It is rare for any of the DMs in Fig. 15.7 to occur below 50 000 hours and more than 80 percent of related failures (on both HRSG and fired power boiler designs) have occurred > 80 000 hours. The situation can be complicated, however, by two factors:

- *The existence of manufacturing defects.* Material or, more commonly, welding and post-weld heat treatment (PWHT) defects can easily cause early life failure but are in many cases, unrelated to the main DMs in Fig. 15.7. Most manufacturing weld defects, for example, result in early fatigue failure at <5000 hours.
- *Maloperation of the HRSG.* The effects of this are influenced by whether or not the HRSG has auxiliary firing. Unfired HRSGs have their heat input predetermined by the exhaust temperature from the GT and a control system which closely controls temperature transients by limitation on loading and unloading 'ramp' rates, purge cooling, and similar. Hence the possibilities of serious maloperation, although not eliminated, are restricted. HRSGs with auxiliary firing, however, are at greater risk of out-of-design

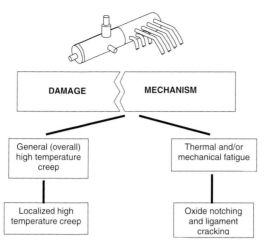

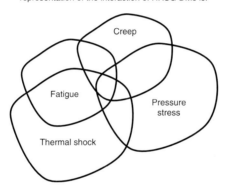

In reality, these DMs are rarely discrete. They act in *combination with* themselves and other stresses, so a better representation of the interaction of HRSG DMs is:

Fig. 15.7   The four main HRSG-header damage mechanisms

temperature and pressure transients. The common maloperation events are:

– out-of-design heating/cooling transient rates;
– excessive pressure cycling;
– poor feed water treatment.

On balance, most cases of maloperation result in a risk of *long-term* rather than short-term (early life) failures.

So are hot-header DMs predictable or unpredictable?

A bit of both. While the individual DMs outlined in Fig. 15.7 are well known and reproducible under laboratory test conditions, the way in which they interact is cloaked in *un*predictability. Individual temperature distributions, accuracy of erection, and operating conditions have a huge effect, to the point where each individual HRSG can be expected to exhibit a character of damage mechanisms almost unique to itself – if not in type, then certainly in severity. The challenge of in-service inspection is therefore to find and identify these damage mechanisms before they progress too far. The procedure starts early in an HRSG's life, i.e. at the first cold shutdown, with a 'benchmarking' exercise, and continues throughout its life. A correctly managed inspection programme then *increases* in frequency as the HRSG gets older and the risk of service DMs increases.

General high-temperature creep

'Generic' creep in HRSG hot headers follows the classical model of high-temperature creep. Creep is a permanent and non-reversible stretching (inelastic strain) caused by sustained exposure to stress at high temperatures. The temperature at which creep starts depends on the material but ranges from 400 °C for plain carbon steels to about 480 °C for Cr/Mo alloys. Creep is a progressive mechanism – the higher the temperature, the worse the creep. The key point about creep is that it occurs at 'as designed' steady temperature conditions. Process temperature upsets make it worse but it does not need them in order to occur. Figure 15.8 shows the basic process.

   The characteristic of general creep is that it occurs uniformally over the body of the hot header. This causes a uniform diametral strain, i.e. the header suffers permanent plastic deformation as a result of creep of the material under the influence of pressure stress. In practice, because of temperature variations along the header, the deformation is rarely uniform – it normally appears as bulges. Eventually, general creep like this can lead to cracking and failure of the header. It is acknowledged that this type of creep has a level of *predictability*. It is commonly designated into three stages: I, II, III, and can be accepted, under close monitoring, as not being a serious risk to integrity, until it has reached stage III. Many petrochemical, refinery, and nuclear plants continue to operate with headers and similar pressure envelope components that exhibit early-stage creep.

Creep replication tests

Witnessing of creep replication tests is a common task for the plant inspector involved in shutdown inspections on power station, refinery,

Creep is:

TIME	TEMPERATURE	STRESS
DEPENDENT		

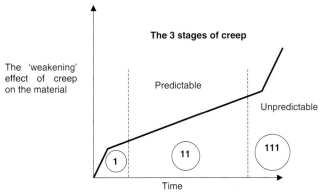

The 3 stages of creep

The 'weakening' effect of creep on the material

Predictable

Unpredictable

1   11   111

Time

Factors that make it worse in HRSGs are:

Temperature upset conditions
Restraint stresses
Excess loadings of any type

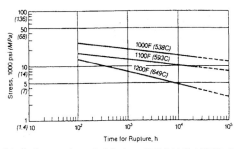

A typical creep characteristic for a P22 2 ¼ Cr HRSG alloy

Fig. 15.8   The basis of creep

or petrochemical plant. The technique is used on plant items operating above approximately 400 °C and its purpose is to identify microstructural creep damage.

Performing replication tests on site during plant shutdowns is more difficult than doing the tests in the laboratory. It is therefore an important witness point for plant inspectors – to make sure the test is carried out properly. Remember that the inspector's role is limited to witnessing and reporting on the test rather than actually doing the test –

this is the job of a trained technician. The site test itself consists of several main stages (see Table 15.4): polishing, cleaning, etching, applying the replica acetate, and 'test viewing' the etched microstructure using a field microscope. The final stage is to view the microstructure photographs after the replica strips have been further processed and photographed in the laboratory. Figures 15.9, 15.10, and 15.11 show the processes.

Choice of test locations

As a plant inspector you should check the planned location of the replication test. The correct location is across a weld (typically a nozzle-to-shell weld as shown in Fig. 15.7). This ensures the test will check the microstructure across the weld metal, heat-affected zone and the parent material – a common area for creep to occur. For large welds it will be necessary to do several separate tests to make sure all these transition areas are covered. Check also that tests are carried out in the parent metal itself (both types if it is a dissimilar material joint). These can provide results that are useful as a comparison to those obtained across the weld area.

Preparation

The area to be tested must have a temperature of less than 40 °C. If it is hotter than this, the replication acetate will 'bubble', giving poor test results. As a guide, if you cannot keep your hand on a component, it is probably too hot. Make sure that the lagging has been removed for at least 50 cm either side of the test area. This is necessary so that the technician can get proper access with grinding and polishing tools.

Grinding

The purpose of grinding is to produce a smooth weld profile to which the acetate replication strip can be applied. Check that the entire weld-cap has been profiled by grinding – normally a 125 mm angle grinder or similar is used.

Polishing

Polishing is carried out using zinc oxide abrasive discs on a smaller electric polishing tool. Check that the technician uses progressively higher grit discs. A normal progression would be 36, 60, 120, 220, 600, and 1200 grit. Each disc is used for a maximum of about 30 s. The final polishing stages use 6-μm followed by 1-μm diamond paste. These should polish the test area to a reflective finish. Check that the finish is absolutely smooth and mirror-like without visible scratches.

Table 15.4 Creep replication tests – inspection guidelines

Step	Activity	Key inspection check points
1	Cool workpiece down before starting work	Replica tests cannot be carried out with a metal temperature > 40° as this causes the acetate 'replica strip' to bubble or melt. There is also a health and safety issue if temperatures are too high
2	Grind weld caps to a smooth profile using angle grinder	Weld caps do not need to be ground flat – just to a sufficiently smooth profile to allow the acetate replica strip to make contact
3	Polish surface to a mirror finish using precision polishing tool	Special discs are used (ZnO or similar) in progressively smoother grit sizes: 36, 60, 120, 220, 600, then 1200 grit followed by 6 μm diamond paste and finally 1 μm diamond paste. Stages may be interspersed with acetate wash.
4	Etch and polish cycle	Multiple etches have been shown to give the best results. Check for a minimum of three, e.g. etch–polish–etch–polish–etch–polish. Etch acid (ferric chloride, nitric acid, or 'Viyellas' or similar is used) – the choice depends on the material
5	Apply acetate replica strip after spraying with acetone	Check the strip is sufficiently large to incorporate parent metal, weld metal, and HAZ regions. It should not bubble or melt when applied – this shows the metal is too hot and the test should be discontinued
6	Spray matt black paint over the replica strip then peel the strip off using double-sided tape. Apply to glass microscope slide for viewing using a field microscope	The black paint reduces reflection when checking the microstructure under a low-magnification ($\times 100$) field microscope. It is not necessary to get a good replication over all the area of the acetate replica strip – but ensure that there is sufficient to view sufficient parts of the parent metal, weld metal, and HAZ
7	View photograph replication in laboratory under high magnification	A magnification of $\times 200$ is adequate for most purposes. Up to $\times 500$ is normally achievable for site replicas and up to $\times 1000$ for those done under laboratory conditions.

Fig. 15.9   1 μm polish for replication test

Etching

The purpose of etching is to make the microstructure of the test material visible. This is achieved by applying an acid which attacks the grain boundaries and other features such as creep voids, so making them visible under an optical microscope. The type of acid etch used varies from material to material and with the preference of the technician performing the test. From the plant inspector's viewpoint it is difficult

Fig. 15.10   Replication test area on high-temperature steam pipe

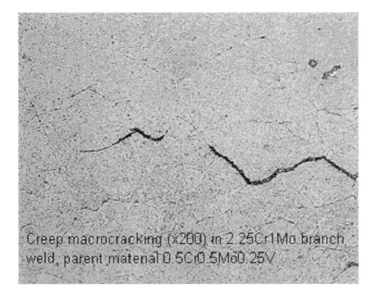

Fig. 15.11   A replica 'macro' showing creep voids and local cracking

to know whether the correct type and strength of acid etch has been used until the microstructure has been viewed. For some materials the etch process has to be done two or three times, with acetone cleaning between each, in order to get the best results.

Apply the acetate replication strip
The replication strip is simply a thin piece of acetate, cut to size to fit the test area. Check that the strip used is long enough to extend over the parent metal, heat-affected zone and weld metal areas. Alternatively, several shorter strips may be used to give the same result. Make sure that the acetate strip has been smoothed firmly down over the test area and that it does not 'bubble'.

View the microstructure
Most technicians will do a quick visual check on the microstructure. This is to make sure that the degree of etch is sufficient to give good photographic results under higher magnification back in the laboratory. The first step is to apply a thin coat of matt black spray paint to the back of the replica strip. This helps reduce reflection under the microscope. A strip of double-sided tape is then applied to stick the replica acetate strip to a glass microscope slide. The slide is then viewed using a field microscope. The magnification is normally $\times 100$. This is low compared with the $\times 200$ to $\times 500$ that will be used for the

422   The Handbook of In-Service Inspection

laboratory photographs but it is sufficient to provide a quick check on whether the etch process is satisfactory. A plant inspector would not normally intervene at this stage – it is up to the technician doing the test to decide whether the view of the microstructure is good enough. One point to check, however, is that the replica strips show enough microstructure in the main areas of interest, i.e. both parent materials, the heat affected zones and the weld metal itself. It is not necessary for 100 percent of each replica strip to show a satisfactory view of the microstructure as long as, together, they provide sufficient coverage of the main areas of interest.

Laboratory microstructure photographs
The next step is for the microstructures to be photographed in the laboratory (see Fig. 15.11). A normal magnification is × 200, or up to × 500 for some specialized material structures. Theoretically, magnification of up to × 1000 is possible but this is rarely achievable for site-prepared samples.

The final stage is to view the microstructure photographs to check if there is any evidence of grain boundary precipitation leading to, in its more advanced state, *creep voids*. Initial grain boundary precipitation starts as thin black lines around the periphery of the grains. This gets progressively more pronounced, forming creep voids, which show as black circular indications, eventually joining up to form creep cracks.

Diagnosis of the extent and criticality of creep is more the job of the metallurgist than the plant inspector. It is necessary, however, for the plant inspector to be able to identify grain boundary precipitation and creep voids in replica photographs, and so be able to report it and seek specialist opinion on the fitness for purpose of the components.

Choice of replication test locations
This is a key issue, particularly for high-temperature headers. The main principles are:

- Six to ten tests should be done per header, in the areas where the highest temperatures are expected.
- The most important locations are across welds, because this is where localized creep effects can be worst, owing to microstructural differences caused by the heat input. Coverage must extend across the parent materials, weld metal, and HAZ.
- At least two tests should cover the parent metal only. It is normal to orientate these in the circumferential and longitudinal directions to allow comparisons to be made.

Localized high-temperature creep

This is a dangerous variation of creep – one which is responsible for many high-temperature header failures. It has slightly contradictory characteristics in that while it is less predictable than general creep, the damage may be more localized, and therefore has a higher chance of being repairable.

The mechanism

Most localized creep occurs in areas of differential strain, i.e. locations where owing to the design, some parts of the material suffer more strain (elongation) than those next to them. The problem is made worse by the existence of welds – these can have a lower ductility than the parent metal, resulting in an increase in stress concentration.

The common locations

Figure 15.12 shows common locations of localized creep in high-temperature HRSG headers. The main problems are with header tube-stub penetrations and their associated seal welds. The area of the header surrounding the tube penetration tends to creep more than the tube stub – this results in differential expansion and encourages cracking. This design feature is one of the inherent design weaknesses of HRSGs. Two other (secondary) effects also have an influence:

- *Header–tube temperature differentials.* During load change, the tube stub cools or heats up more quickly than the header, which has thicker walls. This produces creep and strain differentials in several planes, including in the circumferential 'hoop stress' direction around the header.
- *Nozzle loadings.* The practicalities of large HRSG construction mean that design nozzle loads and moments are sometimes exceeded. There are two problem areas.

    - misalignment of short inter-header pipe runs;
    - wrong adjustment of spring hangers on long vertical pipe runs (e.g. discharge from superheater and reheater outlet headers to the steam turbine). These can have a vertical run of 40 m or more.

These features can place large loads on the nozzle-to-header welds. It is easily possible for stresses due to pipe misalignment or nozzle loading/ restraint to exceed those due to pressure stress.

The end result of localized creep is cracking – either in the welds themselves or in (or near) their HAZ. Cracking in parent material is not unknown but the lower ductility of weld metal, particularly in some of

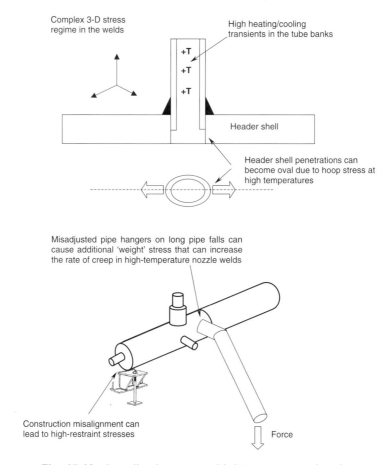

Fig. 15.12   Localized creep on high-temperature headers

the high-alloy header materials, means that the welds usually crack first. Unfortunately, there is no definitive pattern as to which header welds are the most likely to suffer cracking first – tube stub, nozzle, saddle, and attachment welds are all at risk. SA 335-P91 9 percent Cr material has special cracking mechanisms associated with it (see later).

Thermal and mechanical fatigue

As we saw in Fig. 15.7 fatigue due to thermal and mechanical (pressure) effects is a large contributor to high-temperature header damage. HRSG design codes (ASME 1 or similar) do account for resistance to fatigue conditions in their design parameters. This is done by allocating an 'allowable number of cycles' at a certain stress level. This varies from 10 000 to 10^8 and higher, depending on the material. This contrasts sharply with a typical HRSG purchase specification which will rarely require more than 200 start-up cycles per year, divided into cold, warm, and hot starts. In theory, therefore, no HRSG should suffer fatigue damage. The reality of fatigue conditions is somewhat different. Each component of the HRSG is subjected to a complex mix of fatigue conditions. These come from a variety of sources, Fig. 15.13 shows the main ones.

The complexity of these sources means that the exact fatigue condition imposed on any header is largely unknown. For inspection and life assessment purposes, the rational assumption is that HRSG header fatigue is *absolutely unpredictable*. It is, therefore, not wise to ignore possible fatigue effects, even on low-lifetime HRSGs. So:

- HRSG components suffer from complex fatigue conditions. And
- This fatigue is *always unpredictable* (despite what specialist graphs, *S–N* curves and suchlike might infer).

The net effect of all these fatigue mechanisms is either to cause cracking by fatigue only, or to exaggerate the effects of other DMs that are present. Remember that DMs rarely operate discretely; they act *together* to cause failures. Welds suffer the worst effects from fatigue conditions – local stress concentration and geometric discontinuities provide the ideal initiation point for fatigue cracks.

The subject of finding fatigue cracks during HRSG shutdown is a controversial one. Once fatigue cracks have developed sufficiently they can be found by simple visual examination or with the assistance of DP/MPI surface NDT techniques. In their early stages, however, they are much more difficult (sometimes almost impossible) to find with any degree of predictability. The most difficult scenario is if they initiate at a subsurface defect (weld defect or similar). These are not visible on the surface and may not be easily identifiable by simple UT/RT. Hence the absence of early-stage fatigue cracking can never be 100 percent proven under real in-service inspection site conditions. This is one reason why HRSG components sometimes fail, even when they are not expected to.

Oxide notching and ligament cracking

Ligament cracking

This is a damage mechanism that has its origin in the effects of thermal cycling. Ligament cracking is cracking of the area of the header between the tube penetration bore holes, known as the 'ligament'. It has proved a common DM in power boilers, particularly in P22 (and to a lesser extent

---

HRSG headers are subject to fatigue conditions caused by:

- Pressure cycles during loading and unloading.
- Temperature cycles during loading and unloading.
- Gas-flow induced vibration of tube harps during normal operation (this is *high-cycle* fatigue involving a lot of cycles, that can soon reach the theoretical maximum allowable number of stress cycles.
- Induced vibration from header outlet pipework. This is particularly common on superheat/reheat outlet headers which discharge to long piping runs to the steam turbine

---

Fig. 15.13   HRSG header fatigue – the sources

P91) alloy header material. It is therefore one of the main DMs for causing 'retiral' (replacement) of high-temperature headers.

Ligament cracks are *propagating cracks*. They start off as thin star-like cracks on the header inner surface (see Fig. 15.14) then propagate, over time, through the material in a radial direction until they break through the outer surface. The rate of propagation is increased dramatically by thermal cycling and by the differential creep strain effects of radial temperature distributions through the thick-walled header.

Oxide notching

Oxide notching is the most common initiator of ligament cracking. It is caused by high-temperatures acting on the inside diameter of the header tube-stub bore (see Fig. 15.14) to produce a brittle oxide film. As the header is subject to repeated strain due to thermal cycling, the oxide repeatedly cracks off (spalls), exposing the base metal which quickly re-oxidizes. As this procedure repeats itself, the continual spalling and reoxidation produces a pronounced *notch* in the bore hole surface. It is this notch which then acts as a crack initiator for the ligament cracking mechanism.

Detecting ligament cracking

The only practical way to detect early-stage ligament cracking is via borescope examination of the internal surfaces of the header. Access therefore has to be gained either by cutting off a header inspection stub (if fitted) or by cutting open a tube stub near to the area to be examined. Figure 15.15 shows the idea. In practice this can be a difficult exercise.

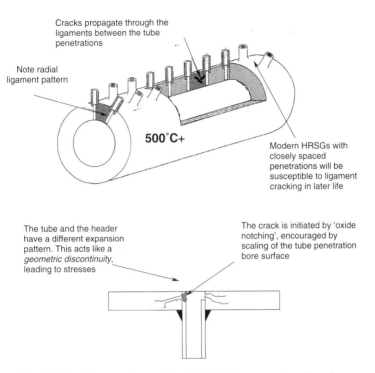

Ligament cracking is a mid-to-late life damage mechanism. It rarely occurs at<50 000 h operation unless there have been serious and prolonged temperature upsets.

Cracks propagate through the ligaments between the tube penetrations

Note radial ligament pattern

500°C+

Modern HRSGs with closely spaced penetrations will be susceptible to ligament cracking in later life

The tube and the header have a different expansion pattern. This acts like a *geometric discontinuity*, leading to stresses

The crack is initiated by 'oxide notching', encouraged by scaling of the tube penetration bore surface

Fig. 15.14   Ligament cracking in high-temperature headers

HRSG steam drums

HRSGs have two, or three, separate steam drums ranging in operating pressure from approximately 8 to 120 barg, depending on the type. They operate at lower temperatures than the HRSG headers and tube harps and so can be constructed of plain-carbon or low-alloy steels (typically SA 299 for HP drums and SA 213 or SA 516-70 for IP and LP drums) without the risk of creep. Despite the thickness of the drum shell and heads (up to 200 mm) the vessel is designed using 'thin-wall membrane stress' assumptions and no insurmountable design problems are caused directly by the use of this thickness of material.

HRSG steam drum damage mechanisms (DMs)

Steam drum DMs are well known. The operating temperatures are outside the creep range, and the absence of serious fatigue conditions

---

Techniques. UT is only effective at finding cracks that have propagated, or have not initiated at the header inner diameter surface. In addition, angle-probe UT is often impossible due to the banks of multiple tube stubs entering the header.

Visibility. Ligament cracks are easily disguised by the oxide film that helped initiate them. In order to increase the possibility of detection, the oxide film has to be removed by remote grinding/polishing. This is a specialist proprietary process.

Classification. Even once ligament cracks are found it is not easy to classify them accurately (i.e. determine their size, depth, etc.). Fluorescent DP techniques can be used to highlight the cracks but precise dimensions of the remaining ligament area etc. are difficult to determine.

Fig. 15.15   Looking for HRSG header ligament cracking – key points

also means that drum DMs are more *predictable* than some other HRSG components. The DMs themselves are rarely life-limiting (within the normal design lifetime of the HRSG) and nearly always *repairable*. They also have the characteristics of occurring later in the HRSG's lifetime, normally > 50 000 hours. The three main DMs are:

- Overall shell metal loss (wall thinning).
- Thermal cracking of welds.
- Tube joint leakage.

Figures 15.16 and 15.17 show some key points.

Overall shell wall thinning
It is almost unknown for shell thinning to be a significant issue for HRSGs in the first 70 percent of their design lifetime. It is caused by corrosion due to either poor feed water treatment (internal metal loss) or corrosion under insulation (external metal loss). In both cases, the situation needs to be extreme before it affects significantly the integrity of the HRSG drum. The drum head and shell wall thickness are designed with a corrosion allowance (> 3 mm) accompanied by possibly additional excess owing to material overthickness, particularly on the drum dished ends (heads). The net result is that overall wall thinning is very rarely a life-limiting DM for HRSG drums. If wall thinning is going to cause problems then it is much more likely to do so on a *localized basis*, i.e. where the thinning corrosion is concentrated in the internal or external high-stress regions around nozzles. Even in this

---

Scope
Most HRSG cold shutdown inspections include:

- Draining and cleaning of the drum.
- Removal of drum furniture (steam separators, baffles, etc.).
- Full visual inspection of drum interiors.
- Sample DP/MPI/UT of critical nozzle welds (from inside the drum).
- Removal of at least 10% of external lagging to access critical nozzle welds.

Benchmark assessment
It is good practice to do a benchmark assessment of HRSG drums during their first cold shutdown. This allows comparisons to be made during subsequent shutdowns. Benchmark records should include:

- Wall thickness assessments.
- Records of which nozzles were subject to UP/MPI/UT and any indications found.

Magnetite layer
Magnetite (Fe_3O_4) is a dark grey/black stable oxide that forms on internal surfaces the drum. Check it is complete and without blistering.

Fig. 15.16   HRSG steam drum inspections – some key points

situation, however, the predominant effect of the corrosion is to initiate conditions of corrosion fatigue or a similar composite DM, rather than cause failure by itself. Figure 15.18 shows some important points on wall thinning inspection.

Thermal cracking of welds

HRSG steam drums can experience several types of weld cracking. The most common cause is thermal fatigue, resulting from excessive cycling of the HRSG. The second most common cause is excessive cooling and heating rates resulting from GT purging (i.e. fast cooling) or fast load-increase rates, respectively. Strictly, the HRSG high-temperature headers are more susceptible than the drums to these high temperature transients but it does not always work exactly like this in practice.

Cracking of longitudinal and circumferential shell welds

This is the *least likely* place to find thermal weld cracking – they are normally high-quality shop submerged arc welds which have been

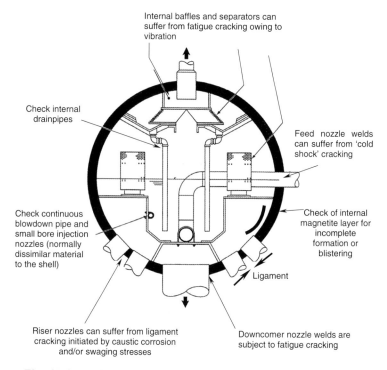

Internal baffles and separators can suffer from fatigue cracking owing to vibration

Check internal drainpipes

Feed nozzle welds can suffer from 'cold shock' cracking

Check continuous blowdown pipe and small bore injection nozzles (normally dissimilar material to the shell)

Check of internal magnetite layer for incomplete formation or blistering

Ligament

Riser nozzles can suffer from ligament cracking initiated by caustic corrosion and/or swaging stresses

Downcomer nozzle welds are subject to fatigue cracking

Fig. 15.17   HRSG steam drum inspection – some key points

subject to 100 percent volumetric and surface NDT. Of the few occurrences of shell weld cracking that have been experienced in HRSG drums, nearly all have been on the head-to-shell circumferential 'finishing' weld (this is an area of high discontinuity stress) and have initiated from pre-existing manufacturing defects. In-service cracking of longitudinal shell welds or shell-to-shell circumferential welds is rare on properly code-compliant drums.

Cracking of nozzle welds
This is much more common but, paradoxically, not so easy to detect. As in any pressure vessel, the nozzles and surrounding areas are subject to a variety of complex stresses. These are easily exaggerated by the effects of thermal cycling – it is fair to say that the fatigue response of these joints is less than 100 percent predictable, so they are high-risk areas. There are several well-documented contributory causes to nozzle weld cracking.

- *Feed nozzle temperature differential.* The HRSG drum feed nozzle has a high temperature differential across it (up to perhaps 150 °C, depending on conditions). It therefore experiences differential

- **Location**: General wall thinning can occur almost anywhere on the head or shell of an HRSG drum. The most common locations are around nozzles (internal) owing to corrosion/erosion and on upper and lower horizontal surfaces [external corrosion under insulation (CUI)], where water has become trapped under lagging.
- **Severity**. Rarely a root cause of failure by itself but it can act as an initiator for fatigue damage. In extreme cases, drum pressure can be de-rated.
- **Inspection**. Wall thinning is measured using standard UT using a $0°$ (compression) probe. In practice, readings are normally accurate to $\pm 0.5\,mm$.

Fig. 15.18   Drum wall thinning – some key points

thermal expansion which encourages thermal cracking. The thermal stress regime can vary greatly between different code-compliant nozzle/compensation designs. It can also differ *between* design codes.

- *Low nozzle weld leg length*. Low leg lengths in critical nozzles and their compensation pad welds result in high stress concentrations. Under such conditions the risk of thermal cracking is increased. One of the worst, often overlooked, scenarios is where the weld profile is marginal or not consistent, resulting in a few small areas of high weld stress concentrations. These are prime areas for in-service thermal cracking.
- *Excess nozzle loading or restraint*. Excess nozzle loading is a common out-of-design issue on large HRSG plant. It is caused by misadjustment of pipe spring hangers (mainly on long pipe runs from the HRSG drum down to ground level) or pipes being forced into alignment with drum nozzles during construction (higher restraint forces result from short, stiff pipe runs at HRSG drum/header level). In such cases it is easy for nozzles stresses, caused by restraint loading, to exceed those due to pressure stress, resulting in unpredictable complex stress levels in nozzle attachment and compensation pad welds. When combined with thermal cycling, this results in a high risk of cracking.

Tube joint leakage
This is a feature of some older HRSG designs that have the downcomer tubes rolled into holes in the lower half of the steam drums. There are also some hybrid types in which the rolled tubes incorporate a seal weld

(not a full penetration weld) on the drum internal or external surface. Repeated thermal cycles can cause the rolled ('expanded') part of the tube to relax, resulting in leakage. The resulting crevice encourages caustic corrosion which leads to further, more rapid deterioration (see Fig. 15.19).

A second problem with steam drum penetrations is ligament cracking. Although the ligament area is normally larger than in HRSG headers, and therefore has lower stress, crack initiation is encouraged by corrosion in the crevice. This greatly increases the risk of ligament cracking.

Rolled-tube joint leakage and ligament cracking can both be detected by standard inspection methods. Leakage is found by visual examination of the outside of the lower half of the drum when the drum is under hydraulic pressure. This can be a time-consuming exercise, however, as the underside may be difficult to access without removing lots of lagging, or even parts of the upper casing in some HRSG designs. An alternative method is to do a visual examination from the inside of the drum. Realistically, however, this is not a very effective method of finding leaks. Of these older HRSGs that use rolled-tube connection into the steam drums, many have been retrofitted with seal welds to reduce the uncertainties of leaking joints.

Steam drum inspection routines

It is easy to waste inspection effort on HRSG steam drums. Although they are affected by some well-known damage mechanisms, these are mainly found on older HRSGs (> 50 000 hours) and even then, the chances of catastrophic failure are very low. Nevertheless, drums do need to be included in the scope of in-service inspection, both from a 'good practice' perspective, and for compliance with statutory requirements, e.g. the PSSRs. Figures 15.20 and 15.21 show drum nozzles exposed for NDT during an inspection.

HRSG tube banks

HRSG tube bank *harps* have perhaps the most straightforward DMs in the HRSG but, paradoxically, are still responsible for the majority of unplanned shutdowns. The main reason for this is simply the large number of tubes and welded joints that are fitted, coupled with the fact that tubes in the same harp can experience significantly different temperature and fatigue conditions depending on the design and operating conditions of the HRSG.

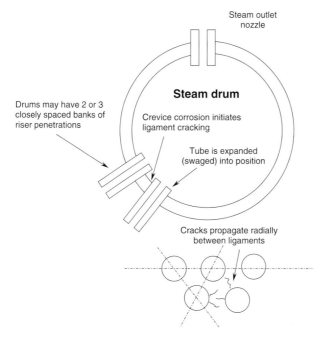

Fig. 15.19 HRSG rolled-tube joint leakage

Tube material is divided into three main groups, chosen to match operating temperature and pressure (see Fig. 15.5). In terms of damage mechanism, however, they divide neatly into two groups: those that are steam cooled (superheater, reheater, and higher-temperature evaporator harps) and those that are water cooled (lower-temperature evaporator and preheater/economizer harps). Each is associated with particular DMs which, on balance, result in broadly similar risks of unplanned outages. Both are similar however in that the harps are *replaceable*.

Oxide scaling

This DM affects mainly the high-temperature steam-cooled tubes. Oxide growth on the inner (and occasionally outer) tube surfaces grows sufficiently thick to form a thermal 'baffle' which can cause catastrophic tube failure. The oxide growth is worst in areas of high temperature, often resulting from uneven temperature distribution across the affected harps. Note that this DM can occur irrespective of the quality of feed water treatment – it is a thermal rather than water chemistry mechanism. In extreme cases the oxide will spall off the surface (exfoliation) and produce significant wall thinning.

Fig. 15.20   HRSG drum nozzles exposed for inspection

Theoretically, oxide scaling occurs later in an HRSG's life (> 50 000 hours), i.e. after prolonged exposure of the tube material to high temperatures. In practice, however, it can occur earlier if process upset conditions or high heating/cooling rates have been experienced.

Fig. 15.21   Lagging sections removed for NDT of HRSG drum upper nozzle welds

Specialist inspection techniques are used to measure the thickness of oxide scaling – they are marketed under proprietary tradenames and are based on a UT technique. Despite the claimed accuracy of such equipment, such tests are often less than conclusive. The main difficulty is in deciding what level of oxide thickness is actually acceptable, and what is not. There are also several different types of oxide scale formation. This means that clear diagnosis is often difficult.

Chemical scaling

This is referred to by several different names. Its main characteristic is that it results from poor feed water treatment rather than any thermal distribution or upset condition. It commonly affects the low-temperature water-cooled tube harps and results in a build-up of scale that reduces flow area and produces crevice conditions that harbour other corrosive effects such as hydrogen corrosion. It is rare on low-lifetime boilers in which feed water quality has been properly managed. It can remain hidden for some time, and steam purity or steam turbine problems often become evident before any HRSG tube problems are noticed. Chemical scaling is easy to detect during HRSG shutdowns by borescope inspection into the harps, accessed from the LP steam drum. Feed water records provide a good clue as to the degree of scaling that can be expected.

Creep

Creep affects those tubes located at the hot ($> 400\,°C$) end of the HRSG, i.e. the steam-cooled superheater and reheater harps. These tubes are made of creep-resistant alloys such as P91 in order to try to keep creep within manageable levels. HRSG tubes react differently to creep than their associated headers that operate at similar temperatures. Creep can cause rupture in tubes (the classical fish-mouth shaped, blunt-edged creep failure) whereas such catastrophic failure is much more rare in headers. Creep is also less predictable in tubes, because of the low wall thickness.

Traditionally, the simplest way to detect creep is by diametrical measurement of the tube in several planes. Most recent HRSG designs have finned tubes, making meaningful measurement impossible, except in the short unfinned area where the tube stubs enter the headers. A further method is by checking for bowing of the tube banks themselves. This can also be inconclusive as most HRSG harps exhibit some bowing due to construction misalignment. In summary, although simple and frequently used, straightforward measurement is not a particularly reliable method of checking for creep of tube harps.

The most reliable way to detect creep in HRSG tubes is to use replication tests. These have been described earlier in relation to headers. The technique is exactly the same for tubes. The main problem is, again, one of access. Access is limited to unfinned areas near the header stubs – it is not usual to remove the fins in other areas of the tubes (they are resistance welded into small helical grooves in the tube outer surface and physical removal could cause mechanical problems). It is also impossible to access most of the tube length without scaffolding. In practice, these limitations often mean that replication tests are concentrated on the high-temperature headers rather than the tube harps themselves. This seems not to introduce any significant increase in risk – creep of headers and their connected tube harps *normally* go together.

Tube inspection routines

Figure 15.22 summarizes the main shutdown inspection points for HRSG tube harps. Inspections are normally limited to a small number of tubes (<5 per cent) and should be concentrated on the areas that have resulted in previous failure on the individual HRSG being inspected. HRSGs harps do exhibit individual failure 'characters' often influenced heavily by the quality of manufacture. It is normal for HRSGs to exhibit tube failure during early life, as manufacturing weld defects become apparent.

Low-temperature headers

These are the headers operating at temperatures (below about 300 °C) located towards the low-temperature chimney stack end of the HRSG. They comprise the main economizer and evaporator banks, and early stages of superheat. The main construction material is low-carbon steel (SA 106B or similar) for the economizer and evaporators, and low-alloy steel such as SA 335P11 ($1\frac{1}{4}$ Cr, 0.5 Mo) for the early stage superheaters. Figure 15.5 shows the typical locations. Construction is almost identical to the high-temperature banks, comprising headers and harps. Despite their relatively low operating temperature, low-temperature HRSG headers suffer from a variety of operational problems. They cannot therefore be ignored during in-service inspections. As a general principle, low-temperature headers suffer from *corrosion-related pro-*

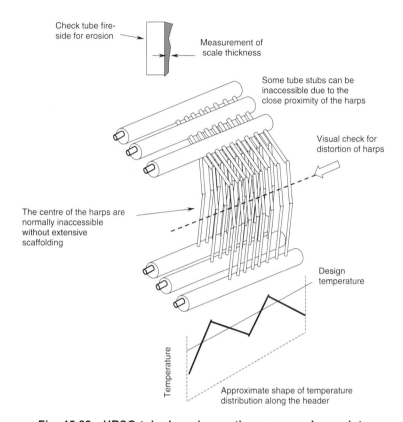

Fig. 15.22   HRSG tube harp inspections – some key points

blems, rather than the more complex DMs that affect the high-temperature end of the HRSG. Figure 15.23 shows the main DMs.

Internal scaling

Low-temperature headers are very susceptible to HRSG feed water quality. There are two reasons for this:

- *Temperature.* The operating temperature range is ideal for the formation of scale-forming salts.
- *Heat flux.* Although temperatures are low, the unit rate of heat transfer (i.e. in W/m^2 of tube surface) is high in the economizer and evaporator sections. This high heat flux encourages the formation of scale.

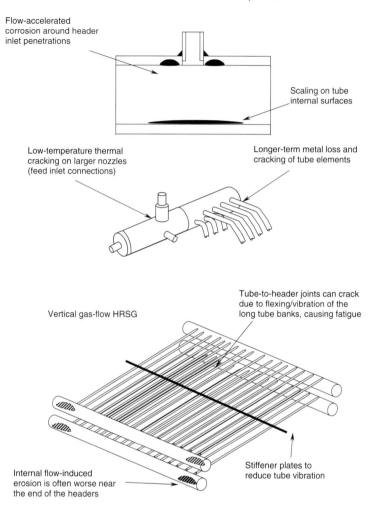

Fig. 15.23   Low-temperature HRSG header damage mechanisms

HRSGs that have operated with high dissolved O_2 levels (originating, for example, from air-cooled condenser leaks) are particularly susceptible to scaling. Similarly, steam systems using seawater-cooled condensers can suffer Cu/Mg salt contamination of the steam circuits from condenser tube leakage – another prolific source of scale-forming salts.

Scaling results in reduced HRSG thermal efficiency, under-scale corrosion, metal loss and eventually, cracking. Hence low-temperature headers are much more susceptible to water-side wall thinning than are high-temperature headers. As HRSGs have long headers, the wall thinning is rarely uniform along the header lengths. It mostly occurs in the hottest centre sections of the headers but, realistically, can appear almost anywhere. Figures 15.24 to 15.27 show HRSG inspection and subsequent repairs in progress.

Flow-accelerated corrosion

This is the result of high flowrates of water (rather than steam) in the economizer and evaporator sections of the HRSG. The severity varies with both the type of HRSG (i.e. whether it is assisted or natural circulation) and with the flow patterns inherent in individual harps. It is, therefore, a rather unpredictable mechanism. One of the worst locations is around the tube penetrations into the header – turbulence here causes differential aeration and the formation of pitting corrosion. Deposits are rapidly scoured away by fluid turbulence, leading to a high annual corrosion rate in excess of published (e.g. NACE) corrosion rates for the particular materials (low-carbon or low-alloy steels) involved. Manufacturing inaccuracies such as joint misalignment (mismatch) or incomplete weld root runs makes things worse. Flow-accelerated corrosion is a *progressive* damage mechanism, most common in the mid to later life of an HRSG, rather than in the first 50 000 hours or so.

Cyclic damage

This is mainly limited to the economizer inlet header. This header suffers from thermal shock as HRSG feed water enters at a low-temperature relative to the bulk metal temperature. The severity varies, depending on the pressure system design. Some triple pressure HRSGs, for example, use a system where the LP drum is continually circulated via the economizer header and tube harps resulting in a lower temperature differential. Other types admit the feed water directly into the low-temperature drum and headers giving temperature differentials of 100–150 °C under some cyclic loading conditions. The results of these cyclic loadings are twofold:

- In the short term, cracking of the tube-to-header seal welds.
- In the longer term, increased risk of ligament cracking propagating from the header bore (inside diameter) surface, radially outwards to the outer surface.

Low-temperature headers – inspection techniques

Inspections of low-temperature headers are mainly visual, and done on a sample basis. On vertical gas flow HRSGs, most of the upper low-

Fig. 15.24   Repair in progress on HRSG header after leak

temperature headers (LP/IP evaporators etc.) are accessible by borescope through their link pipes which penetrate into the lower half of the LP and IP drums. Figure 15.28 shows the arrangement. Several different types of borescope fitting (radial, panoramic, etc.) are used to give a comprehensive coverage of the header internals. Note that, if the

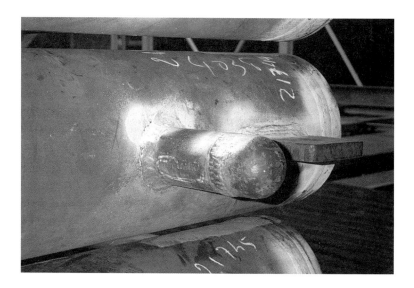

Fig. 15.25   DP of HRSG inspection stub and end-cap welds

Fig. 15.26   HRSG tube-to-header welds

link pipe from the drum is heavily set through the header wall, it can be difficult to achieve a good longitudinal transverse of the borescope head. In this situation, the only solution is to scan the length of the header from multiple entry points. Figure 15.29 shows a borescope in use.

Fig. 15.27   HRSG low-temperature economizer header

Is borescope examination effective?

It is most effective in early-life HRSGs where internal scaling is not well developed. Under these conditions, features such as pitting corrosion show up clearly under borescope examination. Flow-accelerated corrosion shows up particularly well owing to its scoured shiny surface appearance. Once an HRSG gets older and its internal surfaces start to be obliterated by scale, features such as under-scale corrosion and cracking become more difficult to see. In extreme cases, the only solution is to employ special equipment to clean, remotely, the header inside surfaces followed by fluorescent MPI testing, all viewed through the borescope. Similar problems exist with the high-temperature headers, except that the high-temperature oxide is generally more tenacious than low-temperature (water phase) scale, so removal is even more difficult.

Attemporators

Most HRSGs have two 'contact' attemporators: one for the final stage superheat and one for hot reheat steam. Their purpose is steam temperature control rather than actual desuperheating to feed an auxiliary steam range. Temperature control is achieved by spraying condensate-quality feed water into the steam thereby reducing the degree of superheat.

Both HRSG attemporators are normally of almost identical design, the only significant difference being physical size (the reheat attemporator is 20–30 percent smaller owing to its lower operating pressure; 30–40 bar against the 100 bar+ of the superheat attemporator). Material of construction is predominantly SA 335-P91, the same as the high-temperature headers. Although the material is the same, the character of the damage mechanisms are different; more regular temperature distribution along the attemporator means that unpredictable localized creep effects are less of a problem than in headers. Attemporators have also proved themselves much less susceptible to the cycling effects of HRSG load changes.

The main damage mechanism in attemporators is caused by thermal shock from spraying cool condensate into high-temperature steam. The resulting high differential expansion rates cause cracking. A common location for damage is the internal liner and its securing arrangements locating it centrally inside the attemporator shell (see Fig. 15.30). Damage can occur over a short timescale (1–2 years) so this is not a DM

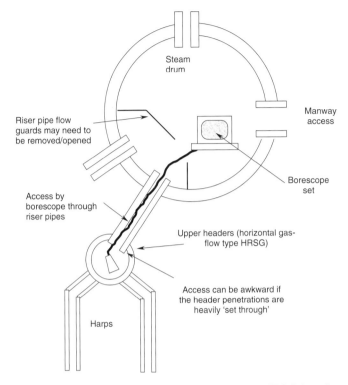

Fig. 15.28   Borescope examination of upper HRSG headers

Fig. 15.29   Borescope head for header inspection

that is limited to later life of the HRSG. Figure 15.31 shows some key points about attemporator cracking.

Shutdown inspections of HRSG attemporators are initially done by borescope through one of the nozzles, normally the small-bore drain connection. This is mainly an inspection for integrity and cracking of the internal components. The spray nozzle assembly is frequently replaced on a periodic basis – most develop cracks after a period of operation of a few years. Replication tests on older units are performed on the shell welds to check for general creep. Theoretically, shell distortion caused by creep can be found by diametrical measurements but, in practice, access and accurate measurement can be difficult.

From a Pressure System Safety Regulations (PSSRs) perspective, HRSG attemporators are high-risk components. This is because they are situated at ground level outside the confines of the HRSG casing, and are therefore in close proximity to site personnel. A similar situation exists with the superheat and hot-reheat steam pipework and external feed heaters (see Fig. 15.32).

HRSG remnant life assessment

Analytical remnant life assessment (RLA) of HRSGs is one of those subjects that inhabit the uncomfortable limbo between academic theory and engineering reality. Like all real engineering equipment, HRSGs do not always comply with the neat world of engineering theory and analysis. Notwithstanding this, RLA techniques are used extensively to decide the future of older HRSGs and other types of power boilers. Chapter 9 gives an explanation of RLA techniques. Figures 15.33 and 15.34 below show some key points when applying these techniques to HRSGs.

RLA techniques (and their validity)

In recent years, mainstream RLA techniques (developed for conventional fossil-fired boilers and nuclear plant) have been applied to HRSGs. The main driving force in this has been the HRSG licensors themselves, particularly in the USA. There is therefore nothing unique about the use of RLA techniques for HRSGs – except perhaps that they are less well proven and validated than for other types of power boilers.

The basic concept
Figure 15.35 shows the basic concept behind HRSG remnant life assessments. The whole thing is an exercise in *prediction* – trying to

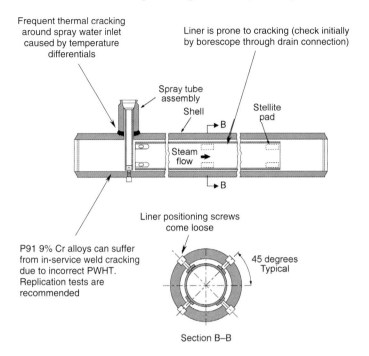

Frequent thermal cracking around spray water inlet caused by temperature differentials

Liner is prone to cracking (check initially by borescope through drain connection)

Spray tube assembly
Shell
Stellite pad

B

Steam flow

B

Liner positioning screws come loose

P91 9% Cr alloys can suffer from in-service weld cracking due to incorrect PWHT. Replication tests are recommended

45 degrees Typical

Section B–B

Fig. 15.30 HRSG attemporator inspection points

anticipate how various components of the HRSG will degrade in service.

What about the sketch at the bottom of Fig. 15.35? This shows the relationship between the overall techniques of RLA in comparison to the technical content of design codes (EN 13445, ASME I/VIII/PD 5500 and similar). Note how much of the content of the RLA technique lies inside the code, but that there is also a lot that lies *outside*. So:

• A remnant life assessment can involve showing that a component that is non-code-compliant is still fit for service.

Within the scope of Fig. 15.35 many of the techniques of RLA are based on assumptions. These are wide-ranging, covering HRSG performance, temperature distributions, time/temperature history, material performance, and a raft of parameters describing the way that cracks in material initiate, grow, and eventually cause failure. Frankly, many of these assumptions are either not well validated or remain a matter of technical opinion as to whether they are 'correct' or not. This is the problem with RLA techniques – they may, or may not, pass a reality check, depending on your point of view. Figure 15.36

- The most common damage mechanism is *cracking* due to thermal shock.
- Initially, the cracks form in the *internal liner* and their attachments. While this is not an immediate pressure envelope integrity issue, it can become one if the cracks are able to propagate into sections welded to the attemporator shell.
- The water *spray nozzle* and surrounding areas can suffer from thermal shock and fatigue cracking.
- Scaling in attemporators is not very common – except in HRSGs where feed water quality control has been badly neglected.

Fig. 15.31   HRSG attemporator damage mechanisms – some key points

shows the situation – note how only the tips of the 'star' represent accurate results. Notwithstanding this, we can look briefly at the three main elements of the RLA technique.

Principal stress prediction

This is the crudest of the family of RLA techniques. It is basically a reassessment of header or tube design against code requirements. The main assessment done is against principal stresses [circumferential (hoop) stress, axial stress, and stresses around main nozzles and manway/inspection openings]. Major simplifying assumptions which are made are:

Fig. 15.32   HRSG feed heaters with lagging section removed for NDT

- **Technique**. RLA techniques applied to HRSGs are *much the same* as used for all other types of high-temperature pressure plant – there are no special techniques involved.
- **Methodology**. Standard analytical techniques such as finite element/finite difference and similar iterative methods are used.
- **Software**. Various proprietary software products are available, each claiming to be better than the rest. All, however, work on similar mathematical principles and suffer the same dependence on the accuracy of the boundary conditions that are used as input.
- **Verification**. Most RLA techniques are probably not very well validated in HRSG applications. This is due to the limited long-term experience with HRSGs in general, the existence of many different HRSG designs, and the use of newer materials such as P91 9% Cr alloys.

Fig. 15.33   Remnant life assessment (RLA) of HRSGs – some fundamental points

- **Header design**. HRSG headers employ a variety of different ligament arrangements and orientations. Hence, ligament *behaviour* is a complex, wide-ranging subject.
- **Operating conditions**. Temperature fluctuations (under steady state and cyclic HRSG loads) have a dramatic effect on material creep and fatigue performance.
- **Material choice**. In particular, SA 335-P91 material has a wide, not entirely consistent, pattern of operational performance in traditional fossil-fired boilers and (more lately) HRSGs. Technical opinions vary on aspects such as PWHT/normalizing/tempering procedure, microstructure constraints, and the acceptability of longitudinal seams.
- **The effects of cycling**. Many HRSGs are designed for a small number of warm/cold starts under the assumption of prolonged base-load operation. In practice, the nature of electricity supply spot-markets means that cycling is often a necessity. This may push an HRSG outside the confidence limits of its design envelope.

Fig. 15.34   Some technical aspects of HRSGs that affect their remnant life assessment

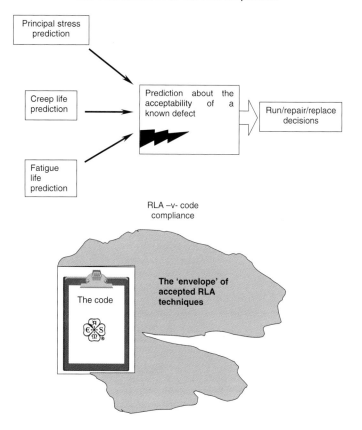

Note how a lot of RLA technique methods and results lie *outside* code compliance

Fig. 15.35   RLA techniques – the concept

- The material behaves as described by the simple uniaxial stress–strain curve.
- Everything exists at a constant temperature and undergoes a constant strain rate.
- Component design uses the elastic (yield) limit as the main design reference point. This changes with prolonged time – temperature exposure (the yield point can reduce by up to 40–50 percent in some alloys).

Fundamentally, the only real purpose of this type of analysis is to 'correct' an original design code assessment to take into account as-built features such as material overthickness, excess weld-cap reinforcement

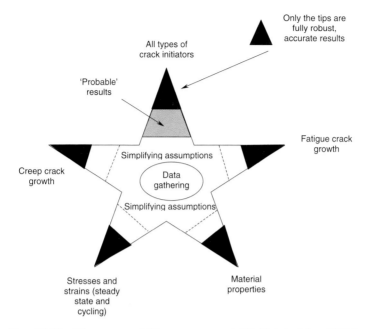

All types of
crack initiators

Only the tips are
fully robust,
accurate results

'Probable'
results

Simplifying assumptions

Creep crack
growth

Fatigue crack
growth

Data
gathering

Simplifying assumptions

Stresses and
strains (steady
state and
cycling)

Material
properties

Fig. 15.36   The remnant life assessment (RLA) 'star' for HRSGs

and similar that might reduce the actual principal stresses in an HRSG component compared with the original design calculation. Surprisingly, many HRSG remnant life assessments do not progress beyond this stage. Figure 15.37 shows a typical example, a principal stress reassessment of misaligned HRSG link pipework.

HRSG components that are commonly subject to this type of assessment are:

- Header link pipework (as in Fig. 15.37).
- Power piping (ANSI B31.1).
- Long headers containing circumferentially welded joints (vertical gas flow HRSG designs).
- Steam drums with excessive head-to-shell misalignment.

Creep life prediction
Creep is a time- and temperature-dependent damage mechanism that reduces the life of the high-temperature components of the HRSG (see earlier). RLA techniques related to creep degradation use well-established theories of how the creep mechanism works, based on a raft of heavy qualifying and simplifying assumptions. Figure 15.38 shows the situation – creep lifetime is divided into three phases as follows:

Stage I creep. The initial phase. This is a largely unpredictable stage in which creep begins. In many cases it is undetectable by standard in-service inspection techniques – its existence being inferred, rather than proved.

Stage II creep. This is the 'progressive' stage of creep. Creep strain (the degree of distortion) is considered proportional to time, i.e. it is a direct linear relationship. It is widely acknowledged that this stage II creep *is predictable*, so the amount of creep damage can be inferred, if the time/temperature history is known. The microstructural damage resulting from stage II creep can be detected by replication test or ultrasonic crack detection, if the damage is more extreme.

Stage III creep. This is the final stage before catastrophic failure of the components (formally termed creep 'rupture'). Like stage I, it is unpredictable, governed more by chaos theory than any predictable mathematical relationship.

Nearly all the content of creep-related HRSG remnant life assessments is based around the stage II creep process. The main part of the assessment is therefore a simple life-fraction calculation based on the estimated 'gradient' of the assumed linear relationship between creep strain and exposure time. While this all sounds fine in theory, there are a number of factors that restrict its accuracy in practice, see Fig. 15.39.

The main alloy used in HRSG high-temperature components, i.e. SA 335-P91, is by design a creep-resistant alloy (it has replaced P22 for final stage superheat and reheat harp tubing, headers, and pipework). It is designed for life of 100 000 + hours at metal temperatures of 540 °C +, having a low stage II creep strain rate and inherent resistance to creep crack initiation (stage I). In practice, however, experience with P91 in 20–30 HRSGs has shown that the material is very sensitive to manufacturing parameters such as post-weld heat treatment and tempering temperatures. Non-optimum post-weld heat treatment has been found to cause significant early-life in-service defects such as *reheat cracking*. This is a microscopic cracking mechanism caused by imperfectly heat-treated welds attempting to normalize themselves (at non-optimum temperatures) when in use in the HRSG.

The end result is that the longer-term creep resilience of P91 may not have the chance to be realized. Manufacturing-induced problems increase the order of magnitude of nearly all known damage mechanisms – resulting in unplanned failure in early or middle life (10 000–40 000 hours). Early-life creep failure is therefore *not unknown* in high temperature HRSG components.

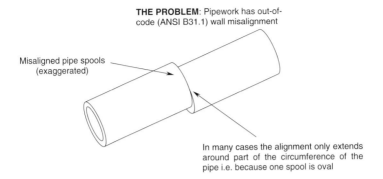

THE PROBLEM: Pipework has out-of-code (ANSI B31.1) wall misalignment

Misaligned pipe spools (exaggerated)

In many cases the alignment only extends around part of the circumference of the pipe i.e. because one spool is oval

THE SOLUTION: In some (not all) cases it is possible to recalculate the allowable misalignment, taking into account the additional reinforcing effect of:

The weld cap 'reinforcement'
The part of the circumference that is not misaligned

Additional effective shell thickness achieved from the weld cap reinforcement

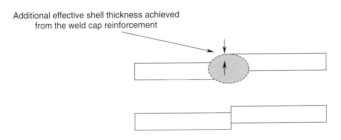

Fig. 15.37   Principal stress assessment – HRSG link pipework

Fatigue life prediction

The main fatigue damage mechanism in HRSGs is low cycle fatigue, caused by thermal and pressure cycling. The main areas affected are the headers (low- and high-temperature ones) and in particular, the tube stub-to-header welded joints. Pressure and thermal cycles are caused both by transients during full-load steady state operation and all start-ups, i.e. cold, warm, and hot starts. All the resulting fatigue effects act together in combination, the chances of failure increasing as time progresses.

Remnant life assessment techniques are based on standard fatigue failure graphs, generally known as *S–N* curves. Chapter 8 shows typical examples. The number of strain cycles N is plotted against the applied stress, S, resulting in a 'failure line', i.e. the material is at risk of failure when the prevailing conditions are above this failure line. As with creep,

The theoretical creep characteristic

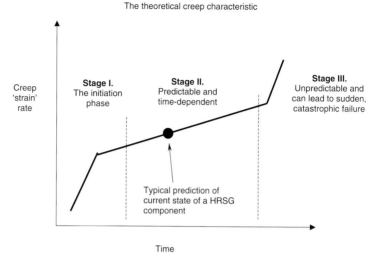

HRSG creep remnant life assessment –
some practical constraints

- **HRSG geometry**: Creep theory is based absolutely on the performance of neat, uniaxial test specimens. Real HRSG component geometry is more complex and 3-dimensional.

- **Temperatures**: HRSG component temperature distributions particularly in headers, vary by up to 100°C in both axial and radial (through-wall) directions even under steady-state HRSG load. Actual time/temperature exposure under a combination of steady-state and cycling conditions is very complex and can never be fully quantified using 'as supplied' HRSG instrumentation.

Fig. 15.38   HRSG creep remnant life assessment – theory and practice

however, this analyses method is not entirely representative of what happens in a real HRSG component. Some difficulties are:

- The *S–N* curve is obtained from tests on accurately machined 'waisted' test specimens rather than a real component with complex three-dimensional geometry.
- *S–N* curves are not available for all materials. For those that are, the fatigue-life line is normally represented by a probability 'band' rather than a single curve. This band is formalized in vessel design codes as a series of quality levels Q1, Q2, up to Q10 (see Fig 9.8 which is taken from BS 7910) giving a wide-ranging set of criteria for acceptability. It is common practice to assume a mid-range quality level Q5 or Q6

- **HRSG geometry**. Creep theory is based absolutely on the performance of neat uniaxial test specimens. Real HRSG component geometry is complex and three-dimensional.
- **Temperatures**. Component temperature distributions (particularly in HRSG headers) vary by up to 100 °C in both axial and radial (through-wall thickness) directions, even under steady state HRSG load. Actual temperature/time exposure under a combination of steady state and cycling conditions is incredibly complex and can never be fully measured using standard HRSG instrumentation.
- **Crack growth model**. The whole concept of creep strain rates is based on a particular crack growth model. Any evidence that this model is actually representative of the behaviour of HRSG materials such as SA 335-P22 and P91 is circumstantial, and a matter of technical opinion.
- **Unavoidable unpredictability**. However robust the background theory, the creep mechanism at work in any specific HRSG will always be subject to a degree of unpredictability. It is *absolutely impossible* to remove this entirely – some will always remain.

Fig. 15.39   Creep RLA – some practical constraints

but there is nothing that unilaterally imposes this – it is left to the decision of the assessor.

The problem of predicting crack initiation

One of the weakest parts of the fatigue element of an RLA is the prediction of crack *initiation*. While it may be possible to predict crack growth once a crack has started, there seems to be almost no technical consensus on how to predict how and when the crack will start. This undermines large parts of the technical rationale of both fatigue and creep considerations of a remnant life assessment – it doesn't destroy it totally, but it certainly places limits on its credibility.

Creep–fatigue interaction

A rational development of the RLA techniques of fatigue prediction is an attempt to predict the combined life-limiting effect of creep and fatigue. As a combination of independent processes, this uses various mathematical techniques such as the Robinsons' life fraction rule and the Miners linear damage rule. These are robust mathematical techniques but are, again, weakened by relying on simplifying assumptions and various parameters that cannot easily be measured.

In terms of validity, RLA results that predict fatigue failure are *less reliable* than those for creep damage mechanisms. The techniques themselves are equally robust but the confidence band is much wider.

Acceptability of known defects

This is similar to the principal stress prediction techniques outlined earlier. The assessment is carried out to investigate whether a pre-existing defect (material lamination, weld slag inclusion, crack, or similar) causes a stress level that is higher than the code allowable stress. The calculation can take two forms:

- A simple re-run of code calculations. Or
- A more detailed 'outside code' FFP assessment using BS 7910 or API 579.

Both BS 7910 and API 579 contain several 'levels' of FFP assessment: Level 1 is the simplest, involving simplifying assumptions about fracture properties of the material under consideration. If a component fails a BS 7910 Level 1 assessment, then a Level 2 assessment is done – this is more detailed, requiring actual impact strengths, fracture toughness data and ratios that can be validated to the component. Level 3 is the same idea, but with even more stringent data validation requirements. Chapter 9 explains the techniques.

Note that FFP assessment methods form *heuristic* (rules of thumb) techniques, rather than absolute requirements. This means that the ultimate decision as to whether a defective component that has, for example, failed a Level 2 assessment (but passed a Level 3 assessment) *is* fit-for-purpose is all a question of technical opinion. The final decision depends on the degree of risk that the person reviewing the assessment is prepared to take.

Implications for the in-service inspector

In most in-service inspection situations the inspector will not be the person called upon to actually perform the RLA. Inspectors do have a role, however, in requesting that an RLA be done and in reviewing the results once it is completed, and then deciding what to do next. The start of the procedure is normally an inspection scheme – a typical extract is shown in Fig. 15.40. Given its technical limitations, an RLA will *almost never* tell you (the inspector) whether to reject an HRSG component or accept it as suitable for further use. The RLA is only part of the picture – you have to temper its output with practical engineering judgement and experience, and *then* decide.

HRSG HEADER HP SHTR 2 INSPECTION SCHEME

NDT SCOPE		COMPONENT	HP SHTR 2 inlet header			
Drg Ref	66054/10047/6	Material	SA335 P91			
KKS Ref	31HAH71AC001	Scheme details	I.R	120	Criticality	8

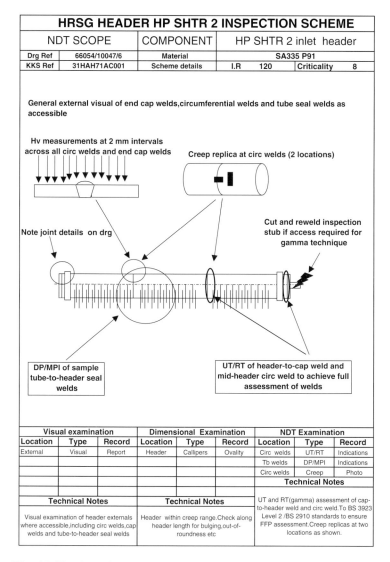

General external visual of end cap welds, circumferential welds and tube seal welds as accessible

Hv measurements at 2 mm intervals across all circ welds and end cap welds

Creep replica at circ welds (2 locations)

Note joint details on drg

Cut and reweld inspection stub if access required for gamma technique

DP/MPI of sample tube-to-header seal welds

UT/RT of header-to-cap weld and mid-header circ weld to achieve full assessment of welds

Visual examination			Dimensional Examination			NDT Examination		
Location	Type	Record	Location	Type	Record	Location	Type	Record
External	Visual	Report	Header	Callipers	Ovality	Circ welds	UT/RT	Indications
						Tb welds	DP/MPI	Indications
						Circ welds	Creep	Photo

Technical Notes	Technical Notes	Technical Notes
Visual examination of header externals where accessible, including circ welds, cap welds and tube-to-header seal welds	Header within creep range. Check along header length for bulging, out-of-roundness etc	UT and RT(gamma) assessment of cap-to-header weld and circ weld. To BS 3923 Level 2 /BS 2910 standards to ensure FFP assessment. Creep replicas at two locations as shown.

Fig. 15.40   A typical, high-temperature header inspection scheme

Thinking Outside the Box – In-Service Inspection of HRSGS

If you do want to find defects	If you have an embarrassingly large inspecting budget but don't want to find (many) defects
Specify replication and hardness tests on high-temperature P91 headers particularly any that have longitudinal seams	DP all the shell welds on the HP/IP drums
Look for ligament cracking on P22 and P91 headers using a borescope	UT all the shell welds on the IP/LP drums (keep going, you're wasting lots of time)
Check for flow acceleration corrosion inside low-temperature economizer headers	UT All the low-temperature carbon steel headers
Do sample UT/DP checks on heavily loaded (or misaligned) set-through drum nozzle welds	Build huge scaffold constructions inside the gas space and take OD measurements on at least 50% of the tube banks
	Cut off all the inspection stubs on P91 headers, power piping that don't have ligaments. Do a borescope inspection. Don't check the P91 stubs for new defects After they've been rewelded
	Do a loosely specified remnant life of fitness for purpose assessment to justify any embarrassing defects that may have been found

Fig. 15.41 TOTB – in-service inspection of HRSGs

Chapter 16

Heat exchangers

Heat exchangers are found in many applications in power, process, and general engineering plants. They range from straightforward shell-and-tube and plate-type heaters and coolers to more complex designs including steam condensers, air-cooled 'fin-fan' coolers, and exchangers with external firing. Most heat exchangers operate at a positive or negative gauge pressure and so are categorized as *pressure equipment*, and subject to periodic in-service inspection.

The design criteria for exchangers are not as straightforward as those for vessels and pipework. It is, for example, rare for an exchanger to be designed and manufactured in compliance with a single design code. The most common situation is that the shell (or headers, depending on the design) is designed to one of the unfired vessel codes (ASME VIII-I, PD 5500, EN 13445, etc.) while the tube plate and tube bundle assembly follow the TEMA (Tubular Equipment Manufacturers Association) standards. TEMA standards cover integrity requirements, resistance to pressure stresses *and* design aspects relating to the heating surface and thermal transfer. Other specific standards on heat exchangers *do* exist (see later) but the overall situation is much more fragmented than for boilers, unfired pressure vessels, and pipework systems.

The variety of different designs and applications of heat exchangers is reflected in the way that they appear in risk-based inspection (RBI) schemes. The 'risk level' ascribed to exchangers varies across a wide range, from *negligible risk* for low-pressure coolers or water heaters through to *high risk* for safety-critical exchangers dealing with high-pressure inflammable, toxic, and corrosive process fluids in refinery and petrochemical plant applications.

Shell-tube exchangers – construction features

Figure 16.1 shows a general exploded view of a shell–tube heat exchanger. The shell side acts as a conventional pressure vessel designed to resist circumferential (hoop) and longitudinal (axial) stresses as well as any stress concentrations that result from the flanges and heads. As with pressure vessels, heat exchanger shells are designed using 'thin-shell assumptions'. As a general rule, however, the factors of safety present in an exchanger shell tend to be *higher* than those used for conventional unfired pressure vessels. This is due more to the wall thickness of commercially available schedule pipe material rather than any inherent design reason. Shell flanges also tend to be overdesigned – sized to accommodate the bolt circle and resultant bolting force, rather than just to resist pressure stress. The small diameter ($< 500\,\text{mm}$) of most heat exchanger shells means that shell and head welds are normally single-sided, because they are often impossible to access from the inside. The various shell and tube configurations are outlined in the TEMA regulations. Figures 16.2 and 16.3 show the main types and terminology used.

Tube bundles

Tube bundles vary in their configuration but their design and construction principles are similar. Tube plates are made of rolled plate material and the tubes are *drawn*, by pulling (extruding) over a lubricated mandrel. There are well-defined manufacturing tolerances on tube straightness, ovality, and minimum bend radius. The tube layout follows either a triangular or square pitch pattern. The square pattern is most commonly used for dirty service conditions – the greater distances between the tubes reduces fouling, and makes mechanical cleaning easier. Tube material can be cupro-nickel, brass alloy, titanium, or plain carbon steel and is chosen to meet the service conditions.

The tubes are expanded (rolled) into the tube plates. Higher-pressure designs are seal-welded, normally on the outer face only. The tube-to-tube-plate joint needs a close tolerance to achieve the correct expansion – the tubes are manufactured to a standard size and the tube plates drilled and reamed to fit. The tube bundle is supported in the shell by baffle plates (see Fig. 16.1) which act both to help direct the flow of the shell-side process fluid and as an antivibration device for the tube bundle. The tubes pass through clearance holes in the plates – there is a small air gap but no bush or sleeve.

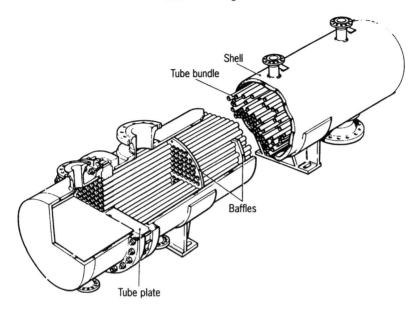

Fig. 16.1 Tube-type exchanger exploded view

Separate *impingement plates* may be fitted in the shell. These prevent the shell-side fluid eroding the tube bundle in areas of high velocity and swirling. The most common location for impingement plates is opposite the exchanger shell-side inlet nozzle.

'Fin-fan' coolers – construction features

Air-cooled tube-nest heat exchangers (known loosely as 'fin-fan' coolers) are in common use for primary cooling purposes in desert areas and inland plant sites (see Fig. 16.4). On a smaller scale, they have multiple uses in chemical and process plants where a self-contained cooling unit is needed – avoiding the complication of connecting every heat sink component to a centralized cooling circuit. In their larger sizes, fin-fan coolers can cover an area of up to 4–$5000\,\mathrm{m}^2$ and often stand up in a shallow angle 'A' configuration. Smaller ones usually stand horizontally resting on a simple structural steel frame. Figure 16.5 shows a basic fin-fan cooler – they vary very little between manufacturers. The main design points are outlined below.

The cooling matrix

This consists of a matrix of extruded carbon steel or stainless steel finned tubes arranged in a complex multi-pass flow path. The matrix is

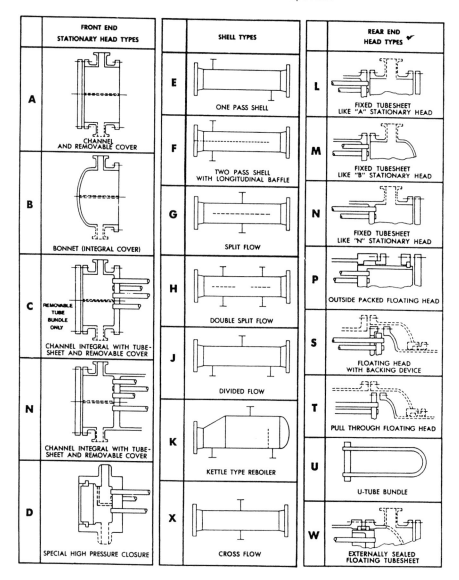

Fig. 16.2   Surface-type exchanger configurations (Source: TEMA)

divided into discrete banks of tubes, extending horizontally between a set of headers. The fins consist of a continuous spiral-wound steel strip, resistance-welded into a thin slot machined in a close helix around the tubes' outer surface. The extended surface of the fins adds significantly to the effective surface area, thereby increasing the overall thermal

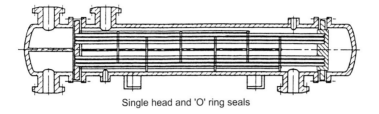

Single head and 'O' ring seals

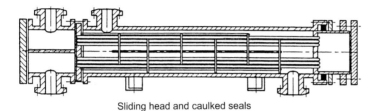

Sliding head and caulked seals

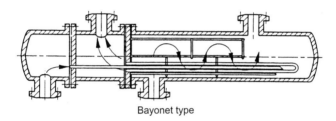

Bayonet type

**Fig. 16.3   Surface-type exchangers – various arrangements
(Source: EN247: 1997)**

transfer. A typical tube bank will be between six and ten tubes 'deep' in order to achieve the necessary heat transfer in as small an area as possible.

The headers

The ends of the tubes are stub-welded into heavy-section cast or welded headers. These contain internal division plates and baffles which give the desired multi-pass pattern through the system. Each header also contains stub pieces and small access hatches for inspection, cleaning, and bleeding off unwanted air during commissioning. In most designs, the headers are designed and built to an accepted pressure vessel standard.

Fig. 16.4   A large 'fin-fan' air-cooled condenser

The air fans

Primary cooling effect is provided by a bank of axial-flow cooling fans which blow air vertically upwards through the tube nest. The fans are generally belt driven for simplicity, and have variable-incidence blades positioned by a pneumatic actuator arrangement. The electric motors are often two-speed, (typically 300 and 600 r/min) to allow operating current and power consumption to be reduced when air temperature is low. In a typical unit, each fan will be located about 2 m off the ground and be protected by an expanded metal safety guard. The tip speed of the fan is kept below about 60 m/s to avoid overstressing the aluminium blades.

The fan running test

Fans are tested prior to installation and after any refurbishment or rebuild work. Figure 16.6 shows a section through a typical fan, the shape of its performance characteristic and the main points to check during a running test. The test does not normally follow any particular technical standard but is simply organized around the task of demonstrating the fan's FFP in use. Specific points are as follows:

- *Static pressure versus blade angle.* The performance of the fan does not keep on improving as blade incidence is increased. There is a well-defined 'cut off point', above which the blades start to become

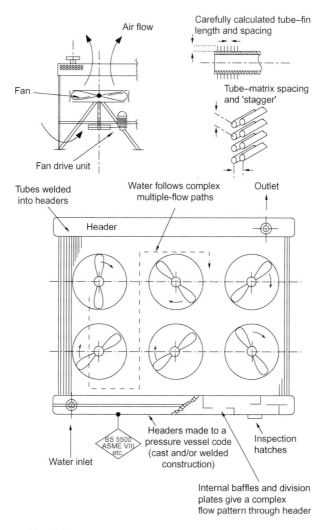

Fig. 16.5 Fin-fan cooler – general arrangement

aerodynamically inefficient and will actually produce less, rather than more, cooling effect.

- *Blade angle versus motor current.* This places a limitation on the FFP of the fan. Maximum motor design currents usually have a design margin of about 30 percent (to keep the cost of the motors down). A properly designed unit should reach full operating current before the static pressure curve levels off.

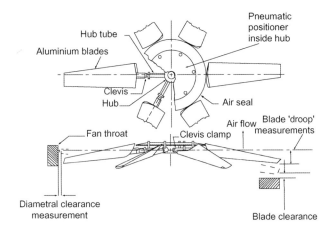

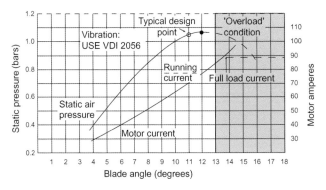

Fig. 16.6   Fin-fan cooler fan – typical performance characteristics

- *Vibration.* Axial fans are smooth running units and rarely experience serious vibration problems. A maximum V_{rms} level of about 2.5 mm/s is acceptable, using the principles of VDI 2056.
- *Mechanical integrity checks*
  - Blade locking arrangements, including the fitted clevis used to locate the blades accurately in position on the hub.
  - The pneumatic positioner and diaphragm that move the blade angle.
  - The blades themselves (usually aluminium): checks for length and obvious mechanical damage.
  - All locknuts and lockwashers fitted to the rotating components.

Technical codes and standards

TEMA standards

The TEMA (Tubular Exchange Manufacturers' Association) standards are published in the USA but used for guidance in most other countries worldwide. Their acceptance is more widespread than other US standards (e.g. ASME code) and they are used in almost all industry sectors, from petrochemical/refinery applications to power, food manufacturing and general engineering plant. TEMA standards cover mainly the design and new construction of heat exchangers but, in doing so, provide useful guidance on some areas of in-service inspection. Figure 16.7 shows the structure of the standards.

Two useful points contained in the TEMA standards are the classification of exchanger *types* and *service classes*.

- *Exchanger type classifications* are shown in Fig. 16.2 Each design is given a three-letter acronym describing its type of front head, shell, and rear head. For example (referring to Fig. 16.2) a type 'AES' denotes a single-pass shell with a channel-type front head (removable cover) and a floating rear head fitted with a backing device. These classifications are shown on exchanger datasheets and drawings and show the type of construction to expect during a shutdown inspection. They can also show when inspection of some parts of the exchanger may *not* be possible, because of the tube plate/head configuration or dismantling clearances.
- *Exchanger service class.* TEMA subdivides exchangers into three service classes: R, C, and B.
  - *R class* is for the most severe applications (mainly refinery and petrochemical industry use) with arduous conditions of pressure, temperature, and corrosive/erosive service fluids.
  - *C class* exchangers are the common type; most general power/ process industry exchangers which operate in moderate service conditions.
  - *B class* are dedicated specifically for chemical process service.

This classification system can be difficult to understand. While R class generally has the highest design criteria and C class the lowest, the B class is a combination of the two. The main differences are in design features such as minimum corrosion allowance, tube pitch and tolerance, and nominal shell thickness. Note that there are many aspects of exchanger design where the requirements for classes R, C, and B are exactly the same. The clause numbers of the TEMA standards

TEMA Section	Subject
1*	Nomenclature (terminology of heat exchanger components)
2	Manufacturing tolerances
3*	General fabrication information
4	Performance and operation
5	Design information
6	Flow-induced vibration
7	Thermodynamics
8	Properties of fluids
9–10*	Recommended good practice

* Most useful parts for inspectors

Fig. 16.7   TEMA standards – content

show this directly. Clauses designated, for example, RCB 1.5, indicate that clause 1.5 is applicable to all three service classes of exchanger: R, C, and B.

API 660   Shell and tube heat exchangers for general refinery service

This follows the model of many of the API 600-series standards in being a general guide to the design and manufacture of shell and tube heat exchangers, without giving specific information to cover all inspection-related areas. The general philosophy of the standard is similar to that of TEMA, being based around refinery and petrochemical plant applications. There is little in this standard that is essential reading for in-service inspectors.

API 662   Plate heat exchanger specifications

This has equivalent coverage to API 660 but applies specifically to plate-type exchangers. These are less complex owing to the absence of tube bundles and are based on pressure vessel code 'intent'. There are no links to TEMA standards. Most of the standard's content deals with design and new construction rather than in-service inspection.

API RP 572   Inspection of pressure vessels

This is a wide-ranging document (not strictly a *standard*) covering technical aspects of the in-service inspection of all types of pressure vessels. It includes several appendices relevant to shell and tube

exchangers, outlining typical corrosion and erosion problems experienced in the refinery/petrochemical industries. In line with the other
API RP (recommended practice) documents, it does not contain any
specific requirements or inspection 'acceptance criteria' that an
inspector can use on specific tasks. It is best used for technical
information only.

Appendix A of API RP 572 covers heat exchangers in general,
including air-cooled and coil types. It also includes checklists and
reporting formats which can be used during in-service inspections.
Figure 16.8 shows an example. Note that these are limited to wall-
thickness records and are not very effective at recording detailed data
about corrosion, erosion, and other degradation mechanisms.

In-service inspection of heat exchangers

Heat exchangers are subject to in-service and cold shutdown inspection
in accordance with a plant's Written Scheme of Examination. As with
some types of pressure vessels, there is little that can be concluded from
an on-stream external visual examination, except perhaps that the
exchanger does not have any visible leaks to atmosphere. The main
inspection activities are therefore done during periodic cold shutdown
inspections.

Preparation and cleaning

It is virtually impossible to do a meaningful shutdown inspection on
shell–tube heat exchangers without removing the tube bundle from the
shell. It is therefore normal to remove some of the external lagging to
allow access to critical external parts of the shell and head. It is not
always necessary to remove *all* the external lagging, unless there are
specific concerns about widespread corrosion and/or wall thinning on
either the internal or external surfaces of the shell.

Lagging removal
Figure 16.9 shows the extent of lagging removal necessary to allow
external visual examination of key areas of the shell. Removal of lagging
is always a controversial issue during shutdown inspections – it is *quite
normal* for plant operators and management to want to remove only the
minimum amount necessary to carry out the external visual inspection.
Note (in Fig. 16.9) the importance of exposing the shell area around the
shell-side fluid inlet and outlet nozzles, particularly if they are larger
than $D/3$ and use compensation pads welded to the shells. These are

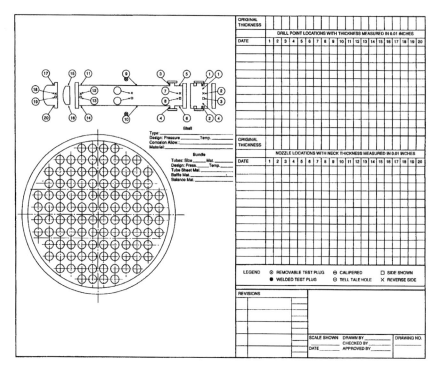

Fig. 16.8   API RP 572 heat exchanger inspection reporting format

likely to be highly stressed areas and possibly be subject to fatigue-induced cracking at weld toes. On larger exchangers located outdoors, visual thermography is used to identify areas where water has entered the protective cladding then remained soaked into the lagging sheets. For exchangers working at an external shell temperature of less than about 120 °C, these areas have a high risk of shell thinning due to corrosion under insulation (CUI). The technical standard API 570 *In-service inspection of pipework*, although not directly targeted at heat exchangers, contains useful guidelines on other conditions that produce a high risk of CUI.

Disassembly

Tube-type heat exchangers should be shut down by gradually isolating the hottest fluid side first, followed by the cold fluid. For exchangers which have a fixed tube plate (see Fig. 16.10) this isolation must be carried out slowly and progressively so as to minimize abrupt differential expansions between the shell and the tube bundle. For the main disassembly, flange bolts should be slackened in diametral

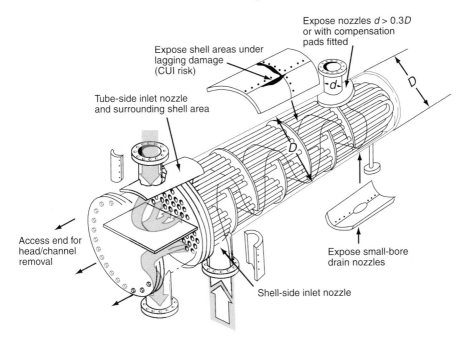

Expose nozzles $d > 0.3D$
or with compensation
pads fitted

Expose shell areas under
lagging damage
(CUI risk)

Tube-side inlet nozzle
and surrounding shell area

Access end for
head/channel
removal

Expose small-bore
drain nozzles

Shell-side inlet nozzle

Fig. 16.9   Lagging removal – key areas

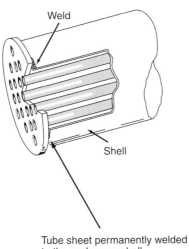

Weld

Shell

Tube sheet permanently welded
to the exchanger shell

Fig. 16.10   Fixed tube sheet arrangement

sequence so that the flanges separate squarely, and so do not experience excessive stress.

For small exchangers, the tube bundle can normally be withdrawn by hand. Larger exchangers (lengths range up to 10 m +) need special lifting tackle. Some designs are fitted with a special jig which fastens onto the head channel or tube plate and then is driven along a horizontal 'slipway' by a hydraulic ram, hence pulling the tube bundle out of the shell.

Tube bundle cleaning

The first step is to visually inspect the tube bundle (inside and outside surfaces of the tubes) *before* any cleaning has taken place. This is so that any loose scale or corrosion product can be recorded, and samples taken, before it is removed. Important points are:

- The *thickness* of scale.
- Whether the scale is *intermittent* or *complete*, and its location.
- The *tenacity* of any scale layer, i.e. whether it is tightly bonded to the tube plate or tube wall (internal or external) or whether it flakes off easily with finger-pressure.

Various methods can be used for cleaning the tube bundles. Scrapers or rotating wire brushes are used for hard scale. Sludge and soft deposits are best removed by a circulating hot wash (water or light distillate at high velocity) or plain high pressure water-jetting. Check for three main precautions:

- Make sure any chemical cleaning compounds used are compatible with the exchanger materials, particularly the tube bundle.
- Take care not to cause damage by excessive mechanical cleaning (chipping, grinding, etc.).
- Do not blow steam through individual tubes. This causes severe strain due to expansion and may either bend the tube or loosen the fit of the tube-to-tube-plate joint.

Visual examination

Perhaps 90 percent of in-service defects in shell-and-tube heat exchangers can be found by visual examination. Figures 16.11 to 16.13 show the most common defects and their possible effect on the fitness for purpose (FFP) of the exchanger. It is important to keep a sense of perspective on FFP issues when considering the results of visual

Component	Inspection findings	FFP implications
Shell	Internal or external shell general wall thinning.	Requires remaining wall thickness calculation to code requirements. Calculate remaining life based on annual corrosion rate.
	Localized shell wall thinning due to internal erosion or external CUI.	As above, but the reason for preferential thinning needs to be identified.
	Flange-face erosion or corrosion.	There are TEMA (Section 2) acceptance criteria for flange-face imperfections, based on hoop stress considerations. Imperfections are assessed by measuring the radial projection length (RPL).
Tube bundle	Tube plate ligament cracking.	Cracking is generally unacceptable. Check remaining ligament width complies with TEMA (Section 5) requirements.
	Leaking tube-to-tube-plate joints.	Leaking joints are unacceptable. Maximum tube projection should be < 50% tube diameter.
	Tube-plate distortion (lack of flatness/ twisting).	In-service distortion is rarely an FFP issue unless the tube plate has been dropped or suffered significant mechanical damage.
	Scaling/corrosion of tubes.	Loose scale is rarely an FFP issue. The main FFP issue is tube or tube-plate thinning underneath tenacious corrosion deposits.
	Distortion of the tube bundle.	Rarely an FFP issue unless it is caused by mechanical damage or excess temperature.

Fig. 16.11   Shell and tube exchangers – in-service inspection points

	Tube cracking.	Can seriously reduce the life of Cu/Ni or Ti tube bundles.
	Cracking of baffles (antivibration plates).	Rarely an FFP issue alone but can be indicative of vibration-induced fatigue which may affect the tube bundle or shell.
Nozzles and fittings	Internal erosion around nozzle-to-shell fitting.	Requires wall thickness/ nozzle compensation calculation to code requirements. Calculate remaining life based on erosion rate.
	Erosion of baffle plate (impingement guards).	Rarely an FFP issue alone but is indicative of erosive conditions/excess fluid velocity.
	Nozzle flange-face distortion.	There are TEMA (Section 7) requirements on flange-face alignment.

Fig. 16.11   (Continued)

inspections of heat exchangers. There are two aspects to this: pressure envelope considerations and process contamination considerations.

Pressure envelope considerations
Inspection findings which constitute a risk to the integrity of the pressure envelope can have serious safety implications, particularly if hazardous hydrocarbon or chemical process fluids are involved. In practice, however, most routine inspection findings *do not* represent an immediate safety risk. They are more likely to have implications on maintenance requirements or expected service lifetime rather than on the immediate risk to the integrity of the pressure envelope. The situation does, however, vary with the exchanger design (see Fig. 16.13). Some designs are such that, for example, defects or failures in the exchanger have a greater chance of resulting in uncontrolled process fluid release to atmosphere, than in others. The type of exchanger head construction and sealing has the greatest influence on this.

Process contamination considerations
Most inspection findings relate to the risk of process contamination (leakage) rather than the integrity of the pressure envelope. Leakage is

not uncommon in heat exchangers – the overt design function of an exchanger in acting as a heater or a cooler means that there will be significant temperature differences across the heat transfer surfaces (the tubes) with the accompanying risk of differential expansion thermal distortions. These can cause leaks. Inspection findings likely to result in process contamination are a softer-focus FFP issue for two reasons:

- They are rarely a safety issue.
- Their effect is not so apparent, at least in the shorter term. Contamination of process fluid with, for example, heating steam or cooling water is likely to result in *long-term* problems, but the current risk may be small, and low on the list of plant priorities.

For the reasons above, the result of heat exchanger inspections can end up being controversial, with varying views on both the validity of findings, and their effect on FFP. All of the findings shown in Figs 16.12 and 16.13 are typical of heat exchangers that have been in use for some

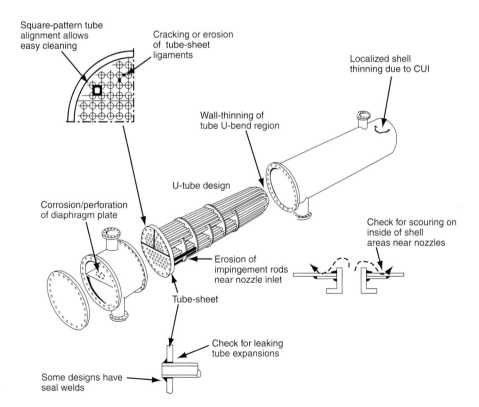

Fig. 16.12 Shell and tube exchangers – inspection points

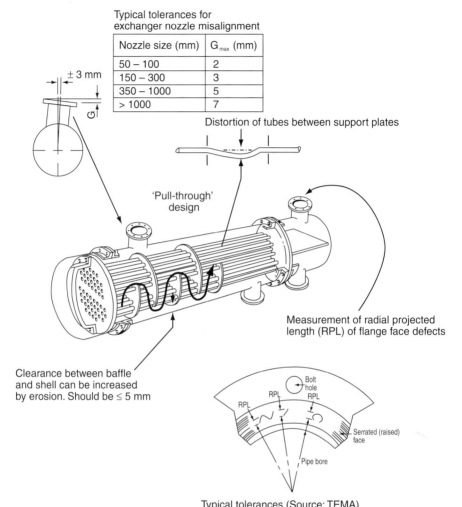

Typical tolerances for
exchanger nozzle misalignment

Nozzle size (mm)	G_{max} (mm)
50 – 100	2
150 – 300	3
350 – 1000	5
> 1000	7

± 3 mm

G

Distortion of tubes between support plates

'Pull-through'
design

Measurement of radial projected
length (RPL) of flange face defects

Clearance between baffle
and shell can be increased
by erosion. Should be ≤ 5 mm

Bolt
hole

RPL

RPL

RPL

Serrated (raised)
face

Pipe bore

Typical tolerances (Source: TEMA)

Nominal pipe size (NPS) (mm)	RPL_{max} of defects ≤ serration depth (mm)	RPL_{max} of defects > serration depth (mm)
125	6.4	3.2
150	6.4	3.2
200	7.9	3.2
250	7.9	4.8
300	7.9	4.8
350	9.5	4.8

Fig. 16.13   Shell and tube exchangers – further inspection points

time. The best way to report them during an inspection is by using a detailed checklist. Owing to the time pressure of a plant cold shutdown, it is not unusual for exchanger visual inspections to be carried out in a too-cursory way, or even for the tube bundle not to be removed at all. Again, the risk is dependent on the design of exchanger, its materials of construction, and the service conditions. We can now look at some specific points.

Common inspection findings

General shell corrosion
Light general shell corrosion on the inside surface of exchanger shells is common. It rarely gets serious enough to cause significant wall thinning, except in particularly corrosive service conditions. Exchanger shells have a corrosion allowance (normally 3 mm +), in addition to the calculated thickness required to resist pressure stresses.

Localized shell corrosion
This is more common and is caused by either localized CUI on the outside or combined erosion/corrosion on the inside. Common internal locations are:

- Opposite the inlet nozzle (particularly if the design fluid velocity is greater than about 10 m/s).
- Around the tubes or antivibration plates that are supported by the tube bundle itself, i.e. are not welded to the inside of the shell. Fluid velocities are high in the gap between the periphery of the plates and the shell surface, giving a scouring action. This increases erosion and can cause shell wall thinning in excess of the corrosion allowance.

Flange-face corrosion (see Fig. 16.14)
Features on shell-and-tube exchangers flange faces caused by corrosive or erosive affects are known by their general name of 'imperfections'. TEMA standards (Section 2) give acceptance limits on the radial projection length (RPL) allowed on the raised flange face. For designs where the tube plate is sandwiched between the main head and shell flanges, limits will apply to all four mating surfaces. Figures 16.13 and 16.15 show some useful points.

Scaling and corrosion of tubes
Scaling is caused by poor process conditions. Excessive temperatures and process transients make the situation worse. Unless it is sufficiently

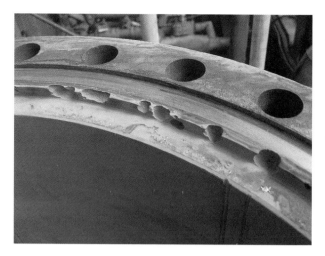

Fig. 16.14   Heat exchanger flange-face erosion

thick and tenacious enough to blanket the tube heat transfer and cause
the tube to overheat, scale itself does not actually *cause* failure. The
main problems with scale is that corrosion can start *underneath* it. Tube
corrosion takes several forms:

- *Galvanic corrosion.* Tubes made of 90/10 cupro-nickel (Cu/Ni) suffer
  bad galvanic microstructural corrosion if they come into contact with
  a more cathodic material – one of the worst is *graphite*. Graphite may
  be present from the tube manufacturing process itself (it is used as a
  mandrel lubricant during tube extrusion) or from contamination by
  graphite-rich grease. Large exchangers are fitted with cathodic
  protection to protect the shell, but this protection is poor at
  protecting tubes against localized galvanic corrosion that occurs
  within the structure of the tube material itself.
- *Microcorrosion of titanium (Ti) tubes.* This happens in longitudinally
  seam-welded Ti tubes that have not been correctly solution-annealed
  after welding during the manufacturing process. It is worst in
  seawater exchangers (for desalination or general process plant). Its
  timescale is unpredictable – it may start soon after commissioning, or
  take several years to develop. The mechanism is that the material
  structure develops multiple microcracks, providing a foothold for
  crevice corrosion – one of the few corrosion mechanisms to which
  titanium is susceptible. Titanium-tubed exchangers, despite their high
  cost (several times that of Cu/Ni), *can* suffer problems – so in-service
  shutdown inspections should not be ignored in the early years of the

Flange design

Heat exchanger flange design follows similar *principles* to those in the general flange design standard ANSI B16.5. Reduction in flange thickness or a change in the radial face area will result in both excessive bending stresses in the flange and poor sealing.

Flange-face corrosion

This is normally an electrochemical corrosion mechanism, made worse by the existence of a tight crevice. Some specific features are:

• Stainless steel flanges are susceptible to stress corrosion cracking (SCC). They show as fine 'hair-like' branched cracks running radially across the raised face and may *only* be visible using DP or MPI. The rest of the flange face is normally unaffected. Actual flange failure *can* occur, but only in extreme cases.
• Overtightening of flange bolts is a common initiating cause of flange-face corrosion. An overtightened spiral-wound gasket will not seal properly, allowing process fluid to contact the flange face and encouraging crevice corrosion.

Flange-face flatness

Check this with a steel straight edge. Any visible bowing indicates serious distortion.

Nozzle flange distortion

This is mainly caused during manufacturing (or repair) rather than by in-service conditions (except in high-temperature heaters using super-heated steam which can suffer twisting). These tolerances (see Fig. 16.13) are important – they can transfer any 'out of design' static load to the nozzle-to-shell welds, causing failure.

Fig. 16.15 Heat exchanger flange face inspection: some technical points

exchanger's life. Figure 16.16 shows some useful inspection-related information.

Tube plugging and sampling

Heat exchanger design allows for overcapacity on tube heat transfer surface. This means that if individual tubes leak or fail, a number of tubes can be plugged without affecting the overall thermal performance of the exchanger. It is not unusual for up to 5–10 percent of the tubes to be plugged in old exchangers, rather than commissioning an extensive re-tubing exercise. Both ends of the tube are normally plugged, with

Tube cleaning

- Weak citric acid solution or hot 2% NaOH solution can remove scale.

General corrosion characteristics

- Ti forms a good corrosion-resistant passive surface film which can easily withstand fluid velocities up to 20 m/s.
- Ti can suffer from crevice corrosion > 130 °C.
- Ti is cathodic to most other heat exchanger materials.

Fatigue limit

- Approximately 50% UTS (in air or seawater).

Erosion resistance

- Negligible erosion in clean seawater at velocities < 20 m/s or water containing sand and grit at < 2 m/s.

Typical Ti tube corrosion/erosion rates

Service	Corrosion/erosion rate
Clean water	1×10^{-8} mm/year
Sulphuric acid 10% @ 25 °C	0. 25 mm/year
Nitric acid 10% @ 100 °C	0.03 mm/year
Chlorine-saturated water @ 100 °C	0.07 mm/year
Sodium chloride solution 10% @ 100 °C	Nil
Hydrochloric acid 10% @ 35 °C	0.76 mm/year

Fig. 16.16   Titanium heat exchanger tubes – useful data

plugs being either tapered or threaded. Special tools are required to install threaded plugs. The technique cannot be used on some designs of heat exchanger that have very high temperature differentials.

Tube sampling is used to assess the condition and assess the remaining life of exchangers that have seen significant service (> 15 years or more). Sample tubes are plugged and cut out during a shutdown inspection and subject to laboratory test. Figure 16.17 shows typical details and Fig. 16.18 shows serious ligament erosion.

NDT techniques

Although most inspection techniques on heat exchangers are visual, several specialized NDT techniques are used for specific tasks.

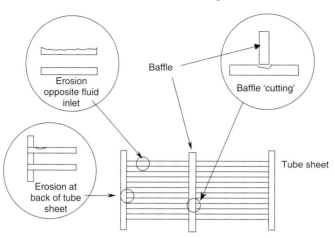

Tube sampling: remove representative sample of tubes

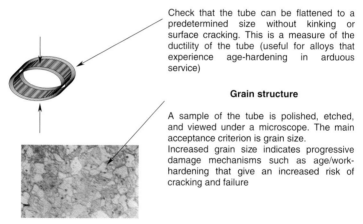

Flattening test

Check that the tube can be flattened to a predetermined size without kinking or surface cracking. This is a measure of the ductility of the tube (useful for alloys that experience age-hardening in arduous service)

Grain structure

A sample of the tube is polished, etched, and viewed under a microscope. The main acceptance criterion is grain size.
Increased grain size indicates progressive damage mechanisms such as age/work-hardening that give an increased risk of cracking and failure

Fig. 16.17   Heat exchanger tube defects and sampling

Borescope inspection of tube internal surface

The simplest way is to use a special 90° illuminated borescope head. This is a small diameter head fitted with a perpendicular mirror so that the tube internal surface can be viewed in a radial direction. Some models give a full 360° panoramic view. The main purpose is to check for microcracking (see Fig. 16.19). This is not a 100 percent reliable detection technique – even if the tube internals are chemically descaled, microcracking cannot always be easily seen. A normal axial-view borescope can be used for a general visual scan down the length of the

Fig. 16.18   An example of heat exchanger tube sheet ligament erosion

tubes. This will detect failure or blockages but little else. It is important that a borescope inspection be targeted in the area of tube where the worst condition is expected – the end 200 mm of the tubes are normally the worst places for erosion and corrosion, because of the turbulent flow regime that exists in these areas. Fewer problems occur near the centre of the tube length.

Tube 'ring' test

This involves 'ringing' each tube by tapping the end with a ball-pein hammer. In theory, a sound tube-to-tube-plate joint gives a firm ringing sound, while a duller sound suggests the joint is loose. Some older industries still use this method, but it is not very common and can cause more problems than it solves.

Heat exchanger eddy current testing

Eddy current testing (ECT) is a specific method used to detect and size defects in heat exchanger tubes (see Fig. 16.20). It is limited to non-ferromagnetic materials such as Cu/Ni, titanium, high-nickel alloys, and non-magnetic stainless steels. It has two main purposes:

- To detect wall thinning caused by erosion or corrosion of the internal surface of the tube.
- To detect cracking (including SCC) in the tube walls.

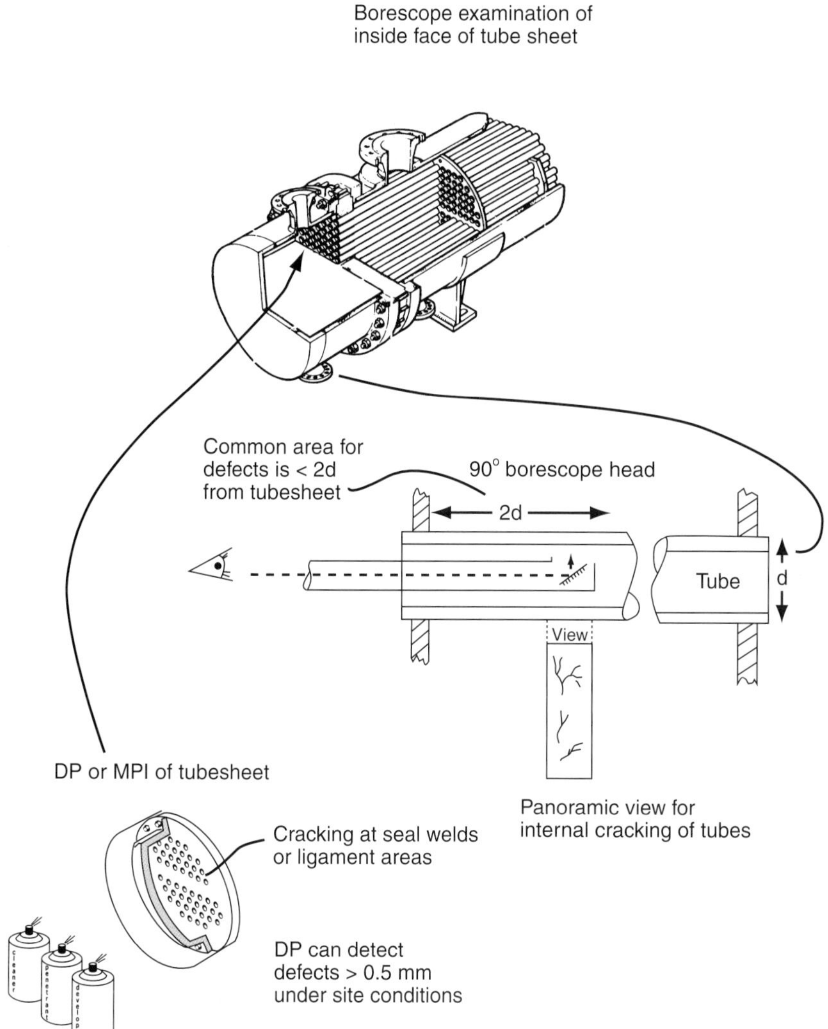
Borescope examination of inside face of tube sheet
Common area for defects is < 2d from tubesheet
90° borescope head
2d
Tube
d
View
DP or MPI of tubesheet
Panoramic view for internal cracking of tubes
Cracking at seal welds or ligament areas
DP can detect defects > 0.5 mm under site conditions
cleaner
penetrant
developer

Fig. 16.19 Heat exchanger visual/NDT inspection

How effective is ECT?

The effectiveness with which ECT can reliably find defects depends on several factors; the degree of scaling or material deposits on the tube, the type of probe used, and operational variables such as current, frequency, speed of traverse, and the degree of signal filtering.

ECT – how it works

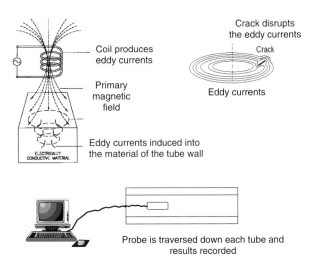

Coil produces eddy currents

Primary magnetic field

Eddy currents induced into the material of the tube wall

ELECTRICALLY CONDUCTIVE MATERIAL

Crack disrupts the eddy currents

Crack

Eddy currents

Probe is traversed down each tube and results recorded

Typical ECT 'tubemap' results showing degrees of wall thinning and defect depth

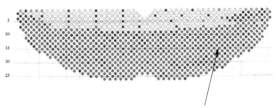

Colours/shading shows corrosion/defect depth

Fig. 16.20   Heat exchanger eddy current testing (ECT)

The technique works on the principle that when an AC current flows in a coil in close proximity to a conducting surface, the magnetic field of the coil will induce circulating (eddy) currents in that surface. Any changes in the surface such as surface-breaking cracks or changes in thickness will change the eddy current flow, affecting the impedance of the coil. By monitoring the voltage across the coil a visual display can be produced, representing the condition of the tube. For use on heat exchangers, the coil is enclosed in a probe which is pushed through the tubes. This is a quick process, making it suitable for testing large numbers of individual tubes. It has limitations however so, like all NDT techniques, it cannot be considered 100 percent reliable. As a rule of thumb, ECT has been shown to have a detection performance of between 85 and 95 percent under test conditions. Its best performance is on titanium tubes – detection falls on Cu/Ni alloys and some stainless steels.

From a practical viewpoint, the main limitations of ECT are:

- It can have difficulty differentiating between defects and metallic deposits.
- It cannot detect cracks orientated parallel to the eddy current path (because only cracks at an angle to the path affect the eddy current flow).
- It has accuracy limits – wall thinning of <5 percent wall thickness is difficult to detect reliably.
- It can only detect cracks that are surface-breaking. Also, the size and character of the crack becomes more difficult to diagnose as its depth of penetration increases. This is due to the inherent mathematical properties of the ECT technique.

Interpretation of ECT inspection results

ECT is carried out by specialist NDT contractors, many of which seem to have their own preferences as to the best way to display the results. Figure 16.20 shows one common graphical method – the display shows a mimic of the exchanger tube sheet with 'defects' shown in four or five different shades or colours. The colours represent *defect depth ranges* – each being chosen to reflect the type of defect expected. For example, an exchanger containing heating steam in the shell side would have the display calibrated to specifically identify expected defects such as external tube thinning (due to steam impingement) and 'cutting' where the tubes pass through the baffle plates (see Fig. 16.17).

Heat exchanger pressure testing

Pressure testing requirements for heat exchangers are governed partly by a mixture of TEMA, the code requirements for the pressure envelope, and non-code tests for leakage between the shell and tube-side process fluids. Figure 16.21 outlines the major points.

Heat exchanger tubes — titanium tube replacements

Titanium (Ti) tubes are used for repairs on condenser, evaporator, and heat exchanger elements where process corrosion has proved to be a problem. Ti has a high resistance to corrosion in sea and estuarine waters, giving a longer lifetime than traditional Cu/Ni alloys. It is a

- A full hydraulic 'body' test of the shell is performed, to code, during manufacture. This is not repeated regularly during shutdown inspections, unless there are issues regarding wall thinning or recertification for changed process conditions.
- For repeat testing during in-service shutdown, the shell and tube side are tested separately so that tube-to-tube-plate expansion/weld joints can be checked for leaks.
- If the higher pressure exists in the tube side, it is usual to leave the tube bundle inside the shell.
- Test time is normally a minimum of 30 min to give leaks time to develop.
- For exchangers that cannot tolerate water contamination, pneumatic tests can be done, as long as the test procedure is fully compliant with code.
- Large exchanger and steam condenser tube bundles may be difficult to get absolutely leak-free. A low-pressure leak test using nitrogen or a similar inert gas can be used to detect leaks (via a hand-held gas-sniffer).

Fig. 16.21   Shell and tube heat exchangers – pressure testing

common in-service role to monitor the quality of replacement tubes to ensure that one set of problems is simply not replaced with another.

Basic technical information
Titanium tubes vary from *unalloyed* grades – almost pure titanium with a restricted number of trace elements – to those known as *alloyed* grades, containing small amounts of nickel or more exotic materials such as palladium. Tubes come in two types:

- Seamless tube: cold drawn to size from a billet.
- Welded tube: this is formed from a flat-rolled product strip then welded using an automatic argon-protected arc process, with no filler rod.

Both types of tube are annealed to relieve welding or work-hardening stresses. The whole process is automatic, all the stages taking place sequentially on a continuous production line.

The problem with titanium is that it is *expensive*. Tube products have to be made by specialized manufacturers who have a continuous production-line facility, which adds to the cost. For these reasons the manufacturing process is subject to closer control than is used for steel tubes. This means that the inspection and test plans (ITPs) used for heat exchanger rebuilds with titanium tubes are quite comprehensive, consisting of a well-defined set of tests and checks. Figure 16.22 shows a typical example – you can use this as a model for checking any of the common grades of welded Titanium heat exchanger tubes up to about 70 mm diameter.

Specifications and standards

The most commonly used standard is the USA and Europe is ASTM B338-02 *Standard specification for seamless and welded titanium alloy tubes for condensers and heat exchangers*. This is targeted specifically at Ti and Ti alloy tubes for condenser and heat exchangers and covers both seamless and welded product forms. Most contractors' specifications reference this standard in one way or another. It specifies a variety of different *grades* of material, both pure and alloyed forms. Note that it is the chemical composition that varies between grades, not the range of product forms (i.e sheet, strip, castings, etc.) that are given their own grade references in some of the ferrous material ASTM standards that you may see. The grade numbers range from 1 to 18, with grades 1, 2, 3, and 7 being the unalloyed materials. This is an important point when inspecting documentation relating titanium tube products to ASTM B338 – a specification must quote the applicable grade number, otherwise the B338 designation becomes meaningless.

Heat exchanger and condenser tubes are normally made out of Grade 2 material. There is nothing overtly special about it – it is simply an easily available alloyed grade with a mid-range tensile strength and good ductility. The typical ITP model shown in Fig. 16.22 is based around this Grade 2 material used for a welded tube. ASTM B338 also defines the set of tests that are done on the finished tube materials. These are mandatory. It is worth looking at some of these in a little more detail – they are often shown as inspection witness points on the repair ITP. Some are quantitative, and have clear pass/fail criteria while others can only be qualitative. They include:

- *Chemical analysis*. This is a straightforward check on the trace elements. Note that they are all specified as maxima.
- *Mechanical properties*. Again, nothing new here, except that there is a maximum allowable value of yield (or proof) strength as well as a specified minimum. The elongation percentage is used as the sole measure of ductility – necessary when there are tube-to-tube plate expansions.
- *Dimensional accuracy*. There are a surprisingly large number of dimensional tolerances applied to titanium tubes. They are important

Activity	Requirement
Material choice	e.g. ASTM B338 unalloyed Ti, Grade 2
Manufacturing method	Cold-rolled strip with continuous automatic arc weld (argon protection)
Heat treatment	Annealed at 600 °C in argon atmosphere
Chemical analysis	N_2 (max) $< 0.03\%$, C (max) $< 0.10\%$, O_2 (max) $<0.25\%$, H_2 (max) $<0.015\%$, F_e (max) $<0.30\%$, Remainder Ti
Mechanical properties (annealed)	$R_m > 345$ MPa $R_e/R_{p0.2} > 275$ MPa but < 450 MPa HV (1 kg) ≤ 200 with maximum variation of 30 around the tube circumference
Dimensional accuracy	Outside diameter (including any ovality and weld-bead reinforcement): for OD < 25 mm: ± 0.10 mm for OD ≥ 25 mm: ± 0.13 mm Wall thickness: $\pm 10\%$ nominal Straightness: ≤ 2 mm per m length Surface: 2.5 μm R_a or better
Tests (1 sample per 1600 m tube length approx.)	Weld macrograph Eddy current test Ultrasonic test Flattening test Reverse flattening test Flaring test Pneumatic test: air at 700 kPa for 5 s Full visual examination

Fig. 16.22   Ti tube heat exchanger repairs – typical ITP content

from a process *accuracy* point of view, i.e. to show that the process is well controlled. Tube wall thickness and surface finish are the most important as they affect the integrity of the tube-to-tube-plate fits.

- *Non-destructive tests.* Welded tubes to ASTM B338 have to be subject to an electromagnetic *and* an ultrasonic test for defects. The standard provides clear guidance on defect acceptance levels. These acceptable levels have to be pre-programmed into the automatic test machines.
- *Destructive tests.* All tubes must undergo a standard flattening test at ambient temperature. In addition, welded tube requires a reverse

flattening test in which a cut section of tube is bent upwards, subjecting the weld to maximum bending stress. The smaller-diameter sizes of tube (welded or seamless) up to 88 mm diameter also need a cold flaring test – a 60° tool is used to flare the end of the tube to a specified increase in diameter (17–22 percent, depending on the grade). For all these three destructive tests, the acceptance criterion is 'no cracking'. Note that this is only assessed visually, without magnification. For good results, however, it is better to use a hand magnifying glass.

- *Pressure test.* Welded and seamless tubes need a hydrostatic or pneumatic test – not both. Hydrostatic test pressure is based on the pressure that produces a stress in the tube wall of 50 percent of R_e – there is a formula in the standard. Check this carefully, it is not uncommon for tubes to be tested at the wrong pressure. They are tested in batches in a special machine, rather than singly. The pneumatic leak test is often done at a fixed pressure of about 0.7 MPa for 5 or 10 s. The best way is to test them under water where leakage can be easily seen. Trying to find leakage by simply looking for gauge pressure drop is difficult when repetitive tests have to be carried out quickly.

One salient point to mention is that of *retests*. Unfortunately, the ASTM and similar standards do not give much guidance. The general principle seems to be that if a tube fails any of its tests then retests are required on 'the batch' from which the tube came. You have to apply a little prudence here – it will be impractical to test them *all*, but you must do sufficient to give representative results. You can use statistical theory to determine the sample size – or just use a common-sense 5 percent, then make a decision on the results that you see.

Documentation

In common with other mass-produced or batch products, you cannot expect a certificate for every tube. The normal way is for the manufacturer to issue one certificate per 'sampled batch' – normally expressed as the manufactured length of tube (1600 m in ASTM B338). ASTM standards do not cross-refer to the EN 10 024 material certification system – they use general references to 'mill test reports' or something similar. Check what these certificates actually *represent* – they should be equivalent to the EN 10 024 Level 3.1B or 3.1A because of the technical complexity (and cost) of the tube manufacture. Level 2.2 certificates of conformity are not really acceptable for this type of product when used in the repair of high-integrity heat exchangers.

Chapter 17

Transportable pressure equipment (TPE)

Transportable pressure equipment (TPE) is the generic name given to the family of cylinders used to contain gas and a wide range of other fluids. Cylinders themselves are known as transportable pressure receptacles (TPRs). This terminology has been driven by the legislative requirements that cover these items rather than by the engineering industry itself. More than 90 percent of TPRs are simple refillable gas cylinders containing propane, butane, oxygen, etc. for domestic and industrial uses.

Cylinders are batch-manufactured items of simple construction which, in terms of design and manufacture, are relatively low-risk items (Fig. 17.1). In use, however, they contain high-pressure gases and hence large amounts of stored energy. The situation is compounded by their actual method of use, i.e. they are used, stored, and refilled by non-specialist users and their use nearly always involves close proximity to people. The net result is that TPRs are classified as high-risk items that require planned in-service inspection throughout their life if they are to remain 'safe'. In practice, TPRs are frequently 'retired', either following the results of inspection or on a simple periodic basis.

TPRs – the statutory background

TPRs are statutory items in most countries of the world. This has been driven by the fact that TPRs are regularly exported and transferred between countries – Asia and India are large manufacturers as well as Europe and the USA. The difficulty with legislative requirements for TPRs is that they are *complex*, and a little confusing. Part of the reason for this comes from the different types of TPE that exist – road- and rail-transported tanks and pressurized cylinders, for example, are covered by different legislation to gas cylinders and lower-pressure items such as fire extinguishers. It is not easy to draw straight technical definitions between these different types of equipment so a certain

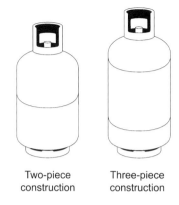

Two-piece
construction

Three-piece
construction

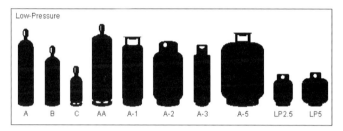

Fig. 17.1   Gas cylinders – general design arrangements

amount is left to historical precedent. This means that the legislative paths taken by different types of equipment are not necessarily consistent (or even, in a few cases, *logical*).

Figure 17.2 gives an overview of the general characteristics of the TPE industry, and highlights some of the difficulties that result when attempting to legislate for their design, manufacture, and use. This figure is not country specific – this situation is much the same in all countries outside the USA (where the legislation framework is much clearer).

Individual countries have responded differently to the TPE characteristics shown in Fig. 17.2. The overall situation is, however, common to most countries – the situation is complex, often overlapping, and occasionally inconsistent. The current situation in Europe, with the issue of European Directives, has attempted to rationalize the situation. The overt objective of EU directives, i.e. to reduce barriers to trade across EU member-state boundaries, fits well with the characteristics of TPE as *transportable items*, likely to be transferred between countries. The Transportable Pressure Equipment Directive (TPED) came into force in 2001 and complements the Pressure Equipment

- **In-service use**. TPE legislation has to cover in-service use as well as design and manufacture.
- **Multiple technical standards**. There are more than 50 different standards covering the manufacture of gas cylinders alone.
- **Diverse duty holders**. The duty holders (i.e. the parties that have a duty to comply with the legislation) comprise designers, manufacturers, third-party bodies, refillers, storers, tranporters, and users of TPE. This is a more complex situation than for non-mobile industrial pressure plant.
- **Straightforward technology**. One of the few simplifying influences is the fact that the technology of TPE is *simple*. Most cylinders are of straightforward seamless or fabricated construction using well-defined design features and materials. This means, thankfully, that the technical aspects of TPE inspection are fairly straightforward, even if the driving legislation is itself complex.

Fig. 17.2   Transportable pressure equipment (TPE) – characteristics that affect their legislation

Directive (PED) and Simple Pressure Vessels Directive which cover static pressure equipment.

The TPED provides for:

- New items of TPE constructed after 1 July 2001, that are placed on the market in EU countries, have to be subjected to a conformity assessment. The technical requirements are set out (or referenced) in the European Agreement on the International Transport of Dangerous Goods by Road (known as 'ADR'), or its rail equivalent (RID).
- Existing items of TPE constructed prior to 1 July 2001 have to have a retrospective assessment of conformity but *only* if they are to be placed on the market. The assessment is against ADR/RID requirements, as for new equipment.
- Periodic inspection is required after the initial conformity assessment or reassessment. This has to be done by either a Notified Body (NB) or Approved Body (AB) as specified (see the TPED for details).
- TPE items are stamped with a 'Pi-mark' (not the CE mark, as for other pressure vessels) to indicate their conformity.

The UK situation

The implementation of the TPED in the UK has caused some controversy and discussion as to the best way to comply. The root of the difficulty is the fact that standards of safety and transport relating to TPE are already high in the UK. Hence the TPED has little to add in terms of *technical effectiveness* to the system that is in place already. While compliance with the TPED is mandatory, rather than an option, the UK is still developing the methodology which will keep existing strengths while maintaining consistency with general agreements across the EU on how the TPED should be implemented.

Current domestic requirements governing design and construction, conformity assessment, and periodic inspection of all TPRs and tanks used for the transport of dangerous goods (i.e. not just those used for transporting gases) are contained in the following regulations.

- *For TPRs*: the Carriage of Dangerous Goods (Classification, Packaging, and Labelling) and Use of Transportable Pressure Receptacles Regulations 1996 (CDGCPL2).
- *For tanks*: the Carriage of Dangerous Goods by Road Regulations 1996 (CDGRoad) and the Carriage of Dangerous Goods by Rail Regulations 1996 (CDGRail2).

These 1996 regulations were part of a package of legal requirements which implemented two directives: Council Directive 94/55EC (the ADR Framework Directive) and Council Directive 96/49EC (the RID Framework Directive). These directives required all EU countries to align their domestic requirements with the requirements governing the international transport of such goods contained in ADR and RID.

The situation is therefore very different from that applying to the PED which was able to be implemented much more freely without contradicting any existing practices. The TPED is implemented in the UK by the Transportable Pressure Vessel Regulations (TPVRs) but it is expected that the actual mechanics of their implementation will take several years to sort out. Realistically, it also seems likely that historical precedent will play a big part in deciding how the TPVRs are interpreted in individual cases.

Consultative document CD163

The UK Health and Safety Commission (HSC) has published a consultative document CD163 setting out proposals for UK imple-

Table 17.1   TPVR* risk categories and 'modules'

TPED risk category	Stored energy (bar litres)	Modules
I	< 300	A1,D1,E1
II	300–1500	H, B + E, B + C1, B1 + F, B1 + D
III	> 1500 and tanks	G, H1, B + D, B + F

*The TPED and TPVRs have the same requirements

mentation of the TPED. This references the way that two separate bodies have authority for related equipment in the UK, i.e:

- The Health and Safety Executive (HSE) for TPRs.
- The Department of the Environment, Transport, and the Regions (DETR) for tanks.

A further function of the document was to consult on the implementation of conformity assessment procedures for tanks and cylinders which lie outside the scope of the TPED, e.g. tanks used for the transport of dangerous goods other than gases. A copy of CD163 is available on the HSE website: www.hse.gov.uk/condocs/index.htm.

The UK TPVRs use a similar 'module structure' as the PED but with a slightly simpler set of risk categories (see Table 17.1). While detailed knowledge of these risk categories is not essential for in-service inspection purposes, knowledge of the risk category can help in making decisions about the severity of inspection findings.

In-service inspection of transportable pressure equipment

TPVR requirements

The TPVRs substantially reproduce the provision of the TPED (Part III, Annex IV) in requiring that 'periodic inspection must be carried out' if equipment is to remain in conformity. This is about as far as the prescriptive technical requirements go – it is then left to the discretion of the appropriate Notified/Approved Bodies to translate these into practical engineering tests and requirements.

Engineering requirements – codes and standards

Although there are many standards covering transportable pressure equipment, they nearly all concentrate on design, construction, and testing of new equipment and virtually ignore the requirements of in-service inspection. Two recent standards that do cover in-service inspection are:

- EN 1968: 2002 *Transportable gas cylinders – periodic inspection and testing of seamless steel gas cylinders (excluding LPG).*
- EN 1803: 2002 *Transportable gas cylinders – periodic inspection and testing of welded steel gas cylinders (excluding LPG).*

These standards superseded BS 5430: 1990 Parts 1 and 2, respectively (BS 5430 Part 3: 1990 covers aluminium cylinders). Although the EN standards have a slightly different approach and inspection format, the technical aspects of BS 5430 remain valid as an indication of good engineering practice for the in-service inspection of transportable pressure receptacles and provide useful technical guidance to follow.

Visual inspection of cylinders

Visual examination remains the core activity of in-service inspection of cylinders. The main in-service defects are normally caused either by mechanical damage or corrosion. Corrosion damage is more prevalent in older cylinders, or those that have been stored in a marine or similar corrosive environment. Tables 17.2 and 17.3 (based on EN 1803: 2002) give sound principles to work to. Use these for guidance only – it is necessary to consult the body with formal responsibility for the integrity of the equipment before re-using a cylinder that has mechanical or corrosion damage.

Preliminary activities required before close visual examination are:

- Remove external coating (loose paint or plastic sheathing) by wire brushing, shot blasting, etc. as appropriate.
- Depressurize to atmospheric pressure, remove the valve and clean the inside of the cylinder by water/steam jet to allow internal inspection by illuminated borescope.
- Check the weight of the cylinder (known as 'tare') to check for corrosion loss. Any cylinder that has lost more than 5 percent of its original tare is generally deemed unfit for further service.
- Check the internal neck threads of the cylinder using a thread gauge.

Most cylinders that are retired in use are done so on the basis of defects found during the visual inspection. The most common defects are those caused by simple mechanical damage from the cylinder being dropped or dented. Cylinders which have been dropped also commonly suffer damage to the neck ring threads. The threads themselves can be chased or re-tapped (within limits).

Hydraulic test

Cylinders are subject to a periodic hydraulic test. Note that this is referred to in standards as a 'hydraulic proof test' – the word 'proof' being used in a non-conventional sense (see Chapter 10 on pressure testing for further information). The test pressure depends heavily on the type of gas and container and *varies* from code to code. Table 17.4 gives some guidelines, but it is important to check the relevant construction code in each individual case. The hydraulic test itself is usually performed on gangs of cylinders held in a special test rig. Some points to note are:

- The absolute maximum pressure that should be applied is the calculated test pressure plus 3 percent or 10 bar, whichever is the lower.
- Make sure the external surface of the cylinder is dry, so that any leaks can be seen.
- The pressure should be held for a minimum of 1 min.
- During the holding time, the cylinders are isolated (by closing the filling valve) and the gauge watched for any sign of pressure drop.
- Hydraulic test pressure should be checked against the marking on the cylinder.

In practice, as with other types of pressure vessels, cylinders rarely fail catastrophically during hydraulic test even if they have significant wall thinning due to corrosion. They are much more likely to leak, particularly from the neck ring threads if they are worn.

Volumetric expansion test

Strictly, the volumetric expansion test is more relevant to testing of new cylinders during construction rather than periodic testing in-service. It is essentially a *combined test* of the design of the cylinder and the ductility of the material of which it is constructed. It is, however, specified in various standards and used as a practical test for the serviceability of used cylinders. In this context it has particular relevance to old cylinders in which repeated filling and discharging may have caused work-hardening of the cylinder material, reducing its ductility.

There are two main methods used for the volumetric expansion test: the water-jacket method and the non-water-jacket method. The water-jacket method involves enclosing the cylinder in a water-filled jacket. The permanent and total volumetric expansion is determined by measuring the volume of water displaced when the cylinder is under pressure, and then again after the pressure is released (see Fig. 17.3).

In the non-water-jacket method the expansions are determined by comparing the volume of water pumped into the cylinder with that rejected back out when the pressure is released. Measurement is done by a calibrated glass burette. Figure 17.4 shows the details. Although this is a more simple procedure than the water-jacket method, calculations must be made to allow for the compressibility of the water. The compressibility varies with temperature. Figure 17.6 shows a specimen example. Cylinders may also be subjected to a burst test to destruction (see Fig. 17.5).

Test periodicities
Test periodicities for visual, hydraulic, and volumetric expansion tests on cylinders vary between countries and with the type of gas that is contained. Table 17.5 shows the periodicities applied by EN 1803: 2002.

Cylinders are subject to a volumetric expansion test during construction and in-service.

- **The objective**: to check the *ductility* of the material and the way in which it interfaces with the design of the cylinder.
- **The principle**: water is used to pressurize the cylinder (at normal test pressure). The volume of water used allows the percentage volumetric expansion under pressure and the accompanying permanent plastic deformation to be calculated
- **Acceptance criteria**: for in-service testing a 'rule-of-thumb' acceptance level (from BS 5430) is:

  - for non-corroded cylinders permanent expansion should not exceed 5% of the total volumetric expansion;
  - for corroded cylinders, permanent expansion should not exceed 2% of the total volumetric expansion.

Fig. 17.3   TPR (cylinders) – the volumetric expansion test

Table 17.2   Typical rejection criteria for welded TPR cylinders

Type of defect	Definition	Rejection limits[a]	Repair or render unserviceable
Bulge	Visible swelling of the cylinder	All cylinders with such a defect	Render unserviceable
Dent	A depression in the cylinder that has neither penetrated nor removed metal and is greater in depth than 1% of the outside diameter	When the depth of the dent exceeds 3% of the external diameter of the cylinder; or	Render unserviceable
		When the diameter of the dent is less than 15 times its depth	Render unserviceable
Cut or gouge	A sharp impression where metal has been removed or redistributed and whose depth exceeds 5% of the cylinder wall thickness	When the depth of the cut or gouge exceeds 10% of the wall thickness; or	Repair if possible [b]
		When the lengths exceeds 25% of the outside diameter of the cylinder; or	Repair if possible [b]
		When the wall thickness is less than the minimum design thickness	Render unserviceable
Crack	A rift or split in the metal	All cylinders with such defects	Render unserviceable
Fire damage	Excessive general or localized heating of a cylinder usually indicated by: (a) partial melting of the cylinder;	All cylinders in categories (a) and (b)	Render unserviceable
	(b) distortion of cylinder; (c) charring or burning of paint; (d) fire damage to valve, melting of plastic guard or date ring.	All cylinders in categories (c) and (d) may be acceptable after inspection and/or testing	Repair if possible

Table 17.2   (Continued).

Type of defect	Definition	Rejection limits[a]	Repair or render unserviceable
Plug or neck inserts	Additional inserts fitted in the cylinder neck, base, or wall	All cylinders unless it can be clearly established that addition is part of approved design	Repair if possible
Stamp marking	Marking by means of a metal punch	All cylinders with illegible, modified, or incorrect or incorrectly modified markings	Render unserviceable
Arc or torch burns	Partial melting of the cylinder, the addition of weld metal or the removal of metal by scarfing or cratering	All cylinders with such defects	Render unserviceable
Suspicious marks	Introduced other than by the cylinder manufacturing process or approved repair	All cylinders with such defects	Continued use possible after additional inspection

[a]When applying the rejection criteria, the conditions of use of the cylinders, the severity of the defect and safety factors in the design shall be taken into consideration. [b]Provided that after repair by a suitable metal removal technique, the remaining wall thickness is equal to the design minimum wall thickness.

[c]If it can be clearly established that the cylinder fully complies with the appropriate specifications, altered operational and modified marking may be acceptable and inadequate markings may be corrected, provided there is no possibility of confusion.

Table 17.3   Rejection criteria for corrosion of the cylinder wall

Type of corrosion	Definition	Rejection limits[a]	Repair or render unserviceable
General corrosion	Loss of wall thickness over an area of more than 20% of the total surface area of the cylinder	If the original surface of the metal is no longer recognizable; Or	Repair if possible
		If the depth of penetration exceeds 10% of original thickness of wall; or[b]	Repair of possible
		If the wall thickness is less than design minimum wall thickness	Render unserviceable
Local corrosion	Loss of wall thickness over an area of less than 20% of the total surface area of the cylinder except for the other types of local corrosion described below	If the depth of penetration exceeds 20% of the original thickness of the cylinder wall;[b] or if the wall thickness is less than design thickness	Repair if possible
Chain pitting or line corrosion	Corrosion forming a narrow longitudinal or circumferential line or strip, or isolated craters or pits which are almost connected	If the total length of corrosion in any direction exceeds the diameter of the cylinder and the depth exceeds 10% of the original wall thickness[b]	Repair if possible[c]
Crevice corrosion	Corrosion taking place in, or immediately around, an aperture	If after thorough cleaning, the depth of penetration exceeds 20% of the original wall thickness	Repair if possible[c]

[a]If the bottom of the defect cannot be seen and if its extent cannot be determined using appropriate equipment, the cylinder shall be rendered unserviceable. [b]If corrosion has reached limits of depth or extent, the remaining wall thickness should be checked with an ultrasonic device. The wall thickness may be less than the minimum, with the acceptance of the inspection body, e.g. small (depth and extent) isolated pits. When applying the rejection criteria given in this table, the conditions of use of the cylinders, the severity of the defect, and safety factors in the design shall be taken into consideration.
[c]Provided that after repair by a suitable metal removal technique, the remaining wall thickness is at least equal to the design minimum wall thickness.

Table 17.4 Gas cylinder test pressure – some guidelines

Formulae for deriving minimum values of test pressure		
Gas classification	Test pressure P_1^*	
Permanent gases in uninsulated containers, the charged pressure not exceeding 300 bar	$P_1 = \dfrac{p}{0.9}$	P_1 shall be the greater of: (a) $\dfrac{p}{0.85}$ or (b) $1.5 \times$ charged pressure at $15\,°C$ except that P_1 shall not exceed $\dfrac{pY}{0.63\,T}$
Permanent gases in uninsulated containers and liquefiable gases in insulated containers	$P_1 = \dfrac{p}{0.9}$ except that P_1 shall be not less than 200 bar for carbon dioxide or nitrous oxide	$P_1 = \dfrac{pY}{0.63\,T}$ except that P_1 shall be not less than 200 bar for carbon dioxide or nitrous oxide
Low- and high-pressure liquefiable gases in uninsulated containers	$P_1 = p$ except that P_1 shall be not less than 200 bar for carbon dioxide or nitrous oxide	P_1 shall be the lower of: (a) $\dfrac{pY}{0.7\,T}$ or (b) $\dfrac{p}{0.85}$ except that P_1 shall be not less than 200 bar for carbon dioxide or nitrous oxide
Dissolved-acetylene containers	$P_1 - 52$ bar	$P_1 = 52$ bar

$P_1 =$ test pressure. $Y =$ minimum specified yield strength, (R_e) (in MN/m^2).
$T =$ minimum specified ultimate tensile strength, (R_m) (in MN/m^2).
$p =$ actual pressure of contents at the pressure reference temperature.
*Treat these test pressures as guidelines only. Consult the relevant *construction code* for the individual cylinder under test.

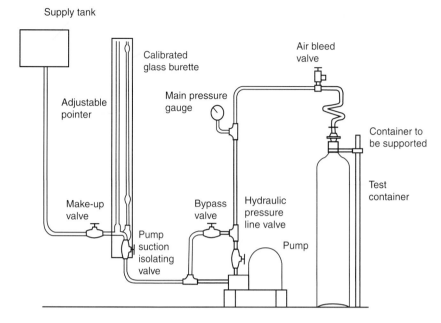

Fig. 17.4   TPR (cylinder) volumetric expansion test – non-water-jacket method

Fig. 17.5   A TPR burst test specimen

The equation for the compressibility, C, of water (in ml) is:

$$C = mP\left(K - \frac{0.6P}{10^5}\right)$$ (1)

where m is mass of water (kg) at test pressure; P is test pressure (barg); and K is a factor from the below table.

Water compressibility factor, K

Temperature °C	K
10	0.048 12
15	0.047 25
20	0.046 54

Example

For a 200 bar test on a 110 kg cylinder:

Test pressure $P = 200\,barg$
Mass of water in cylinder at atmospheric pressure $= 110\,kg$
Temperature of water $= 15\,°C$
Mass of water added to raise pressure to 200 bar $= 1740\,ml = 1.74\,kg$
Total mass of water at 200 bar $= 100\,kg + 1.74\,kg = 101.74\,kg$
Mass of water expelled to depressurize $= 1735\,ml = 1.735\,kg$
Permanent expansion $= 1740 - 1735 = 5\,ml$

Calculating C at 15 °C from equation (1)

$$C = mP\left(K - \frac{0.6P}{10^5}\right) \qquad K = 0.04725 \text{ at } 15\,°C$$

$$C = 101.74\,kg \times 200\left(0.047\,25 - \frac{0.6 \times 200}{10^5}\right)$$

$$C = 937\,ml$$

Hence,
Total volumetric expansion $= 1740 - 937 = 803\,ml$

$$\text{Percentage plastic deformation} = \frac{5 \times 100}{803} = 0.62\%$$

Fig. 17.6   Cylinder volumetric expansion test – specimen calculation

Table 17.5 TPR (gas cylinder) inspection periods (after EN 1503: 2002)

Description	Gas type (examples)[b]	Normative intervals[c] (years)	Informative recommendations for next revision of ADR (years)
Compressed gases	Ar, N_2, He, etc.	10	10
	H_2[d]	10	10
	Air, O_2	10	10
	Self-contained breathing air, O_2, etc.	[e]	5
	Gases for underwater breathing apparatus	[e]	2, 5 (internal visual) and 5 (full)[f]
Liquefied gases	CO[g]	5	5
	CO_2, N_2O, etc.	10	10[h]
Corrosive gases	[i]	3	3 (internal visual) and 5 (full)[j]
Toxic gases	CH_3Br	5	10
Very toxic gases	AsH_3, PH_3, etc.	5	5
Gas mixtures	(a) All mixtures except (b) below	3, 5, or 10 years according to classification	(a) Lowest test period of any component

Table 17.5 (Continued).

Description	Gas type (examples)[b]	Normative intervals[c] (years)	Informative recommendations for next revision of ADR (years)
	(b) Mixtures completely in the gaseous state containing toxic and/or very toxic components	3 years for groups TC, TFC, TOC 5 years for groups T, TF, TO 10 years for groups A, O, F	(b) For such mixtures, if the toxicity of the final mixture is such that $LC_{50} \geq$ volume fraction of 200×10^{-6}, a 10-year period shall apply, and if the toxicity of the final mixture is such that $LC_{50} <$ volume fraction of 200×10^{-6}, a 5-year period shall apply

[a] At all times certain requirements can necessitate a shorter time interval, e.g. the dew point of the gas, polymerization reactions, and decomposition reactions, cylinder design specification, change of gas service.
[b] This list of gases is not exhaustive. A full list of gases can be found in RID/ADR.
[c] These intervals conform to the 1999 edition of RID/ADR.
[d] Pay particular attention to the requirements of Clause 5 and possible additional testing in accordance with EN 1795 for change of service.
[e] Not currently listed in RID/ADR.
[f] For cylinders used for self-contained underwater breathing apparatus in addition to the full retest period of 5 years, an internal visual inspection shall be performed every 2.5 years.
[g] This product requires very dry gas (see EN ISO 11114-1).
[h] This test period may be used provided the dryness of the product and that of the filled cylinder are such that there is no free water, and that this condition is proven and documented within a quality system of the filler. If these conditions cannot be fulfilled, alternative or more frequent testing may be appropriate.
[i] For RID/ADR purposes, corrosivity is with reference to human tissue and not cylinder material, as per Annex I.
[j] For gas mixtures shown to be corrosive for the cylinder material, the time period for single corrosive gases applies.

Chapter 18

Industrial cranes

Introduction

Inspection of lifting equipment accounts for a large proportion of the 'high-volume' in-service inspection market in the UK and other developed countries. Unlike other types of industrial and domestic equipment, most items of lifting equipment are covered by specific statutory regulations. This is also one of the few areas where inspection periodicity is still *prescribed* rather than left to the discretion of a 'Competent Person' organization.

The scope of lifting equipment is very wide, incorporating industrial cranes (see Fig. 18.1), passenger and goods lifts, vehicle lifts and recovery equipment, and almost every other type of appliance and small tackle used for the purpose of lifting or lowering persons or goods. The technology level involved in this range of equipment is also wide – from complex cranes down to simple items such as eyebolts and wire slings.

Lifting equipment is one of the few areas of in-service inspection with sufficient volume of work to warrant specialist inspectors. Insurance companies in particular have retained inspectors who inspect only lifting equipment, and who therefore become very proficient and experienced at this type of work.

The LOLER regulations

In the UK, the in-service inspection of lifting equipment is driven by the existence of the Lifting Operations and Lifting Equipment Regulations (LOLER). These came into force in December 1998 and cover a wide range of equipment items that were previously covered by a variety of separate industry-specific bits of legislation. LOLER applies to all industries and to all situations and premises covered by the Health and Safety at Work Act 1974. It also implements the lifting provision of the

Fig. 18.1   A typical overhead industrial crane

Amending Directive to the Use of Work Equipment Directive (AUWED 95/63/EC).

From an inspection viewpoint, LOLER provides very few loopholes by which equipment can escape inspection requirements. The one weak point is that the regulations do not actually define 'lifting equipment', so there are a very small number of equipment items that are generally considered not to be covered [tractor ploughs and linkages, horizontal conveyor belts, and winders pulling loads along the ground (rather than hoisting them above it) are three examples]. Passenger escalators are exempt from LOLER as they are covered by other specific regulations [Regulation 19 of the Workplace (Health, Safety and Welfare) Regulations 1992)].

LOLER documentation

The LOLER regulations are comprehensively covered in the three documents:

- *Approved Code of Practice (ACoP) L113 (HSE Books, 1998)*. This is the ACoP accompanying the LOLER Regulations. It has similar status to the ACoP to the Pressure System Safety Regulations (PSSRs), see Chapter 5, in that it has been accepted as giving practical advice on how to comply with the law and so has special legal status. The format is the same as other AcoPs, consisting of:

  - The regulations themselves in full.

- ACoP paragraphs giving an interpretation of individual regulations and clauses within them.
- Further informal guidance notes that explain 'best practice' rather than giving specific legal interpretations.

- *SAFeD Guidelines* on LOLER 1998. This is a document published by SAFeD (Safety Assessment Federation). Its stated aim is to assist duty holders (owners, users, etc.) to understand and comply with LOLER and to advise engineering inspectors employer by inspection bodies (SAFeD member companies). The guidelines are robust, but not particularly incisive in providing additional guidance outside the scope of the ACoP.
- *HSE Open Learning Guide* on LOLER: 1999 (HSE Books; ISBN 0-7176-24464-1). This is a comprehensive guide to the content of LOLER *and* is well presented in open-learning format. It includes information on less common and domestic-type lifting equipment as well as industrial items, which are consequently not covered in great technical depth.

LOLER and the in-service inspector

From the viewpoint of the in-service inspector, LOLER sets the *framework* for the scope of inspection activity. It specifies (via LOLER Regulation 9) that lifting equipment has to be thoroughly examined at regular intervals to determine that:

- It is correctly installed and safe to operate.
- It is in a good state of maintenance and repair.
- Any deterioration is identified and remedied in good time.

The above items are consistent with common engineering checks such as visual examination, functional tests, calibration, non-destructive testing, and similar.

One of the characteristics of LOLER is that it also covers lifting operations themselves, as well as the integrity and functionality of the equipment. Control of operations is normally not the responsibility of the in-service engineering inspection – which means that the responsibility of LOLER compliance is shared between several 'Duty Holder' parties. Figure 18.2 shows the situation. Another characteristic of LOLER, (shared with many other statutory regulations), is that engineering details of inspections etc. are not prescribed in detail – instead they are left to the decision of a 'Competent Person' organization to decide. Figure 18.3 outlines some of the key content of LOLER. Figure 18.4 is a transcript of Schedule 1 of LOLER, showing the information that must be contained in an inspection report for a 'thorough examination'.

LOLER periodicities

LOLER is one of the few sets of regulations that still defines statutory inspection periods. Table 18.1 shows a more detailed schedule of UK industry practice. Remember that this is an *interpretation* of LOLER rather than a transcript from the regulations or ACoP.

Chapters 18 to 21 of this book look at in-service inspection aspects of specific lifting equipment, i.e. industrial cranes, mobile cranes, lifts (elevators), and small lifting tackle and accessories. Other items such as fork-lift trucks are not covered – these are more often inspected by specialist lifting equipment inspectors rather than generalist industrial plant inspectors.

Industrial cranes are common inspection items. They are subject to very tight legislation in most countries, which means that they have well-defined certification and inspection requirements.

Crane types and construction

There are many different types and designs of industrial crane. Fortunately they exhibit very similar engineering principles and features, even between types that are visually quite different. Part of the task of the in-service inspector is understanding how the various mechanisms of the crane (see Fig. 18.5) work; the following text describes the construction of a typical overhead crane but the general principles are similar for many other types.

The structure

The largest stressed member is the bridge, which is typically of double-girder construction. The girders are fabricated stiffened box sections, constructed with full-penetration welds. Note the stress regime on these members; it is a straightforward bending case because of the 'simple' supports provided by the crab and bridge wheels. It should be clear that the maximum tensile stress will occur on the lower flange of the girders. A maximum vertical deflection of 1/750 bridge span is allowed when the crane is lifting its safe working load (SWL). Design calculations use the SWL as a reference point. Bridge end-carriages tend to be designed with a relatively small wheelbase (<one-seventh of the crane span) so they are stiffer and deflection is less significant. The crab is also relatively stiff

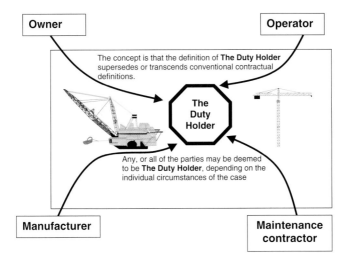

Fig. 18.2   LOLER – the concept of 'The Duty Holder'

but it is subject to significant additional inertia loads. Construction therefore incorporates a lot of cross-bracing and deep sections to resist the superimposed stress regime. This component can contain a lot of load-bearing fillet welds, in addition to the full-penetration welds used for the fabricated base-frame.

The mechanisms

Crane mechanisms, although not in continuous use, are subject to high static loadings *and* an unpredictable set of dynamic conditions due mainly to inertia loads. Figure 18.5 shows the main components.

The most important mechanism is the winding arrangement. The winding drum is normally manufactured to a pressure vessel standard (ASME, EN, etc.) because of the high compressive stress imposed by the rope turns under load. Rope grooves are machined to a carefully designed profile and spacing. A load cell or mechanical cantilever arrangement prevents excessive weights being lifted. There is also a 'hook approach' mechanism to stop the hook being wound too near the drum – it works by limiting the axial position of the rope guide and is adjustable.

Separate electric motors drive the main hoist, auxiliary hoist (a smaller capacity hoist often fitted on the same crab), cross-traverse, and longitudinal travel motors. There is sometimes an additional 'inching' motor on the main hoist to move the load very slowly when lifting, for instance, heavy turbine rotors. All these motors have friction brakes, which are a key safety feature of the crane – those for the hoisting

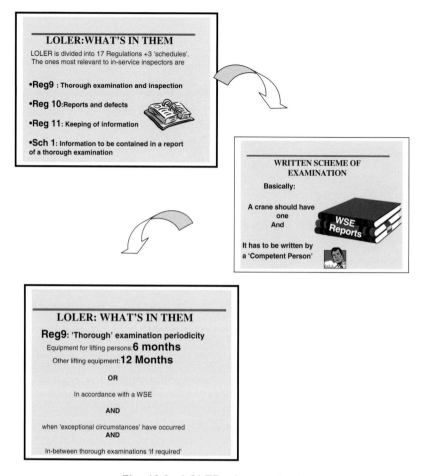

Fig. 18.3   LOLER – key content

motors may be of the centrifugal type. All brakes work on the 'fail-on' principle and are designed to stop the relevant movement in a well-defined braking distance. A range of electrical protection devices are installed on the various electric motors and mechanisms.

Fitness-for-purpose criteria

Statutory regulations normally require that cranes be inspected both during manufacture and in-service. The concept of fitness for purpose (FFP) plays a large part. The three main FFP criteria are:

- Design classification.
- Function and safety.
- Statutory compliance.

INFORMATION TO BE CONTAINED IN A REPORT OF A THOROUGH EXAMINATION

1. The name and address of the employer for whom the thorough examination was made.
2. The address of premises at which the thorough examination was made.
3. Particulars sufficient to identify the lifting equipment including (where known) its date of manufacture.
4. The date of the last thorough examination.
5. The safe working load of the lifting equipment or (where its safe working load depends on the configuration of the lifting equipment) its safe working load for the last configuration in which it was thoroughly examined.
6. In relation to the first thorough examination of lifting equipment after installation or after assembly at a new site or in a new location –

   (a) that it is such a thorough examination;
   (b) (if such be the case) that it has been installed correctly and would be safe to operate.

7. In relation to a thorough examination of lifting equipment other than a thorough examination to which paragraph 6 relates –

   (a) whether it is a thorough examination —

       (i)   within an interval of 6 months under Regulation (9(3)(a)(i);
       (ii) within an interval of 12 months under Regulation (9(3)(a)(ii);
       (iii) in accordance with an examination scheme under Regulation (9(3)(a)(iii); or
       (iv) after the occurrence with an exceptional circumstances under Regulation (9(3)(a)(iv);

   (b) (if such be the case) that lifting equipment would be safe to operate.

8. In relation to every thorough examination of lifting equipment –

   (a) identification of any part found to have a defect which is or could become a danger to persons, and a description of the defect;
   (b) particulars of any repair, removal, or alteration required to remedy a defect found to be a danger to persons;

Fig. 18.4   LOLER Schedule 1 reporting requirements

(c) in the case of a defect which is not yet but could become a danger to persons –
   (i) the time by which it could become a danger;
   (ii) particulars of any repair, renewal or alteration required to remedy it;
(d) the latest date by which the next thorough examination must be carried out;
(e) where the thorough examination included testing, particulars of any test;
(f) the date of the thorough examination.

9. The name, address, and qualifications of the person making the report; that he is self-employed or, if employed, the name and address of his employer.
10. The name and address of a person signing or authenticating the report on behalf of its author.
11. The date of the report.

Fig. 18.4 (Continued)

There is a specific standard covering the FFP of cranes: EN 12644-3 *Cranes – Safety requirements for inspection and use – Part 3: Fitness for purpose.* This is part of the EN 12644 series and while it may not be as prescriptive as some earlier standards, gives sound guidance in this area.

Design classification
Cranes have a well-developed system of design classification that is accepted by ISO and most other standards organizations. Its purpose is to define accurately the anticipated duty of the crane and provide a framework for clear technical agreement between the user and manufacturer. This design classification also becomes a valuable tool for the in-service inspector, particularly where an existing crane has repairs, design changes, or re-rating.

Function and safety
Function is concerned with whether the crane can continue to lift its design loads, at the speeds specified, without excessive deflection or plastic distortion. The various safety features must also be proved to operate correctly. The main ones are the brakes and motion/load-limiting devices.

Table 18.1   Lifting equipment inspection frequencies

Main item	Considered to include	Examination periodicity (months)
Cranes	Jib cranes, static and wheel mounted	12
	Container cranes	12
	Crawler cranes	12
	Derrick cranes	12
	Dockside cranes	12
	Goliath and semi-Goliath cranes	12
	Lorry loaders	12
	Overhead cranes	12
	Pillar jib cranes	12
	Portable jib cranes	12
	Portal cranes	12
	Shipbuilding cranes	12
	Tower cranes	12
	Transporter cranes	12
	Wall jib cranes	12
	Telpher cranes	12
	Concrete pumping boom	12
	Pipe laying machine	12
	Excavator when used for lifting	12
Elevator	Bucket elevator	12
	Elevator	12
	Elevator, man riding	6
Hoists and lifts	Ash/coke/skip hoist	12
	Builders hoist, goods only	12
	Builders hoist, passenger	6
	Passenger hoist/lift	6
	Goods only hoist/lift	12
	Passenger/goods hoist/lift	6
	Inclined material hoist	12
	Service lift	12
	Home lift	6
	Paternoster passenger lift	6
	Paternoster goods only	12
	Scissor lift	6
	Stair lift	6
	Teagle hoist	12
Patient hoist	Patient hoist	6

Table 18.1   *(Continued)*

Main item	Considered to include	Examination periodicity (months)
Winches	Winches, if used for lifting loads	12/6
	Capstans, if used for lifting loads	12/6
Sheer legs	Sheer legs with winch	12/6
Blocks	Rope	12
	Hoist	12
	Manual	12
	Powered	12
	Pulley	12
	Snatch	12
	Chain	12
	Ratchet	12
	Gin wheels	12
	Hook hoist	12
Safety and rescue	Aborealists' equipment	6
equipment for	Bosun's chair	6
supporting, raising and	Mountain rescue sets	6
lowering persons	Fall arrestor/retractable lifeline	6
	Harness with lanyard	6
	Safety belts	6
Miscellaneous items	Anchorage, suspension point	12
(provided for the support	Tracks	12
of lifting equipment)	'A' frame	12
	Overhead gantry	12
	Davits	12
	Gantry	12
	Jib arms	12
	Overhead crane bridges	12
	Runway tracks and beams	12
	Trolleys	12
Lifting accessories	Eyebolts	6
	Cradle	6
	Fork attachments	6
	Fork-lift truck attachments	6
	Girdar clips	6
	Grabs	6
	Hooks	
	Lifting beams/frames	6
	Lifting lug/bar/plate/arm	6
	Magnets	6

Table 18.1 (Continued)

Main item	Considered to include	Examination periodicity (months)
	Plate clips	6
	Rigging screws	6
	Running out block/pole carrier	6
	Shackles	6
	Slings	6
	Vacuum lifting devices	6
Access equipment,	Suspended access equipment	6
suspended	Window cleaning rig	6
Work platforms	Work platforms	6
	Mast climbers	6
	Mast hoists	6
	Bridge maintenance access equipment (where lifting is involved)	6
Platform stacker	Platform stacker	12
Car parking systems	Car parking systems	12/6
Vehicle recovery	Vehicle recovery equipment	12
equipment	'Spectacle' frame	6
Tailboard hoist/lift	Tailboard hoist/lift:	6
	passenger lifting	6
	goods lifting	12
Jacks	Jacks, multi-stage	12
	Jacks, trolley	12
Vehicle lifts	Motorcycle lifts	6
	Vehicle lifts/hoists	6
Vehicle skip hoists	Vehicle skip hoists	12
Road vehicle wheel lifter	Road vehicle wheel lifter	12
Drum, coil, roll lifting device	Drum, coil, roll, lifting devices including where fitted to process plant	12
Fork-lift truck	Fork-lift truck	6/12
Pallet trucks/palletizers	Pallet trucks, manual, electric and hydraulic	12
Order pickers	Order pickers, inclined	6/12
	Order pickers, overhead	6/12
Pick and place robot	Pick and place robot	12
Loading shovel	Loading shovel	12
Front-end loader	As fitted to a tractor	12
Straddle carriers	Straddle carriers	12

Table 18.1   *(Continued)*

Main item	Considered to include	Examination periodicity (months)
Telescopic load handler	Telescopic load handler:	
	hydraulic rams	12
	hydraulic and chains	6
Lighting mast	Ariel mast	12
	Lighting rigs where lifting is involved	12
	Articulated lamp post	12
Balancers	Tool balancers	12
Cable drum raising system	Cable drum lifters	12
Stage equipment hoist	Stage equipment hoist, utilities, scenery, stages:	
	structure	12
	structure and lifting accessory	6
	Camera boom	12
Gymnasium equipment	Gymnasium equipment if used at work (climbing ropes)	12
Piledriver	Piledriver	12
Drilling rigs	Drilling rig where lifting is involved	12
Rail tamping machine	Rail tamping machine with lifting equipment	12
Wagon tipler	Wagon tipler	12
Miscellaneous items	Drop sections	12
	Gangways and bridges	12
	Lock gates	12
	Tidal booms	12
	Service gates	12
	Suspended furnace/process plant doors	12

Statutory compliance

In most countries, legislation requires that cranes are *certificated* by a third party body or Classification Society. Typical areas covered are:

- Integrity of construction.
- Isolation arrangements.
- Emergency stop and control devices.
- Motor rating and tests.
- Insulation rating and condition.

These apply equally to in-service inspection activities, i.e. it is the inspector's job to make sure that the crane has not moved out of compliance as a result of any deterioration, repairs, etc.

Crane design classification

The common system of classification is set out in detail in codes such as ISO 4301-1 and the French Fédération Europiene de la Manutention (FEM) codes. It is also recognized in the more detailed design standards that apply to the structure and mechanisms. The main principle is that the crane *structure*, comprising the fabricated bridge girders and crab frame, has a different set of classes to the crane *mechanisms*: the rotating parts and connected components. There is no imposed direct dependency between the two. Figure 18.6 shows the idea.

Structure class

Typically, structure class (numbered A1 to A8) is determined by the combination of two design factors: the 'utilization' (U1 to U9) and the 'state of loading' (Q1 to Q4). The utilization factor relates to the projected number of operational lifting cycles of the crane structure and ranges in a series of preferred numbers from 3200 cycles (U1) up to 4×10^6 cycles and above (U9). The state-of-loading factor refers to the frequency with which the structure in use will actually experience the SWL: Q1 is infrequently and Q4 is where it lifts it very regularly. Figure 18.6 shows these factors expressed pictorially as a matrix (see individual crane codes for full details).

Mechanism class

This uses a similar principle but different letter designations. Mechanism class (M3 to M8) is determined by the 'utilization' factor, this time termed T1 to T9 and the 'state of loading' (L1 to L4). For mechanisms, the utilization factor is based on projected lifetime operating hours (400 hours for T1 up to 50 000 hours for T9) rather than operating cycles.

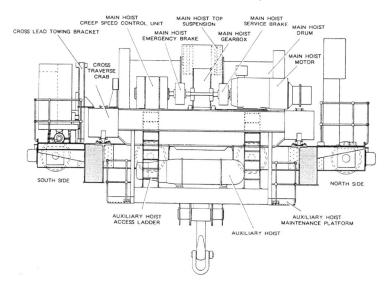

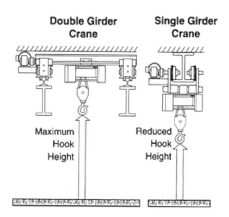

Fig. 18.5   Overhead crane construction details

The state-of-loading factor is on the same basis as that used for structures.

A commonly used standard is FEM *Rules for the design of hoisting appliances*. This is a French standard, known more generally as the 'FEM regulations'. The most relevant part to overhead cranes is Part 1, *Heavy lifting equipment*. The FEM document is predominantly a set of design rules (not unlike BS 2573 Part 1), concentrating on structural aspects.

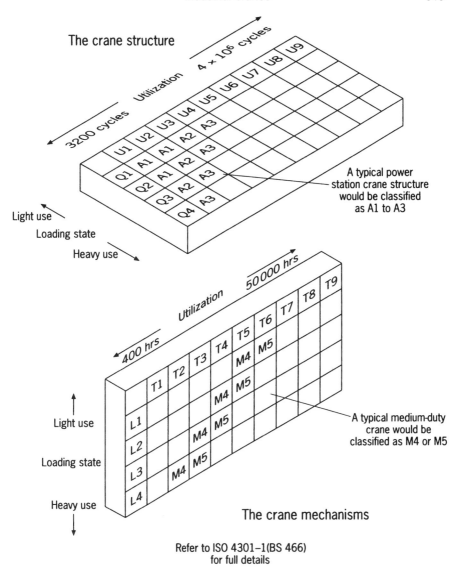

Fig. 18.6 Crane design classifications

Crane test, inspection, and repair

Inspection and test plans (ITPs)

When cranes are repaired, which they frequently are, the need for statutory certification imposes a significant pressure on all the involved parties to perform and monitor the repairs in a controlled way. This is done using an ITP. Important points are:

- The ITP is normally subdivided by *component parts* of the crane. Structural members, drum, and electrical equipment should each have their own section.
- Witness points shown on the ITP must include those relevant to the third-party certification body. Make sure the repair design appraisal activity is shown, as it is likely to be a legislative requirement.
- There will be a number of proprietary (bought-in) items such as the hook, wire ropes, and brake assemblies. These are included in the statutory certification requirements, so provision should be made to specify the correct certification requirements to sub-suppliers.
- Material traceability requirements are high. This will be reflected in the ITP entries showing material identification and marking activities. Some ITPs may specify a separate traceability mechanism for 'batch' materials that are drawn from the manufacturer's stock. Fig. 18.7 shows a typical crane repair/rebuild ITP.

Crane test procedures

With the exception of some very specialized designs, most cranes can be tested effectively on-site during periodic in-service inspections. Because cranes are subject to statutory certification it is normal for site tests to be witnessed by several inspection parties.

Crane visual and dimensional examination
The visual and dimensional examination can be carried out before or after load testing of the crane. In practice, the major measurements are best taken afterwards, as part of the checks for post-test distortion. The minimum checks normally made are as described below.

Roof clearance
An overhead crane, by definition, operates near the roofs of buildings and is designed so that it does not foul the structure or fittings such as ventilation ducts and lights. Check the roof clearance 'envelope' dimensions shown in the general arrangement (GA) drawing and make sure that there are no protuberances on the crane itself that exceed these dimensions.

Pendant length
Once installed, the crane will be controlled from the floor using a suspended pendant control. Check the installed length of pendant cable; it should be long enough to meet the specified pendant height, which will be shown on the general arrangement (GA) drawing.

Component	Check	Reference document	Acceptance standards	Witness M	C	TP
Crane bridge material	100% ultrasonic laminations scan; visual inspection	BS 5996	BS 5996	X		X
Fabrications (girders,	Review of weld procedures;	EN 287/288	EN 287/288	X	X	X
end-carriages, and hoist	Ultrasonic test 100%;	BS PD 5500 (typically)	BS PD 5500 table 5.7(2)	X	X	X
drum)	MPI 100%	BS PD 5500 (typically)	BS PD 5500 table 5.7(3)	X	X	X
	Visual/ dimensional examination	Drawing	Drawing tolerances	X	X	
Wire ropes	Proof test	BS 302	BS 302			X
Hoist components (load-bearing members)	Full document review	Material specification	Specification mechanical properties	X		X
Control panels	Functional test	ISO 4301-1	ISO 4301-1	X		
Motors	Insulation test			X		X
Proprietary items: wheels, rails, shafts, and gears	Full documentation review	Specification and ISO 4301-1	Specification and ISO 4301-1	X		X
Hook	Proof load test	BS 970	BS 970 (or type test certificate)	X		X
	Material tests	BS 970				
	Dimensional check	BS 970				
Crane assembly	Visual/ dimensional inspection	GA drawings	Checklist	X	X	X

Fig. 18.7   A typical crane repair/rebuild ITP

				M	C	TP
	Interim document review			X	X	X
Crane assembly	Light run check	ISO 4301-1		X	X	X
	Functional check	ISO 4301-1		X	X	X
	Overload test	ISO 4301-1	Checklist	X	X	X
	Insulation checks	ISO 4301-1		X	X	X
	Final certification review	ISO 4301-1		X	X	X
Crane assembly	Painting and Packing check	Purchase specification	Referenced paint standard	X	X	

M, manufacturer; C, contractor/purchaser; TP, third party.

Fig. 18.7   (Continued)

Insulation test

The technical standards, and certification organization, require that an insulation test be carried out on all the electrical components fitted to the crane. In practice this is normally performed on the motors and electrical panels only. The test is quite straightforward, a D:C voltage of twice the rated voltage is applied and the resistance to earth is measured. The minimum allowable resistance is $0.5\,M\Omega$. Make sure the test is carried out in dry conditions. You can also make a useful visual examination of the insulation to check for mechanical damage; pay particular attention to the conductors that run the length of the bridge, and to the loops of the pendant control cables that move with the crab.

The light run test

This is a no-load test. The purpose is to check the function of the various operating and safety systems before operating the crane in the loaded condition. Note that this is essentially a *safety procedure*; the function of the systems could be checked, if required, with the crane under load. The test includes the following elements.

Speed checks
The hoist motor, crab traverse motor, and longitudinal travel motor are operated (separately) to check that they are operating correctly and

provide the correct driving speeds. If speeds are not specified, check ISO 4301-1 for acceptable values. Specific points to watch are:

- Check the rotation of motors in both directions, including up and down and at 'inching' speed for the hoist. Use a tape measure and stop-watch to measure all the travel and hoisting speeds.
- There is a normal tolerance of ± 10 percent on allowable speeds. The preferred speed categories given in ISO 4301-1 are quite broad, so small differences are unlikely to be critical. It is preferable for speeds to be too slow, rather than too fast; this is particularly applicable to the long travel motion, where a too-high speed can significantly increase the inertia forces experienced as the crab or bridge is stopped when carrying full SWL.

Hook approach test
The hoist is wound up at slow speed until the hoist limit trip operates, isolating the power to the motor. Note that this is an adjustable limit (it works using an adjustable axial position switch on the hoist rope-guide) so after the test, make sure the setting is sealed to prevent accidental movement. The minimum acceptable hook approach distance will be stated in the crane specification or the technical schedules/data sheets. Some designs have an additional limit arrangement to control the 'lay' of the rope on the drum

Hoisting height
All crane design standards require that two full rope turns on the drum be classed as dead turns. This means there should be two empty grooves (sheaves) when the hook is at minimum approach distance and two full rope turns left on the drum when the hook is fully lowered. Sheaves can be checked for wear using special gauges.

Brake operation
Brakes are fitted on the hoisting, cross-traverse, and longitudinal travel mechanisms. Brake *efficiency* is only proven properly during the loaded test, however the light run test is a good opportunity to test the function of the brake mechanisms. Check for:

- *Fail-safe mode.* The brakes should *fail-on* when the power supply is cut off.
- *Free operation.* There should be no binding of the brakes when they are in the 'off' position. You can detect binding by excessive noise or heating of the brake housing, accompanied by high current reading from the relevant drive motor. Most disc and drum (older type) brakes can be adjusted.

SWL performance test

This is the main proving test of the operation of the crane (see Figs. 18.8 and 18.9). The load is a concrete block in a steel lifting frame or water-filled bags; expect the contractor to have a 'lifting schedule' that shows the weight of the individual components of the lift (frame, shackles, etc.) to demonstrate that the correct overall SWL is being used. A separate smaller weight will be required for the auxiliary hoist. The test consists of the following three elements.

Hoist up and down

The load is hoisted up and down several times, as near to the specified vertical limits of travel as is possible on the test rig. Important points to watch are:

- Use slow hoisting speed first and do not let the load swing too much.
- Check brake operation. Stopping should be smooth in both directions – there should be absolutely no juddering as the load comes to rest. Estimate the stopping distance using a tape measure or a chalk mark on the hoist drum. Pay particular attention to the 'worst case' condition, i.e. when the load is being lowered, and watch for any slipping of the brake. As a guide maximum braking slip should be not more than 7–8 mm per minute of hoisting speed.
- Check brake release. The brakes should release cleanly when the power is restored. Check the motor current again and listen for any undue noise.

The hoisting test is the best time for checking for any obvious problems with the crab – use the same checks as those shown below for the overload test. Note that it is necessary to do the checks *twice*, first during the SWL test and then again during the overload test. It is not good practice only to do them during the overload test – if there is a problem with cracking or yielding it will be useful to know whether it occurred during the SWL or the overload test.

Traverse and longitudinal travel

The main points to check are:

- Make sure that the crab is tested over several complete traverses of the bridge length. A single traverse is not always sufficient to show any problems.
- Check the *tracking* accuracy of the crab and the bridge rails. You can do this using an accurate measuring tape or feeler gauges. It is important to make sure that the clearances are the same at several points along the rail – this will show whether the rail is straight.

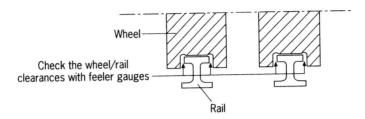

Wheel

Check the wheel/rail
clearances with feeler gauges

Rail

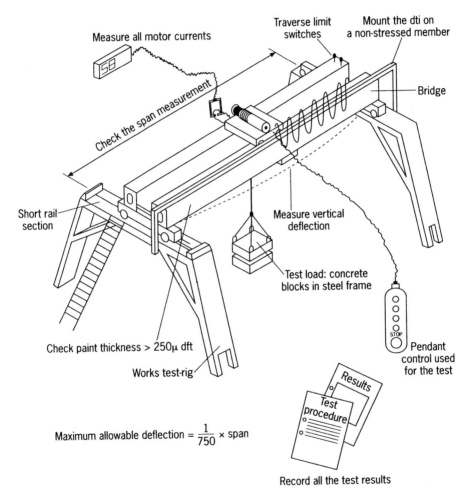

Measure all motor currents

Traverse limit
switches

Mount the dti on
a non-stressed member

Check the span measurement

Bridge

Short rail
section

Measure vertical
deflection

Test load: concrete
blocks in steel frame

Check paint thickness > 250μ dft

Works test-rig

Pendant
control used
for the test

Results

Test
procedure

$$\text{Maximum allowable deflection} = \frac{1}{750} \times \text{span}$$

Record all the test results

Fig. 18.8   The crane SWL performance test

Fig. 18.9   Load test of top-running crane

Acceptable tolerances on rail distortion are quite small because misalignment will cause quite rapid wear in use, causing the crab to become unstable (it will wobble) in some positions, most often at the end of the bridge.

- Check the operation of the crab traverse limit-stops. These are normally a simple microswitch. There should also be physical stops or buffers to stop the crab traverse if the limit switch becomes defective – they should be positioned accurately so the crab will contact both stops at the same time, hence spreading the impact load.
- Watch the crab closely as it stops after a full traverse, you will need to climb into the access gantry for this; it is of little use viewing it from the ground. Check what effect the inertia force has on the crab assembly. There should be no 'jumping' of the crab on the rails as the load swings, or visible movement of the hoist mechanisms bolted to the crab. Everything should be securely dowelled and bolted.

Deflection measurement

It is normal to measure vertical deflection of the bridge under SWL, as required by the design standard. Sometimes it is measured at overload conditions as well. Measure deflection at mid-span on the bridge using a dial test indicator (dti). Figure 18.8 shows the general arrangement. A good test should incorporate the following points:

• The 'undeflected' reading should be taken with the crab positioned at one end of the bridge.
• Do not take the deflection reading immediately after lifting the load, move the load up and down a few times and then wait a minute or so for the structure to settle.
• It is essential to have an accurate fixed datum point on which to mount the dti. The mounting point should not be connected in any way to the stressed structure. If it is, the deflection reading taken will be meaningless (you will get an undermeasurement). Usually, it is best to try and mount the dti on the access gantry or similar. A more difficult way is to measure the distance vertically from the underside of the bridge to some reference point on the floor. Double-check that the crab is at the exact mid-span of the bridge when measuring the bridge deflection. There should be a hard-stamped reference mark. If there isn't, make one before the test so that the test can be repeated accurately.
• Vertical deflection under SWL should be an absolute maximum of 1/750 of the span (longitudinal rail centre-to-centre distance). Anything more is a major non-conformance point.

The overload test

Crane design standards and most contract specifications specify an overload test at 125 percent of SWL. This acts as a proving test for the factors of safety incorporated into the mechanical design, and of the in-service conditions of the structure and mechanisms. Your task as an inspector during the overload test is to look for three things yielding (i.e. *plastic* distortion), cracking; and breakage. Surprisingly, none of these three are easy to find, they will only show themselves as a result of a structured approach to the test.

The overload test consists of lifting a 125 percent SWL weight to impose *static* stresses on the structure. Note that crane standards do not specify directly that the overload test should test for the structure's ability to withstand dynamic (inertia) loads resulting from movement in the overload condition. Dynamic stress concentration factors can be very large so beware of inertia loads – it is acceptable to use slow speeds for hoisting and traverse, and to wait until the load stops swinging before starting the next movement. You can measure vertical deflection during the overload test but this is generally not subject to an acceptance level.

The test should consist of a complete set of hoisting, crab traverse and, if applicable to the test rig, longitudinal travel movements. During and after the test, a full visual inspection of all the stressed components should be made. A brief analysis of the stress case will quickly indicate the areas of highest tensile and shear stresses. For the structural sections of the crab and bridge, areas of *stress concentration* such as fillet weld toes (particularly on stiffeners which are subject to a shear stress, transverse to a web axis), changes of section from thick to thin material, and sharp corners are the most common locations to find cracks. Fillet and gusset-pieces that do not have 'mouse-holes' at the enclosed corner are also common areas. Some useful points to note are:

- Make sure fillet welds are clean before the test so that you can see any subsequent cracking. Very thick paintwork that obliterates the weld is not a good idea; it can and will hide cracks.
- You should ask for a dye penetrant test of critical areas if you have evidence that there is any cracking at all. It is also useful to help detect any movement of the main hoist on its mountings. This is only a stand-by technique, however, and does not replace a proper dye penetrant or magnetic particle examination.
- Yielding (plastic deformation) is more difficult to detect, particularly if it does not progress to the extent where it causes cracking or breakage. The *only* quantitative way to detect this type of yielding is by direct measurement before and after loading the particular component. In practice, because of the size of overhead crane components compared with the accuracy of measurement (using a tape or straight-edge rule) the only yielding that you can identify with any certainty is that related to the bridge girders. The crab is much smaller and stiffer so the resulting small deflections are difficult to measure. For the bridge girders make sure, during the deflection check, that the dti reading returns to zero when the load is removed. This gives some comfort that the bridge has only been loaded within its elastic limit and that yielding has not occurred.
- Drive gear teeth. Gear teeth do not have such a high factor of safety as some of the structural parts of a crane. This is particularly the case if the mechanism design classification is in the M3 or M4 category. You should inspect all gear teeth for the main drives. The objective is

to look for tooth breakage or obvious visible cracks. Normally, if there is a problem with gear teeth you can expect to find teeth completely broken, or chipped due to overstressing or shock load – you are not looking for scuffing or any of the failure mechanisms associated with rotational wear.

- Load display. If this feature is fitted, it should be correctly indicating the overload during the test. Check this carefully; it is an important safety consideration.
- Rope grooves. Inspect the rope grooves on the hoisting drum after the test is finished. Look for any obvious abrasion of the groove lips that would indicate the rope guide mechanism is not adjusted correctly.
- Double-check for flaking paint on all components – this is a sure sign of excessive stress. The only possible exception to this is on the bridge girder, particularly if it is a single box section girder or offset crab type sometimes used on lightweight cranes. These may twist within their elastic limit causing some slight paint flaking on the box section corners around mid-span. This can be acceptable; but check the vertical deflection accurately to make sure that yielding has not occurred. Designs with double box section girder bridges should not twist or deflect enough to cause the paint to flake.

Codes and standards: cranes

ISO 4301-1: 1986	*Cranes and lifting appliances – Classification – Part 1: General.*
ISO 4302: 1981	*Cranes – Wind load assessment.*
ISO 4304: 1987	*Cranes other than mobile and floating cranes – General requirements for stability.*
ISO 4310: 1981	*Cranes – Test code and procedures.* Specifies the tests and procedures to be followed in order to verify that a crane conforms to its operational specifications and is capable of lifting rated loads. Where rated loads are determined by stability, a test procedure and test load are specified that permit stability margins to be easily verified. Defines test procedures such as conformity tests, visual inspection, and load lifting competence testing.
ISO 7363: 1986	*Cranes and lifting appliances – Technical characteristics and acceptance documents.*
ISO 7752-1: 1983	*Lifting appliances – Controls – Layout and characteristics – Part 1: General principles.*

ISO 8686-1: 1989	*Cranes – Design principles for loads and load combinations – Part 1: General.*
ISO 2374: 1983	*Lifting appliances – Range of maximum capacities for basic models.*
ISO 4306-1: 1990	*Cranes – Vocabulary – Part 1: General.*
ISO 7296-1: 1991	*Cranes – Graphic symbols – Part 1: General.*
ISO 11994: 1997	*Cranes – Availability – Vocabulary.*
ISO/TS 15696: 2000	*Cranes – List of equivalent terms.*
ISO 4308-1: 1986	*Cranes and lifting appliances – Selection of wire ropes – Part 1: General.*
ISO 4309: 1990	*Cranes – Wire ropes – Code of practice for examination and discard.*
ISO 9373: 1989	*Cranes and related equipment – Accuracy requirements for measuring parameters during testing.*
ISO 11630: 1997	*Cranes – Measurement of wheel alignment.*
ISO 9926-1: 1990	*Cranes – Training of drivers – Part 1: General.*
ISO 9927-1: 1994	*Cranes – Inspections – Part 1: General.*
ISO 9928-1: 1990	*Cranes – Crane driving manual – Part 1: General.*
ISO 9942-1: 1994	*Cranes – Information labels – Part 1: General.*
ISO 10973: 1995	*Cranes – Spare parts manual.*
ISO 12478-1: 1997	*Cranes – Maintenance manual – Part 1: General.*
ISO 12480-1: 1997	*Cranes – Safe use – Part 1: General.*
ISO 12482-1: 1995	*Cranes – Condition monitoring – Part 1: General.*
ISO 15513: 2000	*Cranes – Competency requirements for crane drivers (operators), slingers, signallers, and assessors.*
ISO 4301-2: 1985	*Lifting appliances – Classification – Part 2: Mobile cranes.*
ISO 4305: 1991	*Mobile cranes – Determination of stability.*
ISO 4306-2: 1994	*Cranes – Vocabulary – Part 2: Mobile cranes.*
ISO 4308-2: 1988	*Cranes and lifting appliances – Selection of wire ropes – Part 2: Mobile cranes – Coefficient of utilization.*
ISO 7296-1: 1991/ Amd 1:1996	
ISO 7296-2: 1996	*Cranes – Graphical symbols – Part 2: Mobile cranes*
ISO 7752-2: 1985	*Lifting appliances – Controls – Layout and characteristics – Part 2: Basic arrangement and requirements for mobile cranes.*
ISO 7752-2: 1985/ Add 1:1986	
ISO 8087: 1985	*Mobile cranes – Drum and sheave sizes.*
ISO 8566-2: 1995	*Cranes – Cabins – Part 2: Mobile cranes.*
ISO 10245-2: 1994	*Cranes – Limiting and indicating devices – Part 2: Mobile cranes.*

ISO 11660-2: 1994	*Cranes – Access, guards and restraints – Part 2: Mobile cranes.*
ISO 11661: 1998	*Mobile cranes – Presentation of rated capacity charts.*
ISO 11662-1: 1995	*Mobile cranes – Experimental determination of crane performance – Part 1: Tipping loads and radii.*
ISO 13200: 1995	*Cranes – Safety signs and hazard pictorials – General principles.*
ISO 4301-3: 1993	*Cranes – Classification – Part 3: Tower crane.*
ISO 4306-3: 1991	*Cranes – Vocabulary – Part 3: Tower cranes.*
ISO 7752-3: 1993	*Cranes – Controls – Layout and characteristics – Part 3: Tower cranes.*
ISO 8566-3: 1992	*Cranes – Cabins – Part 3: Tower cranes.*
ISO 8686-3: 1998	*Cranes – Design principles for loads and load combinations – Part 3: Tower cranes.*
ISO 9374-3: 2002	*Cranes – Information to be provided for enquiries, orders, offers and supply – Part 3: Tower cranes.*
ISO 9942-3: 1999	*Cranes – Information labels – Part 3: Tower cranes.*
ISO 10245-3: 1999	*Cranes – Limiting and indicating devices – Part 3: Tower cranes.*
ISO 11660-1: 1999	*Cranes – Access, guards and restraints – Part 1: General.*
ISO 11660-3: 1999	*Cranes – Access, guards and restraints – Part 3: Tower cranes.*
ISO 12485: 1998	*Tower cranes – Stability requirements.*
ISO 4301-4: 1989	*Cranes and related equipment – Classification – Part 4: Jib cranes.*
ISO 7752-4: 1989	*Cranes – Controls – Layout and characteristics – Part 4: Jib cranes.*
ISO 8566-4: 1998	*Cranes – Cabins – Part 4: Jib cranes.*
ISO 9374-1: 1989	*Cranes – Information to be provided – Part 1: General.*
ISO 9374-4: 1989	*Cranes – Information to be provided – Part 4: Jib cranes.*
ISO 10245-1: 1994	*Cranes – Limiting and indicating devices – Part 1: General.*
ISO 12210-1: 1998	*Cranes – Anchoring devices for in-service and out-of-service conditions – Part 1: General.*
ISO 12210-4: 1998	*Cranes – Anchoring devices for in-service and out-of-service conditions – Part 4: Jib cranes.*
ISO 4301-5: 1991	*Cranes – Classification – Part 5: Overhead travelling and portal bridge cranes.*
ISO 7752-5: 1985	*Lifting appliances – Controls – Layout and characteristics – Part 5: Overhead travelling cranes and portal bridge cranes.*

ISO 8306: 1985	*Cranes – Overhead travelling cranes and portal bridge cranes – Tolerances for cranes and tracks.*
ISO 8566-1: 1992	*Cranes – Cabins – Part 1: General.*
ISO 8566-5: 1992	*Cranes – Cabins – Part 5: Overhead travelling and portal bridge cranes.*
ISO 8686-5: 1992	*Cranes – Design principles for loads and load combinations – Part 5: Overhead travelling and portal bridge cranes.*
ISO 9374-5: 1991	*Cranes – Information to be provided – Part 5: Overhead travelling cranes and portal bridge cranes.*
ISO 10245-5: 1995	*Cranes – Limiting and indicating devices – Part 5: Overhead travelling and portal bridge cranes.*
ISO 10972-1: 1998	*Cranes – Requirements for mechanisms – Part 1: General.*
ISO 11660-5: 2001	*Cranes – Access, guards and restraints – Part 5: Bridge and gantry cranes.*

Chapter 19

Mobile cranes

Mobile cranes are subject to in-service inspection under the requirements of LOLER (Lifting Operations and Lifting Equipment Regulations) and similar legislation in other countries. They are covered by different technical standards to static cranes (overhead cranes, jib cranes, etc.) and are subject to different kinds of tests. As with static cranes the emphasis of in-service inspection and testing focuses on ensuring the *integrity* of the crane and its safety of operation. On balance, mobile cranes are involved in more accidents and fatalities than static cranes. It also seems that the quality of inspection (and inspectors) is more variable, ranging from the comprehensive to the superficial. The situation is better in countries that have in-service inspection legislation (e.g. like LOLER in the UK), but probably can still be improved.

Mobile crane types

Figure 19.1 shows the main types of mobile crane – they divide into wheeled (see Fig. 19.2) and tracked (crawler) types. Rail-mounted cranes are similar but have different legislative coverage. Wheel-mounted cranes may have either single or double-control status. The double-control type can be operated either from the truck driving cab or a separate crane control station that moves with the turntable. The single-control type is operated from the driving cab only. They have axles and rubber tyres. Sizes vary from small road vehicles to large commercial truck size. Crawler cranes have a similar rotating super-structure or turntable mounted on steel crawler tracks for low-speed travel. Crawler types have to be transported by truck when not in use on sites.

Most mobile cranes are of the simple boom type. The boom rotates and raises/lowers, enabling it to lift and swing loads at various radii. The boom may be fixed, telescopic, or extendable by fitting a fly-jib to the upper end. Figure 19.3 shows the main components of a boom-type

Truck-mounted crane: non-telescoping boom

Dragline crawler

Wheeled telescopic boom
crane (single control station)

Wheeled telescopic boom
crane (multiple control)

Truck-mounted crane: telescoping
boom

Fig. 19.1   Mobile crane types

mobile crane, and the common terminology used. It is common for cranes to be fitted with extendable outriggers (or stabilizers) to stabilize them when lifting and/or swinging a load. Figure 19.4 shows the 'work areas'.

For all but very large mobile cranes, a single diesel engine provides both motive power and crane movements (via a hydraulic pump and

Fig. 19.2 Mobile crane in use

distribution system driven, via a clutch, from the engine). For wheel-mounted cranes, the vehicle is normally road-legal with all the necessary lights, safety requirements and test certification.

Load rating

The main design aspect of any mobile crane is its safe *load rating*. This is a figure based on the maximum allowable load the crane should lift in various configurations. It is expressed as a *percentage*, relating to the load it can take without tipping. These figures relate to *static loads only*, i.e. they assume that there is no swinging or jerking of the suspended load, or any other dynamic effects. Table 19.1 shows some indicative values for different types of crane in normal configuration, i.e. without non-standard attachments such as fly-jibs. Use these ratings as a rough

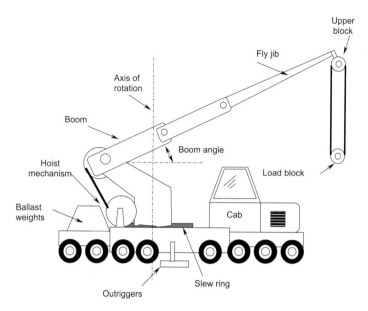

Fig. 19.3 Mobile crane – components and terminology

guide only – they can vary depending on the design standard and specification used.

In all cases the load rating applies to a condition in which the load is in the *least stable* condition relative to its mounting. Further details are provided in published technical standards such as ISO 11662-1: 1995 *Mobile cranes – Determination of crane performance*. Load rating checks are an important part of any in-service/pre-use inspection, particularly if the configuration of the crane has been changed or it has been modified in any way.

Mobile crane tests

The stability test

This is a test of the stability of the vehicle carrying the crane rather than the integrity of the crane structure and mechanical components themselves. Figure 19.5 outlines how it is done. Note that different procedures apply if the crane is designed to lift people rather than goods (e.g. a person-carrying 'cherry-picker'). Note that the least stable load position can vary between vehicle types – it will be shown on the design drawing, or often indicated by a sign or plate on the vehicle itself. It is normal to carry out a stability test for several configuration and loading

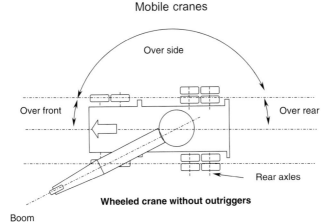

Wheeled crane without outriggers

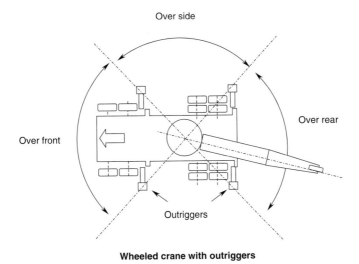

Wheeled crane with outriggers

Fig. 19.4   Mobile crane work areas

conditions, e.g. on wheels, outriggers fully extended, boom retracted, boom extended, etc.

The overload test

This is a test of the integrity of the crane under overload test conditions. Before the test, the crane should be subject to the same preliminary checks as for the stability test. Figure 19.6 shows a test in progress.

The stability test – points to check

- The vehicle should be standing on firm, level ground.
- Check the tyre pressures.

Table 19.1 Mobile crane load ratings

Type of crane	Maximum load rating (allowable load/load to cause tipping) in defined configuration(s)
Wheeled	
Without outriggers, resting on vehicle tyres	75%
Supported on fully extended outriggers, tyres lifted	85%
Crawler	
Without outriggers, resting on crawler tracks	75%
Supported on fully extended outriggers	85%

- If the test is in non-outrigger configuration the road-springs' locks should be on.
- The initial load lift should not exceed 110 percent of the load rating at the corresponding lift radius. This load should be lifted *just clear* of the ground.
- Weights are then added, *or* the radius increased, until the edge of the stability margin is reached. *Do not* increase further, i.e. until the vehicle starts to tip.
- Check that the radius indicator in the cab is reading accurately (check against a physical measurement from the load centre to the rotation axis).

Most overload tests are dynamic, i.e. while loaded, the crane is moved through all its operating movements (rotate, jib raise and lower, etc.) in order to subject the structure and mechanisms to *normal* dynamic stresses. This does not mean overstress caused by excessive speed of rotation, swinging or snatching of the load – cranes are *not* designed with sufficient factors of safety to resist the stresses that would result when overload and excessive dynamic stresses are combined.

Overload tests are only carried out if a proper technical consideration has been made both of the stresses induced during the test and the stability margin of the crane. An outline of the test is shown in Fig. 19.7.

Preliminary safety checks

These are the same as for the stability test, but with some additions:

- Check the manufacturer's loading/radius charts for the crane.

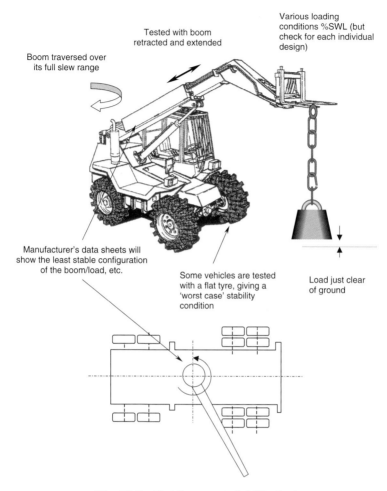

Boom traversed over
its full slew range

Tested with boom
retracted and extended

Various loading
conditions %SWL (but
check for each individual
design)

Manufacturer's data sheets will
show the least stable configuration
of the boom/load, etc.

Some vehicles are tested
with a flat tyre, giving a
'worst case' stability
condition

Load just clear
of ground

Fig. 19.5   Mobile crane stability test

- Set the boom at the maximum radius for the maximum load (do not confuse this with the maximum possible radius that can be achieved).
- With an overload of 110 percent lift the load just clear of the ground.
- Operate the crane through all its motions to ensure the overload is applied to all parts.
- Test that the hoist brake (and, where applicable, the derrick brake) is able to support the overload without slipping.
- Lift the overload just above the ground, using the *minimum* safe working radius of the boom.
- Lift the overload again at the *maximum* safe radius of the boom.

Fig. 19.6   Mobile crane load test

- Repeat the test for each variety of boom lengths and configuration of fly-jibs etc.
- After the test do a close visual and dimensional check for cracks, distortion, or yielding.

Note that the load is checked using a calibrated load cell located between the crane block and the load itself.

Visual/functional check

Periodic statutory in-service inspection of mobile cranes (under, for example, LOLER) includes a large number of structural and general visual checks. These are wide-ranging, being concerned with safety, integrity and function of the crane. While some are specified in relevant legislation, many are not, being left to the inspector's discretion and good engineering practice. Figure 19.8 gives an outline of typical checks that are made. Figure 19.9 is a typical checklist that can be used as a guideline. Remember that the scope of any individual inspection is governed by the content of the written scheme of examination which is written by the Competent Person responsible for the inspection of the crane.

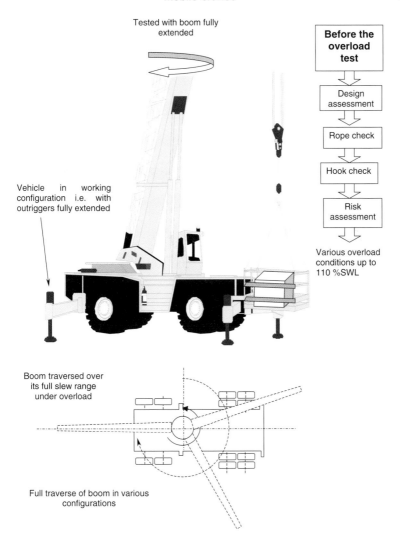

Tested with boom fully
extended

**Before the
overload
test**

Design
assessment

Rope check

Hook check

Risk
assessment

Various overload
conditions up to
110 %SWL

Vehicle in working
configuration i.e. with
outriggers fully extended

Boom traversed over
its full slew range
under overload

Full traverse of boom in various
configurations

Fig. 19.7   Mobile crane stability test

Rope inspection and replacement/rejection criteria

Owing to crane design configurations, sheave diameters, drum diameters, and rope design factors are limited. Because of this, inspection to detect deterioration and organise timely replacement is essential. Periodic inspections should be performed by the Competent Person and should cover the entire length of the rope. Only the surface wires of the rope need to be inspected. No attempt should be made to open the rope.

General

The following items shall be examined as part of all in-service inspections:

- All safety devices for function.
- All hydraulic hoses, and particularly those which flex during normal operation of crane functions.
- Hooks and latches for deformation, chemical damage, cracks, and wear.
- Rope reeving for compliance with manufacturer's specifications.
- Electrical apparatus for malfunctioning or signs of excessive deterioration, dirt, and moisture accumulation.
- Hydraulic system for proper oil level.
- Tyres for recommended inflation pressure, wear, or damage.
- Deformed, cracked, or corroded members in the crane structure and entire boom.
- Loose bolts or rivets.
- Cracked or worn sheaves and drums.
- Worn, cracked, or distorted parts such as pins, bearings, shafts, gears, rollers, and locking devices.
- Excessive wear on brake and clutch, system parts linings, pawls, and ratchets.
- Excessive wear of chain drive sprockets and excessive chain stretch.
- Travel steering, braking, and locking devices for malfunction.
- Crawler tracks for wear or misalignment.
- Signs shall be installed at the operator's station and on the outside of the crane warning that electrocution or serious bodily injury may occur unless a minimum clearance of 3 m is maintained between the crane or the load handled, and energized power lines (up to 50 kV).

Operator control station

- Level gauge(s) to be accurate and easily read by operator.
- All gauges, i.e. RPM, fuel, pressure, temperature, etc., to operate correctly.
- Safety belts to be fitted to all single-operator control stations and in lower station multiple-control station.
- Load charts to be legible and relate to crane by model number and serial number.
- Load movement indicators to be fully operational and reflect the crane load charts. Lock-out systems must operate properly.
- Control levers to be checked for return to neutral position and delay in actuation.

Fig. 19.8   Visual and functional inspection points for mobile cranes

- Cab glazing should be safety-glass material. Windows should be provided in the front and on both sides of the cab or operator's compartment with visibility forward and to either side. Visibility forward should include a vertical range adequate to cover the boom point at all times. The front window may have a section which can be readily removed or held open, if desired. If the section is of the type held in the open position, it should be secured to prevent inadvertent closure. A windscreen wiper should be provided on the front window.
- Cab doors should be restrained from inadvertent opening or closing while travelling or operating the machine. The door adjacent to the operator, if of the swinging type, should open outwards. If it is the sliding type, it should slide rearwards to open.
- A clear passageway should be provided from the operator's station to an exit door on the operator's side.
- Principal walking surfaces should be covered in skid-resistant material.
- Outside platforms should be provided with guard-rails. If platforms are too narrow to use guard-rails, handholds should be provided at convenient points above the platform.
- On all crawler and wheel-mounted cranes, handholds, steps, or both, should be provided as needed to facilitate entrance to and exit from the operator's cab.
- Where necessary for rigging or service requirements, a ladder or steps should be provided to give access to the cab roof.
- Controls for load hoist, boom hoist, swing, and boom telescope (when applicable) should be provided with means for holding in the neutral position, without the use of positive latches.
- Travel distance on hand levers should not be greater than about 350 mm from neutral position on two-way levers and not greater than about 600 mm on one-way levers.

Outriggers

- Means should be provided to hold all outriggers in the retracted position while travelling, and in the extended position when set for operating.
- Power-actuated jacks, where used, should be provided with the means (such as integral load hold check valves on hydraulic cylinders, mechanical locks, etc.) to prevent loss of support under load.
- Each power-operated outrigger should be visible from its actuating location unless the operator is assisted by a signal person.

Fig. 19.8   (Continued)

Boom hoist

- Boom hoist mechanism. The boom hoist may use a rope drum for its drive or hydraulic cylinder(s) and the supporting structure may be gantry or the same hydraulic cylinder(s) used to elevate the boom.
- The boom hoist should be capable of elevating and controlling the boom with its rated load (for rope boom hoists when reeved according to the manufacturer's specifications) and be capable of supporting the boom and rated load without action by the operator.
- In a rope-supporting and elevating arrangement boom lowering should be done only under power control. Free-fall lowering of the boom is normally not permitted.
- The boom hoist drum should have sufficient rope capacity to operate the boom in all positions, from the lowest permissible to the highest recommended when using the manufacturer's recommended reeving and rope size. No less than two full wraps of rope should remain on the drum with the boom point lowered to the level of the crane-supporting surface. The drum end of the rope should be anchored to the drum by an arrangement specified by the crane or rope manufacturer.
- The drum should provide a first layer rope pitch diameter of not less than 15 times the nominal diameter of the rope used.
- On rope boom support machines a braking mechanism and a ratchet and pawl or other locking device should be provided to prevent inadvertent lowering of the boom.
- An integrally mounted holding device (such as a load hold check valve) should be provided with boom support hydraulic cylinder(s) to prevent uncontrolled lowering of the boom in the event of a hydraulic system failure.

Load hoist

- The hoist mechanism may consist of a drum with necessary rope reeving or hydraulic cylinder(s).
- The load hoist drum assemblies should have power and operational characteristics sufficient to perform all load lifting and lowering functions required in crane service when operated under recommended conditions.
- Where brakes and clutches are used to control the motion of the load hoist drums, they should be of a size and thermal capacity sufficient to control all rated crane loads with minimum recommended reeving. Where maximum rated loads are being lowered with near-maximum boom length or operations involving long lowering distances, power-

Fig. 19.8 (Continued)

controlled lowering is usually desirable to reduce demand on the load brake. Brakes and clutches should be provided with adjustments where necessary to compensate for lining wear and to maintain force in springs, where used.
- Drum rotation indicators should be provided.

Load hoist brakes

- When power-operated brakes having no continuous mechanical linkage between actuating and braking means are used for controlling loads, an automatic means should be provided to set the brake to prevent the load from falling in the event of loss of brake control power.
- Foot-operated brake pedals should be constructed so that the operator's feet, when in proper position, will not slip off. A means should also be provided for holding the brakes in the applied position without further action by the operator.

Telescoping boom

- Extension and retraction of boom sections may be accomplished through hydraulic, mechanical, or manual means.
- The powered retract function should be capable of controlling any rated load which can be retracted.
- An integrally mounted holding device (such as a load hold check valve) should be provided with the telescope's hydraulic cylinder(s) to prevent uncontrolled retraction of the boom in the event of a hydraulic system failure (e.g. supply hose).

Swing mechanism

- The swing mechanism should start and stop with controlled acceleration and deceleration.
- A braking means with holding power in both directions should be provided to restrict movement of the rotating superstructure when desired during normal operation. The braking means should be capable of being set in the holding position and remaining so without further action by the operator.
- A device or boom support should be provided to prevent the boom and superstructure from rotating when in transit. It should be constructed to minimize inadvertent engagement or disengagement.

Crane travel

- Travel control: on all crane types with a single control station, the controls for the travel function should be located at the operator's station.

Fig. 19.8   (Continued)

- On wheel-mounted multiple control station cranes, the travel controls should be located in the carrier cab. Auxiliary travel controls may also be provided in the crane cab. If there is an operator in the crane cab when the crane is travelling, communication should be provided between the cabs. Use of audible signalling devices will meet this requirement.

Travel brakes and locks

- On crawler cranes, brakes or other locking means should be provided to hold the machine stationary during working cycles on a level grade or while the machine is standing on the maximum grade recommended for travel. Such brakes or locks must remain in engagement in the event of loss of operating pressure or power.

Reeving accessories

- Eye splices should be made in a manner recommended by the rope or crane manufacturer, and rope thimbles should be used in the eye.
- Wire rope clips should be drop-forged steel of the single-saddle (U-bolt) or double-saddle type clip. Malleable cast iron clips are not normally used.
- Wire rope clips attached with U-bolts have the U-bolt over the dead end of the live rope resting in the clip saddle.

Sheaves

- Sheave grooves should be free of surface defects which could cause rope damage. Flange rims should run true about the axis of rotation.
- Sheaves carrying ropes which can be momentarily unloaded should be provided with close-fitting guards or other devices to guide the rope back into the groove when the load is reapplied.
- The sheaves in the lower load block should be equipped with close-fitting guards that will prevent ropes from becoming fouled when the block is lying on the ground with loose ropes.
- All sheave bearings (except for permanently lubricated bearings) should be provided with means for lubrication.
- Boom-hoisting sheaves should have pitch diameters of not less than 15 times the nominal diameter of the rope used.
- Load-hoisting sheaves should have pitch diameters of not less than 18 times the nominal diameter of the rope used.

Fig. 19.8   (Continued)

Any deterioration resulting in an appreciable loss of original strength should be noted and determination made as to whether further use of the rope would constitute a hazard. Indicative guidelines for rope replacement are as follows:

- In running ropes, six randomly distributed broken wires in one lay or three broken wires in one strand in one lay.
- In rotation-resistant ropes, two randomly distributed broken wires in six-rope diameters or four randomly distributed broken wires in thirty-rope diameters.
- One outer wire broken at the point of contact with the core of the rope which has worked its way out of the rope structure and protrudes or loops out from the rope structure. Additional inspection of this section is required.
- Wear of 33 percent of the original diameter of outside individual wires.
- Kinking, crushing, birdcaging, or any other damage resulting in distortion of the rope structure.
- Evidence of heat damage from any cause.
- Reduction from nominal diameter of more than:
  - 0.4 mm for diameters up to and including 8 mm.
  - 0.8 mm for diameters of 9.5 mm to and including 13 mm.
  - 1.2 mm for diameters of 14.5 mm to and including 19 mm.
  - 1.66 mm for diameters of 22.0 mm to and including 29 mm.
  - 2.4 mm for diameters of 32.0 mm to and including 38 mm.

- In standing ropes, more than two broken wires in one lay in sections beyond end connections or more than one broken wire at an end connection.
- Replacement rope must have a strength rating at least as great as the original rope supplied or recommended by the crane manufacturer.

Inspection checklist

Figure 19.9 gives a typical inspection checklist for a mobile crane.

Common terminology

The following list shows typical specialized terminology used for mobile cranes.

- *Angle indicator (boom)*. An accessory which measures the angle of the boom to the horizontal.

TYRES AND WHEELS

TREAD/
TREADMOUNTING
SIDEWALLS/CRACKS
LUG NUTS
HUB END SEALS
AIR PRESSURE

BRAKES (MECHANICAL)

SWING (FOOT AND HAND)
HOIST (MAIN AND AUX.)

BOOM HOIST

SYSTEM OIL LEAKS
BRAKES(AIR ASSISTED)

HOSES (CONDITION AND LEAKS)
LINE FITTINGS/VALVES
PARKING (MAN.)
ROAD BRAKES

OUTRIGGERS
CYLINDER SEALS (BEAM)
CYLINDER SEALS (JACKS)
HYD. HOSE AND FITTINGS
OUTRIGGER PADS AND RETAINERS
HYDRAULIC OIL LEAKS
OUTRIGGER'S BOX SECTIONS
JACK RETAINER PINS

JACK LOCKING HANDLE
HOLDING VALVES

SUPERSTRUCTURE
SWING MOTION
MOTION CONTROL LEVERS ☐

ELECTROCUTION WARNING ☐
BOOM CRADLE INDICATING SWITCH ☐ ☐
STEERING SYSTEM ☐
ALL LINKAGES ☐ ☐
HOSES AND FITTINGS ☐ ☐
CYLINDERS/VALVES ☐ ☐
REAR WHEEL ALIGNMENT LAMP ☐

BOOM ☐
OPERATION (TELESCOPING) ☐ ☐
OPERATION (RAISE/LOWER) ☐ ☐
MANUAL PIN EXTENSION USE ☐ ☐
SWING AWAY (IN USE) ☐
SWING AWAY (STORED) ☐ ☐

WEAR PADS ☐ ☐
HEEL PIN AND BUSHINGS ☐ ☐
LIFT CYLINDER PINS AND BUSHINGS ☐ ☐
TIP SHEAVES AND SHAFTS ☐
LATTICE TYPE ☐ ☐
MAIN CHORD (TUBULAR) ☐ ☐
MAIN CHORD (ANGLE) ☐ ☐
LACING ☐ ☐

PENDANTS ☐ ☐
BRIDLES ☐ ☐

'A' FRAME (PINS LINKS/ ASSEMBLY) ☐ ☐
GANTRY MAST ☐ ☐
FLY EXTENSION ☐ ☐
BACK STOPS
RELAY SHEAVES/SHAFT ☐
BOOM EXTENSION PINS ☐ ☐
AND COTTERS ☐

LOAD RANGE CHARTS	☐	**HOIST AND BOOM CABLES**	
		MAIN HOIST (REEVED) Y N	☐
CRANE CONTROL VALVES	☐	AUX. HOIST (REEVED)	☐
BOOM RAISE LIMITS	☐	Y N	
		BOOM HOIST	☐
BOOM BACKSTOP TUBES	☐	HOIST UNITS	☐
SWING BEARING AND	☐		
ASSEMBLY		PLANETARY GEAR (FREE	☐
COUNTER WEIGHT	☐	FALL)	
MOUNTING		**LOAD BLOCKS** (MAIN AND	
COUNTER WEIGHT	☐	AUX.)	
FUNCTION		SHEAVES/SHAFTS	☐
CLUTCH FUNCTION	☐	HOOK	
GUARDS	☐	SAFETY LATCH	☐
WELDS	☐	SWIVEL	☐
ELECTRICAL		CROSS-HEAD	☐
GAUGES (INSTRUMENT	☐		
PANEL)		ANCHOR BECKET	☐
WIRING	☐	CABLE DEAD END	☐
LIGHTS/INDICATORS/	☐		
HAZARD		HEADACHE BALL	☐
LIGHTS/REVERSING	☐		
AND AUDIBLE		AUX. HOIST FLAG	☐
SAFETY DEVICES		**GENERAL**	
LOAD INDICATOR	☐	ALL FUNCTIONS	☐
ANGLE INDICATOR	☐	(OPERATION)	
		HYDRAULIC OIL LEVEL	☐
WARNING (AUDIBLE)	☐	ALL PANELS	☐
ANTI TWO-BLOCK SYSTEM	☐	(SUPERSTRUCTURE)	
		DOORS/WINDOWS/	☐
REELING INDICATOR (MAIN	☐	COWLING	
AND AUX.)		FIRE EXTINGUISHER	☐
BOOM OVER REAR ALARM	☐	ENGINE OIL LEAKS	☐
DOOR BACK LATCH	☐	HOUSEKEEPING (CABS)	☐
POSITIVE SWING LOCK	☐	HORN	☐
REAR AXLE LOCKS	☐		
LOAD HOOK TIE BACK	☐		

Fig. 19.9 Inspection check list – mobile crane

- *Anti two-block device*. A device which, when activated, disengages all crane functions whose movement can cause two-blocking.
- *Auxiliary hoist*. A secondary hoist rope system used either in conjunction with, or independently of, the main hoist system.
- *Axis of rotation*. The vertical axis around which the crane super-structure rotates.
- *Ballast*. Weight used to supplement the weight of the machine in providing stability for lifting working loads. The term 'ballast' is normally associated with locomotive cranes.
- *Boom (crane)*. A member hinged to the rotating superstructure and used for supporting the hoisting tackle.
- *Boom angle*. The angle above or below horizontal of the longitudinal axis of the base boom section.
- *Boom hoist mechanism*. Means for supporting the boom and controlling the boom angle.
- *Boom point*. The outer extremity of the crane boom, containing the hoist sheave assembly.
- *Boom point sheave assembly*. An assembly of sheaves and pin built as an integral part of the boom point.
- *Boom stop*. A device used to limit the angle of the boom at the highest recommended position.
- *Brake*. A device used for retarding or stopping motion.
- *Cab*. A housing which covers the rotating superstructure machinery, or the operator or driver's station
- *Cross-over points*. In multiple-layer spooling or rope on a drum, those points of rope contact where the rope crosses the preceding rope layer.
- *Drum*. The cylindrical member around which a rope is wound for lifting and lowering the load or boom.
- *Dynamic (loading)*. Loads introduced into the machine or its components due to accelerating or decelerating forces.
- *Flange point*. A point of contact between rope and drum flange where the rope changes layers.
- *Hoist mechanism*. A hoist drum and rope reeving system used for lifting and lowering loads.
- *Jib*. An extension attached to the boom point to provide added boom length for lifting specified loads. The job may be in line with the boom or offset to various angles in the vertical plane of the boom.
- *Jib backstop*. A device which will restrain the jib from turning over backward.

- *Load block lower.* The assembly of hook or shackle, swivel, sheaves, pins, and frame suspended by the hoisting ropes.
- *Load block upper.* The assembly of shackle, swivel, sheaves, pins, and frame suspended from the boom point.
- *Load indicator.* A device that measures the weight of load.
- *Load ratings.* Crane rating in kilogrammes established by the manufacturer.
- *Mast (boom).* A frame hinged at or near the boom hinge for use in connection with supporting a boom. The head of the mast is usually supported and raised or lowered by the boom hoist ropes.
- *Mast (jib).* A frame hinged at or near the boom point for use in connection with supporting a jib in normal operating conditions.
- *Outriggers.* Extendable or fixed members attached to the mounting base which rest on supports at the outer ends used to support the crane.
- *Reeving.* A rope system in which the rope travels around drums and sheaves.
- *Rotation-resistant rope.* A wire rope consisting of an inner layer of strand laid in one direction covered by a layer of strand laid in the opposite direction. This has the effect of counteracting torque by reducing the tendency of the finished rope to rotate.
- *Running rope.* A rope which traces around sheaves or drums.
- *Side loading.* A load applied to an angle to the vertical plane of the boom.
- *Stabilizer.* Stabilizers are extendable or fixed members attached to the mounting base to increase the stability of the crane, but which may not have the capability of relieving all of the weight from wheels or tracks.
- *Swing.* Rotation of the superstructure for movement of loads in a horizontal direction about the axis of rotation.
- *Swing mechanism.* The machinery involved in providing rotation of the superstructure.
- *Two-blocking.* The condition in which the lower load block or hook assembly comes in contact with the upper load block or boom point assembly.

Mobile crane standards

ISO 11662-1: 1995   *Mobile cranes – Experimental determination of crane performance – Part 1: Tipping loads and radii.*

ISO 11660-2: 1994	*Cranes – Access, guards and restraints – Part 2: Mobile cranes.*
SAE J 1257	*Rating chart for cantilevered boom cranes.*
SAE J 103	*Cantilevered boom crane structures – Method of test.*
ISO 4301-2: 1985	*Lifting appliances – Classification – Part 2: Mobile cranes.*
ISO 7752-2: 1985	*Lifting appliances – Controls – Layout and characteristics – Part 2: Basic arrangement and requirements for mobile cranes.*
PREN 13000:	*Cranes – Safety – Mobile cranes.*
DIN EN 13000	*Cranes – Safety – Mobile cranes*; German version of prEN 13000 1997.
DEF-1422-A	*Cranes, road mobile, fully slewing (2 to 6 tons inclusive maximum free load) (04.66).*
SAE J 987	*Rope supported lattice-type boom crane structures – Method of test.*
SAE J 881	*Lifting crane sheave and drum sizes.*
SAE J 959	*Lifting crane, wire-rope strength factors.*
ISO 11661: 1998	*Mobile cranes – Presentations of rated capacity charts.*

Chapter 20

Passenger and goods lifts

All lifts (elevators) provided for use in a work situation are covered by statutory regulations such as the Lifting Operations and Lifting Equipment Regulations (LOLER). This applies irrespective of whether the equipment is designed to carry people or goods – and also to derived equipment such as hoists, mobile platforms (e.g. for window cleaning), and similar. The responsibilities of the 'Duty Holder' (a formal term introduced in LOLER) include:

- Using a 'Competent Person' or organization to inspect and report on the equipment.
- Maintaining the lift so that it remains safe to use.
- Ensuring that the lift is examined at statutory intervals in accordance with a written scheme of examination (WSE) drawn up by a Competent Person.
- Ensuring that various documentation complies with the regulations.
- Keeping the correct records.

Figure 20.1 outlines key requirements.

Lift types

Figures 20.2 and 20.3 show the two common types of lifts in use. There are many different variations within each type but the fundamental principles are much the same. The key inspection-related issues relate to the mechanisms and safety devices installed on the equipment such as:

- Car/landing doors and their interlocks.
- Gearbox and drive-system components.
- Suspension ropes or chains.
- Overload detection devices.
- Electrical devices (earth bonding, fuses, etc.)
- Braking systems (including buffers and overspeed devices).
- Hydraulic and governor systems.

The thorough examination

The law requires that all lifts when in use should be thoroughly examined:

- After substantial and significant changes have been made.
- At least every 6 months if the lift is used at any time to carry people, every 12 months if it only carries loads, or in accordance with an examination scheme; and
- Following 'exceptional circumstances' such as damage to, or failure of, the lift, long periods out of use or a major change in operating conditions which is likely to affect the integrity of the equipment.

Note: when first installed, new lifts do not require any initial thorough examination as long as they have been manufactured and installed in accordance with the Lifts Regulations 1997 and have a current declaration of conformity, i.e. made not more than 12 months before.

Examination schemes

As an alternative to thorough examinations at statutory intervals, the Competent Person may draw up an 'examination scheme'. The scheme may specify periods which are different from statutory intervals, but this must be based on a rigorous assessment of the risks. An examination scheme may be particularly appropriate if a lift is used infrequently for light goods.

Action following notification of defects	Documentation
The Competent Person is legally required to notify the duty Holder as soon as possible following a thorough examination of any defects which are, or could soon become, dangerous.	The Competent Person is legally required to send a written and signed report of the thorough examination as soon as practicable. This should normally be within 28 days, but if there is a serious defect which needs to be addressed it should be sent much sooner.
Once a serious and significant defect has been reported the Duty Holder should immediately take the lift out of service until the fault has been addressed.	
The Competent Person may also notify of defects which need to be made good within a certain time scale. In this case, the Duty Holder should take steps to have the defective equipment repaired or replaced within the specified time, and not use the lift after that time unless the defect has been satisfactorily remedied.	If the Competent Person identifies a defect which presents an 'existing or imminent risk of serious personal injury' they are also legally required to send a copy of the report to the enforcing authority.
	By law, the report must contain certain information, specified in Schedule 1 of LOLER (see Fig 18.4).

Fig. 20.1   Passenger and goods lifts – some key LOLER points

These areas are responsible for perhaps 99 percent of failures and accidents involving standard passenger and goods lifts. It is rare that structural failure of fabricated items is the root cause of a defect or accident. The situation is different, however, for special hoists, mobile platforms, etc. which rely heavily on welding and structural joints for their integrity.

Lift inspections

The easiest way to perform in-service inspection of lifts is to work from a comprehensive checklist. Figures 20.4 to 20.6 show typical lists – these cover all the critical inspection points for hydraulic and electrohydraulic types. Figure 20.7 shows a typical inspection schedule summary from a written scheme of examination.

Bibliography

Further information

BS 5655: 1986: Part 10#Specification for the Testing and Inspection of Electric and Hydraulic Lifts. British Standards are available from BSI Customer Services, 389 Chiswick High Road, London W4 4AL.

BS 5655 (EN 81) Lifts:

Part 1 *Safety Rules for Construction and Installation of Electric Lifts.*
Part 2 *Safety Rules for Construction and Installation of Hydraulic Lifts.*
Part 3 *Specification for Electric Service Lifts.*
Part 4 Not issued.
Part 5 *Specification for Dimensions of Standard Lift Arrangements.*
Part 6 *Code of Practice for Selection and Installation.*
Part 7 *Specification for Manual Control Devices, Indicators and Additional Fittings.*
Part 8 *Specification for Eye Bolts for Lift Suspension.*
Part 9 *Specification for Guide Rails.*
Part 10 *Specification for The Testing and Inspection of Electric/ Hydraulic Lifts.*
Part 11 *Recommendation for the Installation of New, and The Modernization of, Electric Lifts in Existing Buildings.*
Part 12 *Recommendation for the Installation of New, and The Modernization of,* Hydraulic Lifts in Existing Buildings.

BS 5776 *Stair Lifts Domestic Powered.*

BS5900 *Powered Domestic Lifts (Subject of Revision Consideration).*

BS 5965 *Home Lifts Manually Operated.*

BS 6440 *Powered Lifting Platforms, for Disabled People.*

BS 7255 *Safe Working on Lifts (Appendix 1).*

Guidelines on the Thorough Examination and Testing of Lifts (SAFed) lifts guidelines LG1, 1998 (Safety Assessment Federation, (ISBN 190-1212-35-1). Available from Safety Assessment Federation Limited, Nutmeg House, 60 Gainsford Street, Butlers Wharf, London, SE1 2NY.

The Health and Safety at Work Act 1974.

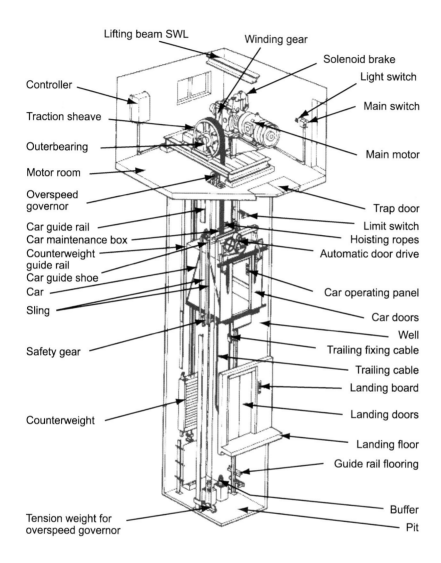

Fig. 20.2   Electric traction lift

Safe Use of Lifting Equipment: The Lifting Operations and Lifting Equipment Regulations 1998. Approved Code of Practice and Evidence, 1998 L113 (HSE Books, ISBN 0-7176-1628-2).

Safe Use of Work Equipment: The Provision and Use of Work Equipment Regulations 1998. Approved Code of Practice and Guidance, L22, 1998 Second edition (HSE Books, ISBN 0-7176-1626-6).

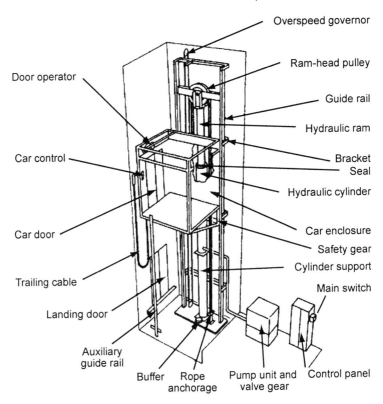

Overspeed governor

Ram-head pulley

Door operator

Guide rail

Hydraulic ram

Car control

Bracket
Seal

Hydraulic cylinder

Car door

Car enclosure

Safety gear

Cylinder support

Trailing cable

Main switch

Landing door

Auxiliary
guide rail

Buffer   Rope   Pump unit and   Control panel
           anchorage   valve gear

Fig. 20.3   Hydraulic-type lift

Motor vehicle repair equipment

Motor vehicle repair (MVR) equipment is covered by a combination of LOLER and PUWER (Provision and Use of Work Equipment Regulations) 1998. All tools and machinery used at work are covered by PUWER while only that equipment to lift a load above the ground is covered by LOLER. Table 20.1 summarizes the situation for common MVR equipment. Note that the information in this figure is based on standard equipment – where equipment is modified or used for special or unusual purposes, an individual risk assessment has to be carried out. Key points on various pieces of MVR equipment are given below.

Vehicle hoists/lifts

Under LOLER, vehicle hoists/lifts are 'lifting equipment' and subject to Regulation 9 thorough examinations. Because people will almost

A typical examination procedure for manual lifts is as follows:

1. Check shaft/enclosure for adequately secured cladding.
2. Check pit for water, rubbish and debris.
3. Check car, counter weight guides, shoes (as applicable) for wear and damage, and verify fitment and security of clamps, fixing bolts, and fittings.
4. Check car sling for damage, distortions, and cracks, and ensure security of all parts and fitment of bolts.
5. Check car and counterweight structures for damage and deterioration and ensure correct fitment and security of parts.
6. Check landing doors and car doors if fitted for adequate protection, wear, damage, or deterioration. Ensure freedom of movement of locking arrangement.
7. Check ropes/chains for wear, damage, or deterioration, and terminations for correct fitment and security of parts.
8. Check drive arrangement and sustaining mechanism for correct operation, wear, damage, distortion, or deterioration, and ensure security of attachments.

Fig. 20.4 Electro-hydraulic lift – inspection checklist

1.0 **Motor room**
1.1 The access door should be substantial and secure.
1.2 Warning notices are correctly worded and the specified dimensions.
1.3 Isolation switches are conveniently positioned and clearly marked.
1.4 Fuses/circuit breakers are rated for function.
1.5 Lighting is adequate.
1.6 Overhead beams are correctly positioned and secure.
1.7 Emergency lowering/recovery instructions are present and legible together with any tools necessary for emergency purposes including landing door release key.

2.0 **Controller**
2.1 Fuses/circuit breakers are rated for function.
2.2 Overloads for setting and operation.
2.3 Operation and setting of phase failure device.
2.4 Correct operation of all mechanical parts/linkage.

Fig. 20.5 Hydraulic lift – inspection checklist

2.5   Controller switches and relays for worn/burnt contacts and tips caused by arcing.
2.6   Deterioration of flexible wiring to switches/moving parts.
2.7   Resistances, transformers, and timers for condition.
2.8   Confirm the existence and operation of a moveable link on the up/down contractors.
2.9   All fixed power wiring (field) is in good condition, all earth connections are sound and any VIR wiring remains fit for purpose.

3.0   **Governor**
3.1   Confirm data plate against rope fitted.
3.2   All linkages are free to operate if provided.
3.3   Check pulley grooves for excessive wear if provided.
3.4   Correct operation of all electrical switches.
3.5   Security and effectiveness of gripping devices/centrifugal switches.

4.0   **Hydraulic circuitry**
4.1   Check for correct operation of hydraulic system emergency lowering valve, rupture/low-pressure switch, pump, and hoses.
4.2   Check valves and fitting for correct reservoir level.
4.3   Check for leakage damage deterioration wear and security and fitment of parts and ensure correct operation.
4.4   For indirect acting type lift. Check divertor sheaves, bearings, shafts, and slack rope switch for wear, damage deterioration security/fitment of parts and ensure correct operation.

5.0   **Lift shaft**
5.1   Correct operation of all stop switches.
5.2   Top/bottom overall limits.
5.3   Emergency hatch electrical/mechanical interlock if fitted.
5.4   Car top control and shaft and car lighting modern lifts usually have the shaft light switch fitted in the lift pit.
5.5   Electric/mechanical interlock on access door to pit.
5.6   Check landing door interlocks for correct operation/adjustment and check condition of electrical wiring.
5.7   Check landing door frame fixings/door rollers/bottom shoes for damage/wear.
5.8   Check clearance between door panels and frames for trap hazard.

Fig. 20.5   (Continued)

5.9   Check for damage to mid-bars and pickets in lattice gates and efficient operation of closers.

5.10   Ensure mechanical connection of slave door (two-panel doors). Recommend fitting of slave door safety switch if none fitted.

5.11   Check integrity of emergency release device (should be tamper-proof) anti-kick rollers, bottom track sills, and toe guards.

5.12   Check shaft electrical wiring and trailing flexes for wear damage and deterioration.

5.13   Check guide rails for alignment and security of fixings.

6.0   **Lift car**

6.1   Check car structure and sling/frame for damage, distortion, cracks, and security of parts/ fitment of bolts.

6.2   Check guide shoes for wear and any adjustments necessary.

6.3   Check for adequate clearances between car and shaft equipment.

7.0   **Safety gear**

7.1   Check safety gear governor/safety ropes for wear, corrosion, and tension. Check anchorages and verify return pulley and electrical switch operation.

7.2   Check safety gear/operating linkage for damage, adjustment, and ease of operation.

7.3   Check operation and adjustment of safety gear switch.

8.0   **Suspension ropes or chains**

8.1   For indirect acting type lifts, check suspension ropes for wear, broken wires, corrosion, and equal tension. Check anchorage and ensure correct fittings and number of rope grips.

9.0   For direct acting type lifts, check ram fixing to car for damage and security/fitment of parts.

10.0   Check cylinder, piston, seals, and hoses for deformation damage, scores, leaks, and deterioration. Check security/ fitment of parts/anchorage and fitment of non-return valve.

11.0   **Pit**

11.1   Check for ingress of water and rubbish and that shaft lights are working.

11.2   Compensating ropes (if fitted) and tension switch.

11.3   Adequate clearance (run-by) under car.

Fig. 20.5   (Continued)

11.4	Operation of buffers switches and load switches (if fitted to car).
11.5	Position and operation of emergency stop switch (stop and lock-off type).
11.6	Provision of access ladder where appropriate.
11.7	Provision and operation of shaft lighting switch and fittings.
11.8	Provision of a pit prop or guide lock pins.
12.0	**Observe operation of**
12.1	Lift, place a call at every landing served in both directions of travel.
12.2	Floor levelling accuracy and re-levelling feature if fitted.
12.3	Door operation, landing and car.
12.4	Door protective devices safety edges and proximity detectors.
12.5	Door timing.
12.6	Push buttons.
12.7	Indicator lights.
12.8	Alarm.
12.9	Telephone.
12.10	Collective group systems.
12.11	Car lighting and emergency lighting (if fitted).
12.12	Check load plate against car floor area.
12.13	For goods-only lifts which are not intended to carry passengers and have not controls within the lift cage, ensure notices are displayed prohibiting persons to ride in the lift.

Fig. 20.5   (Continued)

invariably either work beneath or be inside a vehicle raised on a lift, a 6-monthly period for thorough examination is appropriate, in accordance with a written scheme or examination prepared by the Competent Person.

Body alignment jigs

Many alignment jigs used in body repairs have features similar to those of vehicle lifts. Similar considerations will apply, though persons may not always or regularly work underneath particular jigs. If this is the case, then a 12-monthly period for thorough examination is sometimes used.

1.0   **Motor room**
1.1   The access door should be substantial and secure.
1.2   Warning notices are correctly worded and the specified dimensions.
1.3   Isolation switches are conveniently positioned and clearly marked.
1.4   Fuses/circuits breakers are rated for function.
1.5   Lighting is adequate.
1.6   Moving parts are coloured yellow and where necessary guarded.
1.7   Divertor pulleys are fit for purpose.
1.8   Trap doors are robust and all holes are protected with upstanding edging to prevent items falling down the hoistway.
1.9   Overhead beams are correctly positioned and secure.
1.10   Emergency lowering/recovery instructions are present and legible together with any tools necessary for emergency purposes including a door release key.

2.0   **Controller/selector**
2.1   Fuses/circuit breakers are rated for function.
2.2   Overloads for setting and operation, and where appropriate dashpot oil levels.
2.3   Operation and setting of phase failure device.
2.4   Correct operation of all mechanical parts/linkage.
2.5   Controller switches and relays for worn/burnt contacts and tips caused by arcing.
2.6   Deterioration of flexible wiring to switches/moving parts.
2.7   Resistances, transformers, and timers for condition.
2.8   Confirm the existence and operation of a moveable link on the up/down contactors.
2.9   All fixed power wiring (field) is in good condition, all earth connections are sound and any VIR wiring remains fit for purpose.

3.0   **Floor selector**
3.1   Confirm effective operation.
3.2   Adjustment and integrity of any chain, tape, or wire including return sheave and drum bearings.
3.3   Brushes and contacts for wear/pitting.

Fig. 20.6   Electric traction lift – inspection checklist

4.0   Governor
4.1   Confirm data plate against rope fitted.
4.2   All linkages are free to operate.
4.3   Check pulley grooves for excessive wear.

5.0   Drive machinery
5.1   Gearbox gearing for wear, excessive free movement (backlash), and thrust in the worn shaft.
5.2   Check for oil loss and correct level.
5.3   Check security and integrity of worm wheel to shaft securing bolts and traction sheave bolts.
5.4   Integrity of keys and key stops.
5.5   Look for developing cracks in sheaves, spider, key-ways and shaft steps in diameter.
5.6   Traction sheaves for wear in the groove.
5.7   Bottoming of the ropes on the traction sheave grooves and rope slip.
5.8   Bottoming of the ropes on the divertor sheave grooves.
5.9   For drum lifts, security of drum rope anchorage.
5.10   Presence and operation of slack rope switch (drum lift).
5.11   Adequate rope left on drum when car is on the pit buffers.

6.0   Brakes
6.1   Check friction lining wear.
6.2   Oil/carbon black contamination on the friction linings.
6.3   Presence and safe operation of all linkage and clevis pin arrangement.
6.4   Springs for wear and that only compression springs are fitted.
6.5   Radial centralizing bolts are adjusted to prevent trailing on the drum.
6.6   No residual magnetism is present after power-off condition is selected.
6.7   Check security of holding down bolts.

7.0   Drive motors
7.1   Check for bearing noise and vibration.
7.2   Oil levels and contamination of the rotor/stator; field coils/armature windings.
7.3   Brushgear/commutators (micron level)/sliprings for sparking/thrown solder/arc burns/tracking.
7.4   Drive belt tension/condition.

Fig. 20.6   (Continued)

7.5   Hand winding wheel is smooth (no cranked levers) direction up and down clearly marked.

7.6   If tacho-generator fitted securely, belt drive wear and function of belt failure switch.

8.0   **Lift shaft**

8.1   Correct operation and type of all stop switches (stop and lock-off types).

8.2   Top and bottom overtravel switches.

8.3   Emergency hatch electrical/mechanical interlock.

8.4   Car top controls and lighting conform to required standards.

8.5   Effective operation of all electric/mechanical interlocks fitted to any access door to the shaft.

8.6   Check integrity of landing door fixings, rollers, shoes/spuds for damage/wear.

8.7   Landing door air cords for wear, splintering, and adjustment.

8.8   Clearance at door frame/panels for trapping hazards.

8.9   Damage to mid-bars and pickets in lattice gates.

8.10   Effective door closers to eliminate the possibility of landing doors being inadvertently left open.

8.11   Confirm the integrity of emergency door release mechanism, anti-kick rollers and bottom track sills and toe guards.

8.12   Suspension ropes for wear, broken wires, corrosion, equal tensions and the number of and correct fitting of rope grips.

8.13   Governor and safety ropes for wear, corrosion, tension, anchorage, and wear at the return sheave.

8.14   Check guide rails for tightness, alignment, security lubrication, and shoe/insert wear.

8.15   Shaft electrical wiring and trailing cables for wear/damage.

8.16   Check enclosure for gaps.

9.0   **Pit**

9.1   Check for ingress of water and build up of rubbish.

9.2   Compensating ropes and tension switch (if fitted).

9.3   Adequate (run-by) clearances over and under car and counterweight.

9.4   Operation of buffer switches and load switches if provided.

9.5   Position and operation of emergency stop switch.

9.6   Access ladder if provided.

10.0   **Counterweight**

10.1   Security of filler weights.

Fig. 20.6   (Continued)

10.2   Guide shoes for wear/adjustment.
10.3   Adequate running clearance between counterweight, car and shaft equipment.
10.4   Wear on rope pulleys, bearings, and rope grooves.
10.5   Rope attachment clips.

11.0   Safety gear
11.1   Identify the type and operational parameters
11.2   Prove the safety gear linkage for stiffness and distortion.
11.3   Confirm operation and adjustment of safety gear switch and ultimate limit switch if provided.
11.4   Check operating rope and pulley for corrosion, lubrication and freedom to operate.

12.0   Lift car
12.1   Check car structure and sling/frame for damage distortion, cracks, and security of parts/fitment of bolts.
12.2   Guide shoes for wear and any adjustments necessary.
12.3   Check for adequate clearance between car and shaft equipment.

13.0   General observations
13.1   Run test the lift before and after the examination.
13.2   Check the floor levelling accuracy: ±6 mm for modern lifts; ±15 mm for old single-speed lifts.
13.3   Door operation for smooth and effective operation.
13.4   Landing door closing device when the car is away from the landing.
13.5   Door protective devices, i.e. light beam, sensitive edges, and infra-red detection zones.
13.6   Door timing device for duration and advanced door opener feature.
13.7   Landing push buttons.
13.8   Car push buttons including car top station.
13.9   Indicator lights.
13.10   Alarm, community care system, or telephone.
13.11   Group system, i.e. simplex, group, etc. is fully operational.
13.12   Car lighting including emergency lighting if provided.
13.13   Load plate and identification number.
13.14   For goods-only lifts, with no controls in the car, that a notice warning 'No Passengers Goods Only' prohibits passengers from riding in the car as these lifts are unlikely to be fitted with safety gear.

Fig. 20.6   (Continued)

Examination/test	1 Year E	5 Years E	10 Years E
Earth continuity			◇
Electrical safety device	◇	◇	◇
Terminal speed reduction system			◇
Landing door interlocks	◇	◇	◇
Geared machines			
Shafts and plain bearings			◇
Overspeed governors			
Calibration		◇	◇
Actuation		◇	◇
Safety gear			
Reduced speed		◇	
Rated speed or 1.6 m/s			◇
Over speeding of ascending car			◇
Energy dissipation buffers			
Slow speed	◇	◇	◇
Rated speed or 1.6 m/s			
Car overload detection devices	◇	◇	◇
Hydraulic systems			
Hydraulic rupture/restrictor			
Valves			
Hydraulic cylinders			
Electrical anti-creep device			
Mechanical anti-creep device			
Low-pressure detection devices			
Traction brake and levelling			
Brake condition	◇	◇	◇
Brake dynamic			◇
Levelling			◇

Fig. 20.7   Typical written scheme content for an electro-hydraulic lift

Table 20.1   Summary of recommended inspection/thorough examination requirements for equipment used in motor vehicle repair

Work equipment	LOLER Regulation 9 'Thorough examination'	PUWER Regulation 6 'Inspection'
Axle stands	—	12 months
Body alignment jigs	12 months	12 months
Chain blocks	12 months	
Cab/body tilt mechanism	—	12 months
Cranes:		
Lorry mounted	12 months	
Mobile crane	12 months	—
Engine hoist	12 months	—
Engine stands	—	12 months
Engine lifting brackets (if accessory for lifting)	6 months	—
Eye bolts	6 months	—
Fork-lift truck (FLT)	6 or 12 months	—
Fork extensions	6 months	—
Working platform for FLT	6 months	—
Gearbox lifting table	12 months	
Hydraulic press	—	12 months
Jacks:		
Bottle jacks	12 months	—
Trolley jacks	12 months	
Lifting slings: chain/webbing	6 months	—
Pallets	—	No
Ramps		12 months
Recovery truck:		
'A'-frame crane	12 months	12 months
Jib crane	12 months	
Spectacle lift	12 months	
Towing dollies	12 months	
Rolling road brake tester	—	12 months
Skips	No	No
Stillages	No	No
Tail lifts	6 months	—
Tipper rams	—	12 months
Tow ropes	—	6 months
Towing bars	—	6 months
Vehicle hoists/lifts	6 months	—
Wheel lifters	12 months	—
Winch	—	12 months

Vehicle trolley and bottle jacks

When used as part of garage equipment, jacks should be regarded as lifting equipment. Jacks provided as part of the equipment of a motor vehicle will not normally be subject to LOLER unless they were intended to be used as part of garage equipment.

Engine lifting brackets

Where lifting brackets on vehicle engines are permanently attached to the engine, they will be regarded as part of the load and not as lifting equipment. Those brackets which are kept as part of the garage equipment are treated as 'accessories for lifting' and are therefore subject to LOLER.

Lifting equipment on vehicles

LOLER applies to lifting equipment mounted on vehicles, such as loader cranes fitted to assist with delivery of goods and materials. British Standard BS 7121: 1997 *Code of practice for the safe use of cranes, Part 4 – Lorry loaders*, and the Association of Lorry Loader Manufacturers and Importers of Great Britain: *Code of practice for the installation, application and operation of lorry loaders* provide further information.

Equipment which lifts part of a vehicle but not a load, for example cab tilt and tipper body mechanisms, is not subject to LOLER but a risk assessment should be carried out to determine whether failure could cause injury and if regular inspection under PUWER is required.

Handling aids

Handling aids such as wheel lifters, gearbox lifting tables, and similar devices are considered as lifting equipment under LOLER. Often, however, most will be at the low end of the risk scale and suitable candidates for schemes of thorough in-house examination by suitably independent and impartial Competent Persons.

Pallets, stillages, and skips

These are regarded as parts of loads and not accessories for lifting. Although they will not be subject to LOLER Regulation 9 thorough examinations, Regulation 4 requires that they should be of adequate strength.

Recovery vehicles and associated equipment

Recovery vehicles that have lifting devices which are very obviously cranes should be treated as such under LOLER. Moveable beds and ramps will not normally be subject to LOLER, nor will winches, as their main purpose is dragging a casualty vehicle over level ground rather then lifting it up.

Spectacle frames used for recovering cars and light commercial vehicles are considered subject to LOLER if they lift the vehicle and support it while towing. Towing dollies are not subject to LOLER as the weight is supported on a pair of small wheels. British Standard BS 7121: 1999 *Code of practice for safe use of cranes, Part 12 – Recovery vehicles and equipment* provides further information.

Fork-lift trucks

Fork-lift trucks (FLTs) used in motor vehicle repair are classed as both mobile work equipment and 'lifting equipment'. There are specific requirements in PUWER for rider-operated 'mobile work equipment', particularly to minimize the risks from its rolling over. New or hired FLTs must be provided with a roll-over protective structure and seat restraints, or other suitable devices such as enclosed cabs.

Thorough examination requirements of LOLER cover the parts of an FLT that are lifting equipment (including the mast, forks, and chains) and will depend upon both usage and application. As a general rule, FLTs operating *more than* 40 hours per week, or being used to lift people, or which have a side shift or attachments fitted need to be examined every 6 months. The frequency for trucks operating for *up to* 40 hours which do not have a side shift or attachment fitted will normally be 12-monthly.

Interchangeable equipment used on FLTs such as jibs, fork extensions, and working platforms are treated as 'accessories for lifting', and thoroughly examined every 6 months.

Bibliography

Freight Containers (Safety Convention) Regulations 1984.
Management of health and safety at work: Management of Health and Safety at Work Regulations 1998 Approved Code of Practice and Guidance L21.
Safe Use of Lifting Equipment: The *Lifting Operations and Lifting Equipment Regulations 1998* Approved Code of Practice and Guidance L113, 1998 (MSE Books, ISBN 0-7176-1628-2).

Safe Use of Work equipment: The *Provision and Use of Work Equipment Regulations 1998*. Approved Code of Practice and Guidance L22, 1998 (MSE Books, ISBN 0-7176-1626-6).

Chapter 21

Small industrial lifting tackle

'Small lifting tackle' is a generic term used to describe those industrial items that are used as *accessories* during a lifting operation, i.e. excluding cranes. The main items of equipment are as shown in Fig. 21.1. This is not an exhaustive list – there are probably a hundred or so separate items and variations that could be correctly included. This type of equipment and its use is heavily covered by a raft of statutory requirements, regulations, and technical standards. The origin of these lies with the fact that, historically, lifting tackle has been responsible for a large proportion of industrial accidents. There are two points worthy of note about this regulatory background:

- It has all developed organically, over time. So, like any other part of the legal system that has developed like this (i.e. all of it), it is not *entirely* consistent, and contains repetitions, overlaps, and even the occasional inconsistency.
- Of all the regulations and standards that apply, some are more technically useful than others. Some offer a useful benchmark against which to structure in-service inspection activity while others are simply part of the overall legal background.

On balance, however, the regulatory regime that covers small industrial lifting tackle is very comprehensive. It is also more *prescriptive* than for many other types of engineering plant and, to date remains more resistant to self-regulating approaches such as risk-based inspection (RBI) (see Chapter 7) when deciding inspection periodicities.

Lifting appliances	Lifting accessories
Cranes	Wire rope slings
Fork-lift trucks	Chain slings
Lifts	Man-made fibre slings
Suspended cradles	Hooks and fittings
Powered hoists	Swivels
Manual hoists	Shackles
Lever hoists	Eyebolts
Rope hoists	Rigging screws
Beam trolleys	Wedge sockets
Beam clamps	Plate clamp
Sheave blocks	
Winches	
Runway beams	

Fig. 21.1   Lifting equipment and accessories – summary

Statutory requirements

The main statutory instrument is:

- SI 1998: *The Lifting Operations and Lifting Equipment Regulation 1998 (LOLER)*. This is supported by an approved code of practice (ACoP) as follows:
- LOLER Approved Code of Practice and Guidance L113, 1998 (HSE Books, ISBN 0-7176-1628-2).

LOLER requirements – what are they?

The overall requirements of LOLER have a major influence on the role of the in-service inspector involved in lifting tackle and equipment. Figure 18.3 shows an outline. Note two points:

- Figure 18.3 is an outline, not a full statement of statutory requirements (these are contained in the LOLER regulations themselves and the accompanying ACoP).
- The requirements of Fig. 18.3 apply to both lifting tackle/accessories and lifting appliances (cranes).

Figure 21.2 summarizes the situation regarding the requirements for testing and thorough examinations. (See also the overall LOLER inspection periodicity table in Chapter 20, Table 20.1).

1. All equipment should be of good construction, sound material, adequate strength, and free from defects.
2. Equipment should be designed so that it is *safe* when used.
3. All equipment should have the SWL clearly marked on it, or in the case of wire ropes and chains, a suitable means of identifying the SWL.
4. Such equipment shall be suitably marked so that it can be uniquely identified.
5. Equipment should be maintained in compliance with points 1, 2, 3, and 4, and records kept of such maintenance.
6. Prior to being used for the first time, the equipment should be thoroughly examined by a Competent Person and certified to this effect. A similar examination is to be carried out on the equipment. This certificate should be kept with other such certificates and listing of all other lifting equipment on the site (often referred to as the 'General Register').
7. Ensure that the equipment is certified on a periodic basis as required by statute or as required by a Competent Person (manufacturers or codes of practice may specify more frequent intervals than those laid down by statute).
8. All lifting equipment must be risk-assessed and procedures put in place to reduce the risks associated with its use. Such procedures must be documented in the company safety statement.
9. Ensure employees using the equipment have been trained in its correct use, and the possible hazards associated with its use and misuse. Employees should also be made aware of their obligation to report defects in any such equipment without unreasonable delay.
10. A failure of a load-bearing structure or container is reportable to the Health and Safety Executive (HSE).

Fig. 21.2 Summary of statutory requirements – lifting tackle and accessories

The role of the in-service inspectors

When inspecting small industrial lifting tackle (or any type of lifting equipment) an inspector can fulfil two possible roles:

- Inspector as '*inspector*' (perhaps for pure commercial, technical or quality assurance (QA) reasons).
- Inspector as '*Competent Person*'.

There are important legal differences between the two – as well as some similarities.

The inspector as 'inspector'

This is the simplest type of inspection role in which the purpose of the inspection is a straightforward check for inventory, commercial quality (i.e. compliance with a specification) or as part of a manufacturing QA programme. The key issue is that the inspection is sufficiently *remote* from the activity of 'putting into use' of the equipment that it can be considered unrelated to it. By implication, this means that further inspections will be expected (and necessary) before the equipment is used.

Theoretically, in such situations, an inspector is not bound by the various reporting and documentation requirements of LOLER. In reality, however, if there were to be a subsequent failure or accident involving the equipment, an inspector may be asked to justify any actions (or lack of them) that were non-compliant with LOLER, on the basis that LOLER represents accepted good practice. Hence it is therefore wise to assume that *any inspection* of lifting equipment should be carried out *with due regard to the requirements of LOLER*, whether or not they are considered to apply to the actual inspection performed.

The inspector as 'Competent Person'

This is a clear-cut role where an inspector is commissioned to provide pre-use or in-service inspection under the requirements of LOLER. The essence of the regulations are that:

- *Pre-use inspection.* All lifting equipment must be inspected and certified by a Competent Person prior to being used for the first time.
- *In-service inspection.* All lifting equipment must be inspected and certified by a Competent Person after any repair to its load-bearing structure and at periodic intervals throughout its life.
- *Competent Persons.* The statutes do not define absolutely what a Competent Person actually is. In practice it is generally understood as, someone who has:

  - appropriate training/qualification (relevant to lifting equipment;
  - sufficient practical experience (relevant to lifting equipment);
  - familiarity with relevant codes of practice and standards for lifting equipment;
  - the ability to decide the correct inspection/testing procedures necessary to certify the equipment.

- *Certification.* This involves a visual examination for signs of damage that may compromise the safe use of the equipment. It may also involve checking that the recommended maintenance has been properly done. In some instances additional non-destructive testing (NDT) may be required. Once the inspection and testing are complete, the Competent Person fills out the necessary forms and certificates. The minimum content of these is given in the LOLER ACoP and its referenced documents.

Notwithstanding these extensive statutory requirements, the inspection of small lifting tackle is not *that* complicated. It is nevertheless still an area where inspectors should be *thorough*. In some industries, items of small lifting tackle are frequently ignored, badly stored, or have been subjected to mechanical damage or overload. It is the role of the inspector to find damage when it exists and to make sure the equipment is safe to use.

The following sections show some guidelines for the inspection of small tackle items.

Sheave blocks

Sheave blocks are available in the form of single sheave blocks or multi-sheave blocks with a selection of head fittings, the most popular being the swivel shackle or swivel oval eye (see Fig. 21.3). The main function of a single-sheave block is to change the direction of the hoisting or pulling rope whereas multi-sheave blocks reduce the necessary pull required on the lead rope, to lift the load, i.e. the pull required reduces as the number of falls (sheaves) increases. The main consideration when selecting single-sheave blocks is the load to be lifted and the resultant load on the head fitting, i.e. the load plus the line pull (which will be slightly greater then the load due to friction in the sheaves). The resultant load increases as the angle between the ropes decreases. Figure 21.4 shows outline guidance for sheave block inspection.

Winches

Winches can be of the electric, pneumatic, or the simple manually operated type (Fig. 21.5). One of the most important technical issues is selecting the correct winch for the job, but this is an equipment selection issue rather than, strictly, the role of the in-service inspector. Figure 21.6 outlines some technical checks required during pre-use and in-service inspections.

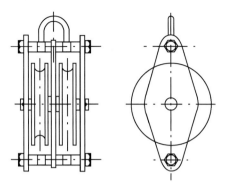

Fig. 21.3   A sheave block

- Check that the SWL is adequate for the load.
- Check that the colour coding (where applicable) is current and the block has a plant number/ID mark.
- Examine sheaves for wear in the rope groove, or any cracks or distortion.
- Try to lift the sheaves to check the bearings/bushes for wear.
- Spin the sheaves to check bearing/bushes and ensure smoothness of operation.
- Ensure all grease ports are clean and unblocked and the machine is well lubricated.
- Examine swivel head fittings and check for wear/stretch.
- Examine (if possible) the thrust bearing/washer and ensure smoothness of operation.
- Examine upper load pin/spigots and check for wear/distortion.
- If the head fitting is of the shank type, check the security of shank and nut and examine for stretch/distortion. Examine the cross-head for wear.
- Examine the head fitting shackle/eye, checking for wear, stretch, or cracking.
- Examine the side plate/straps and check for distortion, wear, or cracking (especially around the main load pin hole and top suspension hole).
- Ensure there are no sharp edges or burrs in the side plates which may be detrimental to the wire rope.
- Check all spacers and tie bolts and ensure that they are not deformed.
- Using a sheave gauge, check the rope groove for compatibility with the winch rope.

Fig. 21.4   Sheave block inspection – areas to check

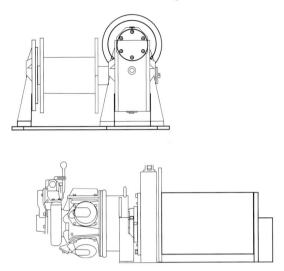

Fig. 21.5   A small winch

Hydraulic rams and jacks

One feature of hydraulic rams and jacks is that they may not have an SWL as such as they are not always subjected to a proof load before use (Fig. 21.7). They do, however, have a 'nominal lifting capacity' which is marked on the equipment. Figure 21.8 outlines some inspection points.

Wire slings

Slings are used in single or multi-leg lifting arrangements (Fig. 21.9). The wire ropes themselves can be of several types (number of strands, types of core, etc.) and standard tables are available showing how the strength of each wire design is affected by its diameter, construction and the tensile strength of the individual wire strands. Figure 21.10 outlines some inspection activities for wire that has been assembled into slings.

Eyebolts

There are several different designs of eyebolts (Fig. 21.11) governed mainly by the material, size and shape of the eye and whether or not it has a collar. Eyebolts frequently suffer damage in use. Figure 21.12 outline some inspection points.

Table 21.1 gives a summary of LOLER requirements for lifting tackle and lifting appliances.

Prior to operating the winch, the following checks should be made:

- The SWL is adequate for the load.
- The colour coding (where applicable) is current and the winch has a plant number/ID mark.
- Examine the rope guard and ensure there is no damage/distortion which may obstruct and/or abrade the winch rope.
- Where possible, examine the winch drum and check for wear, distortion, or cracks.
- Examine the brake bands and drums and check for wear.
- Ensure the bands and drums are clean and free from contamination.
- Where fitted, examine the automatic brake and check springs, link arms, and pins.
- Examine the exposed portion of the piston rod and check for corrosion.
- With power disconnected, check that all the operating levers return to neutral when released.
- Ensure that directional arrows/markings are in place and clearly visible.
- Check the oil level.
- Examine the winch base and check for cracked welds, cracks around bolt holes, distortion, or impact damage.
- Anchorage. Ensure the hold down bolts/welding/clamping are adequate and that the support steelwork has no indications of deterioration.
- Visually examine the winch wire and check for:

  - wear and corrosion;
  - abrasion;
  - mechanical damage (e.g. crushing) and broken wires.

Fig. 21.6   Winch inspection – areas to check

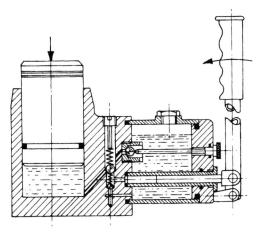

Fig. 21.7   A hydraulic jack

The following checks should be carried out:

- The capacity is adequate for the load.
- The colour coding is current and the cylinder/jack has a plant number/ID mark.
- Examine the body of the cylinder/jack and check for impact damage, cracks, and oil leaks. (With cylinders, examine inlet/outlet couplings and check for leakage.)
- Operate the cylinder/jack, pumping the ram to its full stroke.
- Examine the ram and check for scoring and corrosion.
- If the ram is threaded externally and fitted with a locking collar, examine the threads and check for stretch.
- Examine the seals and check for oil leakage.
- Turn the valve to lower/release and ensure the ram goes down smoothly (jerkiness could indicate distortion to the ram).
- Examine all hoses and fittings and ensure they are not perished, cut, or in any way damaged.
- With the ram in the lowered position check the oil level of the jack/pump unit.
- Function test the pump and ensure the valve does not leak when closed and under pressure.
- Where claw attachments are fitted, examine for distortion/cracks.
- Where gauges are fitted, check for leaks, function test and ensure needles return to zero.

Fig. 21.8   Hydraulic ram and jack inspection – areas to check

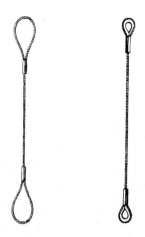

Fig. 21.9   Wire slings

The following checks should be made:

- The SWL is adequate for the load.
- The colour coding (where applicable) is current and the sling has a plant number/ID mark.
- Examine each individual leg along its entire length and check for wear, corrosion, abrasion, mechanical damage, and broken wires.
- Examine each ferrule and ensure the correct size of ferrule has been fitted.
- Check that the end of the loop does not terminate inside the ferrule (i.e. the rope end should protrude no more than one-third of the diameter) unless the ferrule is of the longer tapered design which has an internal step.
- The ferrule should be free from cracks or other deformities.
- Examine each thimble and check for correct fitting, snagging damage, and elongation (stretched thimbles/eyes could indicate possible overload).
- Examine the wire rope around the thimbles as it is often abraded due to the sling being dragged over rough surfaces.
- If fitted with hooks, check for wear, corrosion, and cracking and ensure the safety latch functions.

Fig. 21.10   Wire sling inspection – areas to check

Reference standards

ISO/DIS 1834	*Short link chain for lifting purposes – General conditions of acceptance* (Revision of ISO 1834: 1980).
ISO 3266: 1984	*Eyebolts for general lifting purposes.*
ISO 4309: 1990	*Cranes – Wires – Code of practice for examination and discard.*
ISO 2374: 1983	*Lifting Appliances – Range of maximum capacities for basic models.*
ISO 4301: 1986	*Cranes and Lifting Appliances – Classification – Part 1: General.*
ISO 4305: 1991	*Mobile Cranes – Determination of stability.*
ISO 11530: 1993	*Road Vehicles – Hydraulic jacks – Specification.*
ISO/DIS 15386	*Mobil Cranes – Operational specifications and rated load lifting capability – Verification tests.*
BS 7255: 1989	*Code of practice for safe working of lifts.*
BS 6570: 1986	*Code of practice for the selection, care and maintenance of steel wire ropes.*
BS 1290: 1983	*Specification for wire rope slings and sling legs for general lifting purposes.*
BS 3144: 1959	*Specification for alloy steel chain, grade 80, polished short link calibrated load chain for pulley blocks.*
BS 6210: 1983	*Code of practice for the safe use of wire ropes slings for general lifting purposes.*
BS AU 161: Part 1b: 1983	*Vehicle lifts. Specification for fixed lifts.*
BS AU 161: Part 2: 1989	*Vehicle lifts. Specification for mobile lifts.*
BS 4465:	*Specification for design and construction of electric hoists for both passengers and materials.*
BS 6109: Part 1: 1981	*Tail lifts, mobile lifts and ramps associated with vehicles. Code of practice for tail lifts.*
BS 2903: 1980	*Specification for tensile steel hooks for chains, slings, blocks, and general engineering purposes.*

Chapter 22

Diesel engines

It is by no means easy to inspect a diesel engine properly. There are many different types of engine which, particularly in the large sizes, can display a bewildering array of complex design features. Under nearly all legislation around the world, land-based diesel engines are *not* classed as statutory equipment and therefore are not subject to the requirements of a written scheme of examination. In Europe they are specifically exempt from the requirements of the PED and the PSSRs as they are considered a 'prime mover', even though the engine cylinders are, in reality, pressure vessels with high-integrity requirements. In-service inspection of diesel engines is required, however, on ships; this falls under the control of marine classification societies. They may also be subject to emission control legislation, both on ships and in land installations. Most in-service inspections of diesels involve either emission tests (while the engine is operating) or examination during stripdown, either after a brake test run, or as part of the rebuild procedure, possibly after damage.

Diesel engine FFP criteria

The two main FFP criteria for diesel engines can be clearly defined. They are integrity and emission control. Note that engine performance, while important, is mainly a design/new engine works inspection issue rather than an in-service one.

Diesel engine integrity

As with the other prime movers, the core notion of *integrity* incorporates the mechanical integrity of the mechanical components and the effective operation of the engine's safety and protection features. Because of the design features unique to reciprocating machinery, we must also recognize the importance of correct *running clearances* in ensuring correct operation of the engine. The stripdown examination details given later will look at this in more detail.

Specifications and standards

Technical standards for diesel engines divide into two well-defined categories. The first; manufacturing standards, tends to concentrate on the major forged components of the engine: the crankshaft, connecting rods, camshaft and major gears; the rest of the standards relate to *testing* of the engine. The most accepted standard is ISO 3046: Parts 1 to 7 *Reciprocating internal combustion engines – performance* (identical in all respects to BS 5514 Parts 1 to 7) This is a wide-ranging and comprehensive standard covering engine fuel, performance, brake testing, etc., but does little to address the particular issues of in-service inspection.

Emission measurements

Environmental emission limits are becoming a standard feature of the in-service inspection schedule for land- and marine-based diesel engines above about 750 kW. Manufacturers compete with each other on the claimed environmental performance of their engines when burning standard 'reference' fuel. Emission levels are a function of the fuel specification and the design and adjustment of the engine itself. There are wide differences in claimed emission performance both between slow and medium-speed engines, and within each category. As a general rule, medium-speed engines exhibit the largest variation in limits, mainly because they have more design variants (speed, brake mean effective pressure, combustion regimes, etc.) than do slow-speed types.

There are often problems with an engine's continuing ability to meet its site environmental guarantees. Factory 'proving' tests are generally performed using diesel oil (containing about 1 percent sulphur) which does not need preheating and keeps the engine in a clean condition. Most land-based diesel engines, however, are specified to burn medium to heavy oil (1500–3500 RSec) containing up to 4 percent sulphur. The analysis of some other constituents which affect mainly particulate loading of the exhaust gas are also different and may vary through the engine's operational life. The situation is also variable for NO_x levels – the fuel specification has a significant effect, producing large variations.

Table 22.1 show the outline test methods based on the US Environmental Protection Agency (EPA) standard.

Table 22.1   Emission test methods

Emission	Test method* 1 [reference USA (EPA)]	Typical emission for slow-medium speed engine (mg/m³)
SO$_2$	Method 6C	1900–2500 (with 3500 RSec heavy fuel)
NO$_2$	Methods 7E and 20	1300–1400
	Method 10	160–450
Total suspended particulates (TSP)	Method 5 and gravimetric analysis	100–150
Unburnt hydro-carbons or 'non-methane hydrocarbons (NMH)'	Method 25A	100–150
Volumetric flowrate	Methods 1 and 2	—
Moisture (H$_2$0) content	Method 4	—

*Test methods refer to 15% O$_2$ v/v reference level.

Analysis techniques

Many sampling and analysis techniques used for emission measurement are covered by published technical standards. They are all quantitative techniques, with a robust scientific background. There are two main sets: those from the European ISO series and those using the Environmental Protection Agency (EPA) standards from the USA. Although differing in some aspects of methodology and technical detail, the objectives and general philosophy of the two approaches are very similar. It is not unusual to see a mixture of the two used, although it would not be wise to specify both methods for a single gas constituent – there would be some contradictions.

All the techniques rely on obtaining a representative sample of exhaust gas from a location somewhere between the engine exhaust manifold and the test stack. The exact location varies – it is not specified accurately in the published standards. The gas sample is cooled and subjected to an automatic analysis process to determine the level of individual constituents. All these tests are 'on-line' with the exception of the total suspended particulates (TSP) test – for this, filter papers which have been exposed to the gas stream need to undergo laboratory gravimetric tests to work out the exact TSP loading. An important reference parameter for all the tests is the volume flowrate of the sampled gas – there is a separate methodology specified for this to ensure the accuracy of the test results.

Witnessing the emission tests

The techniques and methods of data analysis for these tests are complicated. They involve multiple data streams and detailed computer analysis incorporating empirical values, theoretical constants, and calculations. While it is *just about* possible to check the calculations manually, you need to be familiar with the methodologies, and have plenty of time, to do it properly. In practice, you will probably have neither, unless you are a specialist. This raises a problem with verifying the integrity of the test results – such integrity is all-important if the results are to be suitable for adjudication in, perhaps, a commercial dispute about whether an engine has met its emission limits. This is part of the whole FFP rationale of works inspection. Expect, for emission level measurement, to find this a bit difficult.

The best way to approach the problem is by using technical *guidelines*. These are basically a checklist of relevant points about the inspection – all of which are 'checkable', without becoming too deeply involved in the complicated technology of the tests. The checklist points are shown in Fig. 22.1 and are much the same for all diesel engine types. Remember that the purpose of these checks is to contribute towards the amount of confidence in the test results – they cannot pretend to be an accurate analysis. The tests are just too complex for that. The other alternative, of course, is to use a specialist to witness the tests.

Reporting

There is a temptation to fill emission test inspection reports with reams of data and computer printouts. There are always plenty of these data available as the final levels are arrived at by an averaging procedure from readings taken at perhaps 20- or 30-s intervals during the test. Clients will rarely read these data. It is much better to make your inspection report a *summary* of the test, mentioning the techniques used, whether they comply with published standards, and any features of the test equipment or techniques that give you cause for concern. Make sure that you summarize the results so that your client can understand them – and say whether the results complied with the relevant emission limits or not. There is little point providing long and confusing descriptions of the tests and then leaving your clients to interpret the results themselves.

- **The fuel specification**. Is it the contract specification (or something else)? Check the sulphur and particulate levels.
- **Engine adjustments**. Check the fuel pump and valve timing.
- **Test methods**. Make sure these follow recognized technical standards.
- **Equipment calibration**. All test equipment must be calibrated (check for valid certificates).
- **Sampling period and frequency**. As defined in the technical standards.

Fig. 22.1   Witnessing emission tests – some checklist points

Turbocharger tests

Large slow-speed diesel engines are generally constant-pressure turbocharged and have a single (or occasionally two) large free-standing exhaust gas turbocharger supplemented by an electrical auxiliary. Medium-speed engines are more commonly pulse-turbocharged by two smaller units mounted on the engine. By their function, turbochargers are inherently high-risk components, mainly because of their high rotational speed (up to 70 000 r/min for smaller models) and the corresponding implications for bearing design, lubrication, and shaft balance. Practically, however, their maintenance is well-controlled by separate specialist companies, which actually reduces the technical risk to below that of many of the other engine components.

Most turbocharger problems occur later in the life of an engine. The main in-service inspection points are as follows (see Fig. 22.2):

- *Temperature distributions.* Check for the correct temperature differentials (under steady state conditions) on the exhaust gas and charge-air sides.
- *Charge air pressure.* Compare this with the engine datasheet. Note that this can be difficult to measure accurately due to pressure fluctuations in the charge-air manifold. A fluid-filled gauge is normally needed.
- *Turbocharger speed.* This is separately monitored from a sensor located on the air-side end of the shaft. Check that it is constant to within about ± 2 percent and that the turbocharger is not surging, particularly at the 100 percent MCR condition. For a low-running-hours engine, surging is an indication of incorrect cylinder balance

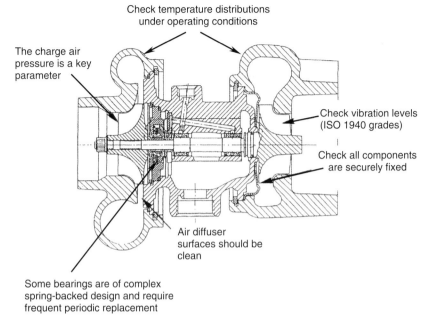

Check temperature distributions
under operating conditions

The charge air
pressure is a key
parameter

Check vibration levels
(ISO 1940 grades)

Check all components
are securely fixed

Air diffuser
surfaces should be
clean

Some bearings are of complex
spring-backed design and require
frequent periodic replacement

Fig. 22.2 Turbocharger inspection points

(injection or valve timing), or of more fundamental fouling problems, probably in the exhaust gas manifold train.

- *Vibration levels.* These need to be closely controlled, often down to a level of 1 mm/s rms. Make sure that the sensors *have been installed* for the load test and that they are providing the correct type of read-out (VDI 2056 is the standard for these 'housing' vibration levels). Look at the turbocharger manufacturer's datasheet to double-check the maximum acceptable V_{rms} level. The most common source of excessive vibration is misalignment of the turbocharger bearings during maintenance – they are either complex concentric spring-backed roller bearings, or precision journal type, both of which can be difficult to install accurately. These bearings are replaceable in a few hours – this is normally the first step taken if vibration problems show themselves during an engine load test.

Unless there are serious problems, there is not usually a turbocharger stripdown after a load test. Some inspectors request a borescope examination of the gas side if the test was conducted using heavy fuel oil.

Vibration

The vibration characteristics of a diesel engine are very different from those associated with turbomachinery and similar rotating plant. The existence of large reciprocating masses means that some of the mass effects simply cannot be balanced. There are two different types of vibration to be considered. Radial and axial 'housing' vibration (in the x, y, z planes) can be measured during engine testing using relatively simple techniques. Diesels also exhibit *torsional* vibration which is a different, specialized subject. In-service inspection activities are normally limited to the measurement and assessment of radial and axial vibration – you are unlikely to become involved much with torsional vibrations when witnessing engine tests on-site

The balancing of diesel engines is also a difficult issue. The rotating element is the crankshaft and flywheel assembly, together with the rotating parts of the connecting rod bottom end bearings. This assembly has complex geometry and there is often uncertainty as to whether it should be classed as a 'rigid rotor', to allow an analysis under ISO 1940/1, or whether it is considered a 'flexible rotor', in which case the specific standards ISO 5343 and ISO 5406 could apply. It is also not possible to link balance grades with vibration performance due to the effect of resonant frequencies. In practice, the balance grade of engine crankshafts is a design issue rather than an in-service inspection one. You should not need to get too involved.

The levels of 'housing' vibration that you will see in diesel engines will be several orders of magnitude *higher* than in turbomachinery. Typical vibration velocity (rms) levels can be up to 50 mm/s during normal running with possibly up to 500 mm/s at critical (reasonant) speeds. Large engines will pass through one, possibly two, reasonant speeds between starting and reaching full speed. It is not easy to find definitive guidance or acceptance levels in technical standards – the best one to use is probably VDI 2056 (the same one used for turbomachinery) which specifies reciprocating engines into its group D and S machines. It does not however provide any graphs or data that you can use to calculate acceptable vibration limits for these groups. Note that the concept of relative shaft vibration (ISO 7919-1) has little relevance to reciprocating engines.

Stripdown inspections

The stripdown inspection is the most important part of an in-service inspection on a diesel engine. A comprehensive stripdown examination is an excellent way to prove running integrity, which is a key FFP criterion. It is also the best way to evaluate the particular lubrication and future wear regime of an individual engine. Stripdown inspections help to build up your inspection experience. You will find that evaluating and diagnosing the condition of components is not a simple pass or fail exercise – it is more about interpretation and experience

The examination normally involves dismantling one complete cylinder liner/piston assembly (sometimes termed an 'upper-line strip') and the corresponding large-end bearings and main crankshaft journal bearings (the 'bottom-end strip'). Together, you may hear these referred to as a 'complete one-line strip'. Some organizations perform a borescope inspection on all cylinders before deciding which cylinder to strip down while others seem to choose one almost at random. Cylinders with high exhaust temperatures and in which corresponding crank webs have 'on-the-limit' deflection are good candidates for stripdown.

Figures 22.3 and 22.4 show which areas to examine and what to look for. Note the various measurement and record sheets that should be filled in – you can structure your checklist around these. Important points are:

- Any *active* wear mechanism such as scoring or scuffing of the sliding surfaces of the piston, piston rings, and liner. This signifies that the lubrication regime is inadequate. You must be careful not to confuse this with mild 'bedding-in wear' which is normal and actually desirable in developing a good lubrication regime.
- Exhaust valve burning or erosion of the valve seat or stem.
- Poor 'bedding' of any of the hydrodynamic rotating bearings to the extent that the white metal backing layer is exposed. This is indicative of serious misalignment.
- Signs of deep scoring or gouging of *any* of the bearing surfaces. This is a sure sign of lubricating oil contamination of some sort. It is a sorry fact that numerous diesel engines have failed on site because of this.
- Look particularly at the gudgeon pin and its bush. Gudgeon pins are a very highly loaded component with tight clearances and only intermittent hydrodynamic lubrication. A lot of potential problems with oil quality or distribution will first show themselves here. Treat gudgeon pin and bush wear seriously, with less tolerance for marginal cases than you would the bottom-end or main journal bearings.

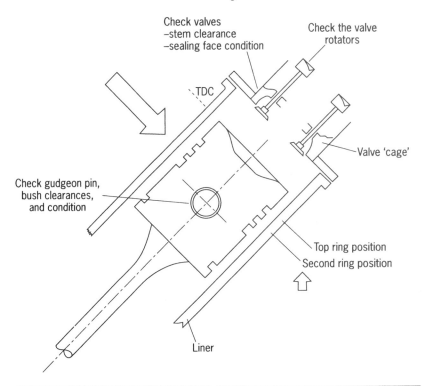

Check valves
–stem clearance
–sealing face condition

Check the valve
rotators

TDC

Valve 'cage'

Check gudgeon pin,
bush clearances,
and condition

Top ring position
Second ring position

Liner

Piston, ring and liner observations	Measure liner wear (to 0.01mm) at
Record any:	A - top ring (axial) B - top ring (transverse) C - 2nd ring (axial) D - 2nd ring (transverse) Mean dia. = (A+B+C+D)/4 Record the mean wear
• scratches • microseizure • active microseizure • wear ridges • scuffing • 'cloverleaf' wear	During the stripdown: • take photographs of the running surfaces • describe accurately what you see
ALSO	
• check the ring gaps • measure ring clearances • check the ring action – watch for sticking	

Fig. 22.3   Diesel engine upper line strip

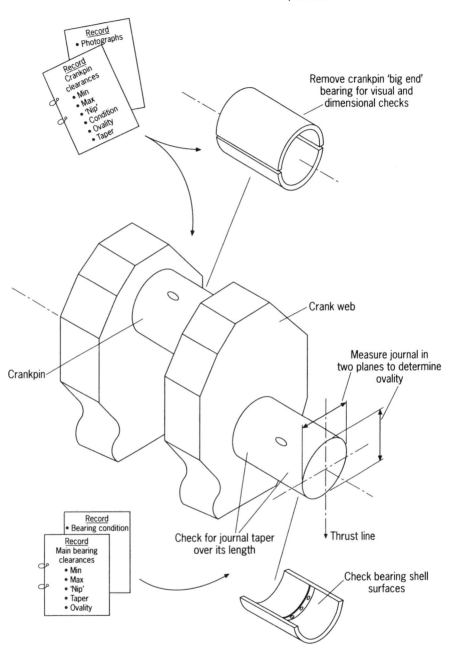

Fig. 22.4   Diesel engine bottom end strip

In common with other rotating equipment, diesel engine failures in service are relatively common. If this happens, the in-service inspection reports become an important baseline to work from when trying to diagnose the reason for the failure.

Inspecting the valves

It is usual to inspect the cylinder valves during the one-line strip procedure. Most large two-stroke engine designs now have a single exhaust valve in the cylinder head, combustion air being admitted lower down the cylinder through radial scavenge ports. In both medium- and slow-speed engines the valves are supported in a water-cooled cage which seals into the cylinder head – hence they are easily stripped for inspection once the head has been lifted off. The state of an engine's exhaust valves is a useful measure of the health of fuel quality, injection efficiency, and the combustion regime.

Preparation

Correct preparation for the inspection is important. Make sure you inspect each valve in its *uncleaned* condition (i.e. the state in which it was removed from the engine) before it has been water-washed or cleaned with wooden scrapers or emery cloth. It should be laid horizontally on an inspection stand or set of vee-blocks, rather than stood on its end in a dark corner of the shop floor. Some of the possible problems can be difficult to see, so make sure that there is good lighting for your visual inspection. You will also need a magnifying glass for looking closely at the valve cone (seating face).

The accuracy check

Figure 22.5 shows the accuracy checks – these are quick and easy but can provide you with a measure of comfort (or otherwise) about the condition of the engine components. The critical aspects are:

- *Straightness* of the polished part of the valve stem. The tolerance varies between about 0.03 and 0.05 mm, depending on design (check the component drawing).
- *Radial run-out* of the valve cone – the tolerance is around 0.03 mm.
- *Surface finish* of the valve stem. This should be a mirror-like lapped finish of about 1.6 μm R_a (N7) or better.

Running condition

The physical condition of the valves after a period of use is referred to, broadly, as the *running condition*. Figure 22.5 shows the main problems to look for:

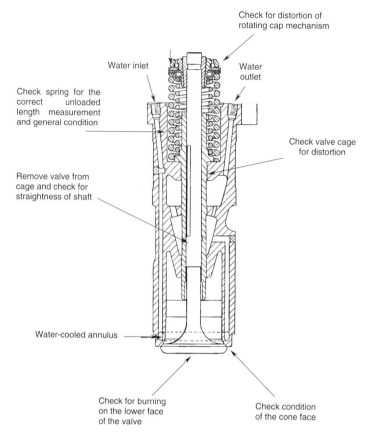

Fig. 22.5   Diesel engine valve condition

- *Cone face condition.* The cone face should be absolutely free of cracks visible using a magnifying glass. Check also for 'blowholes' or erosion channels extending radially across the face.
- *Lower surfaces.* Look for pitting in the lower areas of the valve head. This can either be isolated, or arranged in a semi-continuous pattern – looking a bit like 'paving stones'.
- *Valve head wear.* Heavy fuel engines can suffer severe corrosion of the valve head diameter, d, caused by erosive particles in the fuel. The operational wear limit on a well-used valve is about $0.02d$, but it should be less than about one-tenth of this for a low-running-hours engine. Greater wear than this suggests a problem with valve material, test fuel quality, or both.

Alignments

Alignment checks are particularly important for large slow-speed engines but are also performed for the smaller types. Alignment checks require a full engine stripdown, so are normally limited to inspection following a major rebuild. There are three key alignment parameters to be measured: bed plate, crank shaft, and reciprocating parts alignment.

Bedplate alignment

This is the first alignment operation to be carried out. Its purpose is to check absolute concentricity of the main bearings about the crankshaft centreline after reassembly of the bedplate sections – larger engines often have the fabricated bedplate manufactured in two or three sections for ease of fabrication and transport. This is a key inspection point – accurate alignment at this stage will greatly reduce the risk of later problems during the engine assembly operations.

The simplest method used is a combination of water-level and 'taut-wire' measuring techniques or the new modern method using lasers. Figure 22.6 shows the principles for a two-section bedplate of a large slow-speed engine. Note that there are *two sets of measurements* to be taken: those from wire A will determine the relative concentricity of each main bearing about a datum centreline while those from wires B and C will determine the accuracy of the bedplate in the vertical plane. The test steps are as follows:

- The bedplate sections are assembled on chocks and their relative positions adjusted until they are level. This is checked by filling the channels machined longitudinally along the upper faces with water. Final adjustment is then made to achieve an equal depth of water over the length of the channels.
- The centreline wire A is positioned along the line of the crankshaft axis, passing between the saddles that will hold the main bearing keeps. Measurements are taken from the wire to the milled abutment faces on each side of the saddle (these are the faces that locate the main bearing keeps accurately in position relative to the bedplate). These measurements therefore indicate the accuracy of location of the bearing centreline in relation to wire A (see Fig. 22.6). Note that alignment is required in both the x and y planes to give the correct 'truth' of the main bearing axis.
- Wires B and C are positioned longitudinally above the bedplate flange and across the diagonals as shown. Accurate *vertical* measurements are taken from the wire to the machined horizontal datum faces of the

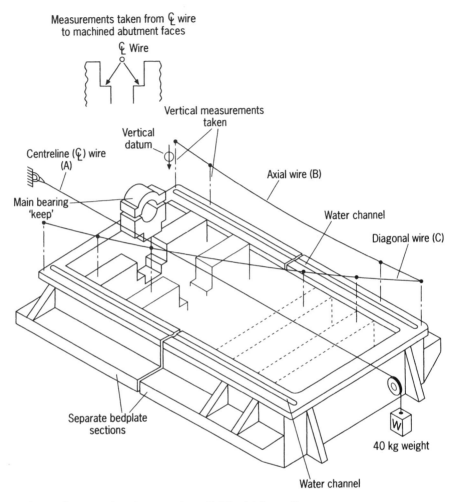

Measurements taken from ₵ wire
to machined abutment faces

₵ Wire

Vertical measurements
taken

Vertical
datum

Centreline (₵) wire
(A)

Axial wire (B)

Main bearing
'keep'

Water channel

Diagonal wire (C)

Separate bedplate
sections

40 kg weight

Water channel

- Large slow speed engines may have 2–3 bedplate sections
- The objective is absolute concentricity of the bearings shells around the crankshaft centreline
- Measurement accuracy is ($\pm$) 0.01mm
- Vertical measurements need to be corrected to compensate for the wire catenary. Tables are available

Fig. 22.6   Diesel engine bed plate alignment

bedplate. Variations in these dimensions indicate whether further skim-milling of these faces is required in order to achieve an accurate flat surface. If these surfaces are not flat, they will pull the cylinder axes, and hence the reciprocating 'running gear' parts, out of alignment once the crankcase and entablature are assembled. This will result in severe wear and probably failure.

Any taut-wire measuring technique suffers from sag of the wire between its suspension points. This is a well-understood phenomenon and is documented in catenary tables which list the corrections that must be added to the vertical dimensions (depending on the overall length of the wire and the position along its length from which a vertical dimension is taken). Dimensions taken during the measurement checks will therefore need to be *corrected* before they will give a true representation of the alignment. The acceptable tolerances for mis-alignment depend on the physical size and stiffness of the bedplate and the type of engine. All diesel engine licensors and manufacturers include a bedplate alignment drawing as part of their design package – compare the test results carefully with the stated tolerances on this drawing.

Laser alignment techniques
Many engine manufacture and rebuild companies use laser equipment in preference to the traditional taut-wire techniques. There are several different types of equipment in common use but all use the same principle – the use of a beam of visible red laser light as a reference. It is a quick and precise method of bedplate and journal alignment but does need accurate setting-up and adjustment.

The equipment consists of a battery-powered laser transmitter and receiver, each mounted on a delta-shaped fixture which locates within a bearing journal. The fixture arms are fitted with pressure-ground rollers to ensure an accurate and repeatable location with the journal surface. The transmitter is installed in the front journal and the receiver initially in the rear journal at the opposite end of the bedplate. A reference position of the laser beam path is taken at this position. The receiver is then installed in each journal in turn and a reading is taken of the relative position of the beam. In practice, this is done by centring the beam 'spot' using a graticule, recording the amount of adjustment required relative to the reference position. The results give the offset of the centreline of each journal in the x and y planes.

Laser alignment can give very accurate results (± 0.001 mm) and has the advantage of providing a direct reading, without the need for catenary corrections as in the taut-wire method. It is worth checking the following points when you witness the procedure:

• Check the rear journal reference position carefully – this is often where errors creep in.

- There should be *no need* for large adjustments while the receiver is in position in each bearing location. If there is, then the fixture is not locating properly (check it is the correct type for the job).
- Accurate and repeatable location of the receiver 'delta' fixture is essential if the results are to be to the required accuracy. Check that the adjusters that extend the fixture rollers into contact with each journal surface are only moved *by hand*. Using spanners can given variable contact pressures and cause measurement inaccuracies.
- Check that the data print-out (most laser devices are connected to a PC/datalogger of some sort) are the same as those you saw on the screen at each journal measurement position. Data errors do sometimes occur.

Crankshaft alignment (deflections)

The purpose of measuring crankshaft deflections is to check the 'as-installed' alignment of the main journal bearings. The check is performed after installation of the crankshaft and after rebuilds, suspected movement of the engine foundations, and similar. Figure 22.7 shows the technique. A dial test indicator (dti) is placed horizontally between the crank webs directly opposite the crank pin. The crankshaft is slowly rotated through one complete revolution and gauge readings taken at the five points shown. This is repeated for each crank web pair (i.e. one set of readings per cylinder) and the results tabulated. Figure 22.7 also shows the sign conventions that are normally used when referring to deflection readings. Note particularly the ± convention relating to the direction of distortion of the crankshaft webs and the way in which the sides of the diesel engine are usually referred to: i.e. exhaust (E) side and camshaft (C) side.

The measured values of interest are the vertical misalignment (shown as T–B) and horizontal misalignment (shown as C–E). There will be clear acceptance limits for deflection readings set by the engine licensor – they may be shown in the form of a graph or a table. The figure shows a typical set of deflection results; note how the values are interpreted to diagnose which bearings are misaligned. Vertical misalignment can be corrected (to a limited extent) by using shims to adjust the position of the bearing keeps.

There are a few key points to bear in mind when witnessing deflection checks:

- Before taking deflections, check the main bearing clearances using feeler gauges. The bottom clearances directly underneath the journals

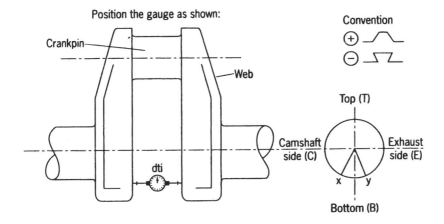

Position the gauge as shown:

Crankpin

Web

dti

Convention

Top (T)

Camshaft
side (C)

Exhaust
side (E)

x   y

Bottom (B)

1. It is convention to set the gauge
   using x = 0 as a datum.
2. Don't confuse the algebra:
   e.g. (− 2) − (− 4) = + 2
3. Diagnosis. In this example the main
   bearing between cylinders 3 and 4 is <u>high</u>.

Typical readings (in 0.01mm) for a 6-cylinder engine

Crank position	Cylinder number					
	1	2	3	4	5	6
x	0	0	0	0	0	0
C	+4	+1	+3	-6	-2	+1
T	+8	+3	+10	-12	-6	+3
E	+4	+2	+5	-6	-4	+2
y	-2	+2	-2	0	0	-2
B=(x+y)/2	-1	+1	-1	0	0	-1
VM=T-B	+9	+2	+11	-12	-6	+4
HM=C-E	0	-1	-2	0	+2	-1

VM - Vertical misalignment
HM - Horizontal misalignment

Fig. 22.7   Crankshaft web deflections

should be zero. If they are not, then the bearing train is *definitely* out of alignment.

- Because of the connecting rod, it is not possible to take a deflection reading when the crank is at bottom dead centre. The accepted way of getting a reading here is to take readings as near as possible either side (positions *x* and *y*) and then take an average.
- The engine staybolts should be tightened to their design torque; slack staybolts will cause errors in the deflection readings.
- Make sure that before each reading is taken, the turning gear is reversed and 'backed-off' slightly to unload the flywheel gear teeth. This helps to avoid inaccuracies.

Reciprocating parts alignment

For large engines it is necessary to check various dimensions that indicate the alignment of the pistons and other reciprocating parts. These dimensions include: piston inclination (axial and transverse), cross-head clearances and cross-head pin clearances, all taken in several piston positions to indicate the truth of the engine assembly. They are important dimensions which must be carefully taken in defined positions. Ask to review the engine manufacturer's original alignment procedure document – this is the best (and only) document to use for guidance.

Inspection of bearing shells

Inspection of tri-metal bearing shells

These are the traditional type of bearing shells used on diesel engines. Most slow-speed designs use them for both the main and big-end bearings, while on some medium-speed designs they are slowly being superseded by aluminium-grooved types for the big ends. Tri-metal bearings consist of a steel backing coated with three layers of material: a lead-bronze cast layer, a copper or nickel interlayer, and a plated overlay, normally a lead/tin/copper white-metal alloy (see Fig. 22.8). The nickel interlayer is very thin (0.002–0.003 mm), its main purpose being to provide a 'key' between the lead–bronze cast layer and the white-metal overlay.

One characteristic of tri-metal shells is that they 'bed-in' during use, and so produce well-documented wear patterns. These are useful in that they help the diagnosis of engine problems. Wear patterns are obviously more pronounced after prolonged running of the engine; however, even after relatively low running hours it is possible to observe the wear pattern starting to develop and draw useful 'fitness-for-purpose'

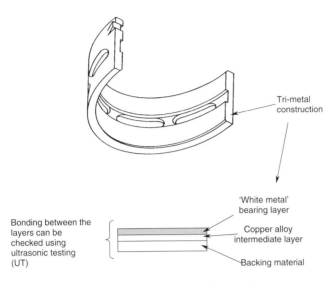

Tri-metal
construction

'White metal'
bearing layer

Bonding between the
layers can be
checked using
ultrasonic testing
(UT)

Copper alloy
intermediate layer

Backing material

Fig. 22.8   Tri-metal bearing shells

conclusions. This is another area where precise reporting is required
during the stripdown inspection. Figure 22.9 shows the common types
of wear pattern and gives brief commentary on their likely cause.

Inspection of aluminium-grooved bearing shells

Aluminium-grooved big-end bearings are being increasingly used on
medium- and high-speed diesel engines because of their high-load-
bearing capability. The bearing surface consists of a series of thin
aluminium ridges, interposed with grooves of approximately $200\,\mu$m
wide, fitted with an electroplated overlay (see Fig. 22.10). In use, most of
the wear is designed to occur in these electroplated areas, leaving the
aluminium ridges almost intact. A new bearing has an overlay:ridge
ratio of about 75:25 – this ratio being used to assess the wear regime
(and therefore serviceability) of the shell after periods of use.

Although hard wearing under optimum conditions, aluminium-
grooved bearings are still susceptible to problems with alignment and
oil quality and should be inspected as part of the one-line stripdown
procedure. Figure 22.10 shows typical wear as it would be discovered
after a running test following a rebuild. The shell wear can be
considered 'acceptable' if the 75:25 surface ratio remains basically
unchanged – note that in practice, both surfaces (aluminium ridge and
plated overlay) will wear slightly, but at the same time). Overall wear up

1. **Uniform dull grey surface**
   This is an indication of a well-adjusted and bedded-in bearing. The main load-carrying area of white metal may be slightly more polished and you can expect to see very minute circumferential *running marks* caused by small impurities in the oil.

2. **Edge glossiness**
   Glossy white-metal areas are indicative of overloading. On a low-running-hours engine this edge overloading is most probably due to incomplete bedding-in. It can be removed using scraping and is generally not a major problem – as long as the white-metal overlay is still fully intact.

3. **Single-edge overloading**
   This is where the lead–bronze layer is visible through a worn white-metal layer. It is normally indicative of bearing housing misalignment and is unacceptable. This type of wear is nearly always accompanied by out-of-limit crankshaft deflection measurements. You should not recommend continued use of an engine that shows this type of wear of any of the bearings – it will soon lead to failure.

4. **Scratches**
   You can expect to find shallow scratches in tri-metal bearing shells after even a short period of running. They will be continuous, running circumferentially around the shell. They are only a cause for concern if they are heavily concentrated, or reach into the bronze backing (you will not see the nickel interlayer because it is so thin). Even with the best flushing of the oil system it is still common for small impurity particles to remain in the oil circuit during the early life of the engine.

5. **Overall white-metal wear**
   Overall *even* wear of the white-metal overlay is a normal feature of well-used bearings. If you see it during the early inspections, particularly if the worn areas have well-defined (non-'feathered') edges, it is a concern. On big-end bearings it is more likely to be caused by roughness of the crankpin surfaces rather than misalignment. For main bearings, it could also be a result of misalignment (check the deflection readings again).

Fig. 22.9   Diesel engine bearings – types of wear pattern

6. **Parting line glossiness**
Very large, flexible bearing shells often show glossy areas near the parting line. It can be particularly pronounced below the 'relief area' that is incorporated into some designs. The glossiness is the result of local pressure overload, most likely caused by something restricting correct expansion of the assembled shell. Check the fit and the size of any shims fitted.

7. **Cavitation depressions**
These are found mainly in big-end bearings. Clear depressions are visible in the white metal – they are generally kidney-shaped with well-defined stepped edges to the damaged area. The outline of the depression is quite irregular. The depressions themselves do not affect significantly the life of the bearing, as long as they do not reach through into the lead–bronze layer. It is, however, important to check the *cause* – it is indicative of oil with the incorrect ISO Vg viscosity being used. This is not an uncommon occurrence, particularly on medium-speed trunk engines where oil specifications are sometimes compromised so that the same grade can be used in both the engine sump and the gearbox. Continued cavitation of bearings over the running lifetime of an engine is undesirable.

8. **Diagonal pressure marks**
This is where heavy pressure marks can be seen on diagonally opposite corners of a bearing shell. Assuming the crankshaft web deflections are within limits, this is caused by a slight misalignment of the shells or bearing cap in an individual bearing, rather than being a general problem with all the bearings in an engine. It will lead to early failure if not rectified. The usual remedy is to fit a new shell (with careful assembly checking of the alignment and any shims).

9. **Smeared overlay on centreline**
Heavy smearing of the white-metal overlay, exposing the lead–bronze along the transverse centreline of the bearing is termed 'galling'. It results from a lack of oil supply, causing overheating and contraction of the bearing shells. This is a serious fitness-for-purpose issue – the only solution is to check all the main big-end and top-end bearings to find the extent of the oil supply

Fig. 22.9   (Continued)

problem. If the supply pumps are working correctly, then there is probably a blockage in one or more of the cooling channels.

10. **Joint face fretting**
    This shows as fretting or 'working-marks' on the joint face of the shells. It is caused by incorrect tightening/pre-stressing of the bearing bolts and is a simple assembly problem rather than a fitness-for-purpose issue.

Fig. 22.9   (Continued)

to about 0.01 mm is quite common during the early test running life. Unacceptable wear will show itself as:

- 'Dragging' or smearing of the overlay over the aluminium ridges.
- Foreign particles embedded in the overlay.
- Rupture of the overlay – to the point where it is sometimes fully 'washed away' out of its grooves.
- Extensive wear of the aluminium ridges to the point where the overlay: ridge area ratio approaches 1:1.

This last category (commonly referred to as '1:1 wear') is the most common type of unacceptable wear regime. Most engine manufacturers that use grooved bearings specify acceptance limits of 1:1. If you see bearings that are getting near this condition there is clearly something seriously wrong. The appearance of any early 1:1 wear *at all* is a cause for concern and warrants investigation before the engine can be subject to further regular use.

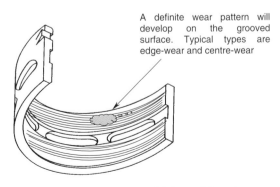

A definite wear pattern will develop on the grooved surface. Typical types are edge-wear and centre-wear

Fig. 22.10   Aluminum-grooved big-end bearing wear

Diesel engine inspection reporting

Owing to the variety of different findings, wear patterns, etc. that are possible during diesel engine inspections, reports should include *photographs*. It is also good practice to annotate the photographs to make the findings clear – wear regions are particularly difficult to describe and are easily misinterpreted. As with other types of inspection reports, conclusions about FFP are the most important part. Diesel engine stripdown is a time-consuming, labour-intensive procedure which tends to be carried out as infrequently as possible, so when a stripdown does occur it is important to draw the best conclusions possible.

Appendices

Appendix A1

The CSWIP plant inspector certification scheme

The Certification Scheme for Welding and Inspection Personnel (CSWIP) plant inspector certification scheme is a comprehensive scheme that provides for the training examination and certification of individuals seeking to demonstrate their knowledge and/or competence in the field of plant inspection. It was introduced in 2001. The scheme is applicable to both 'works' and in-service inspection. It covers both statutory and non-statutory equipment and is universal, i.e. it is not tied to one specific plant inspection code. Various sector- and plant-specific topics are covered at the appropriate level and depth (e.g. for power generation, offshore, and chemical process/refineries sectors).

The scheme allows for increasing specialization at Level 2 and Level 3, with Level 1 as the general entry level. Training for statutory plant (pressure and lifting equipment) is given special significance by the addition of specialized endorsement certificates at Levels 2 and 3. Figures A1.1 and A1.2 show the general scope of the scheme.

Inspection personnel registering for the CSWIP plant inspector scheme normally come from the main disciplines of plant operations, non-destructive testing, and mechanical inspection. It is also draws candidates who are suitably experienced and qualified with related quality assurance and quality control backgrounds.

The scheme is designed to suit plant inspectors working for commercial works (vendor) inspection organizations, independent third-party inspection organizations, classification societies, and insurance companies. Special consideration is given to the responsibility and competence requirements of inspectors working in a statutory 'Competent Person' role under the Pressure System Safety Regulations (PSSRs) or Lifting Operations and Lifting Equipment Regulations (LOLER), or as a 'user inspectorate' in power, petrochemical, or offshore plant.

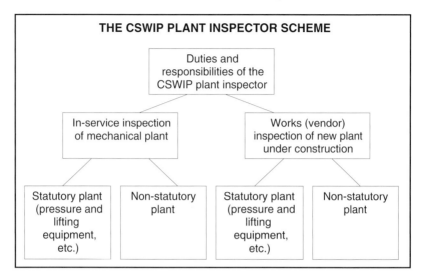

THE CSWIP PLANT INSPECTOR SCHEME

Duties and responsibilities of the CSWIP plant inspector

In-service inspection of mechanical plant

Works (vendor) inspection of new plant under construction

Statutory plant (pressure and lifting equipment, etc.)

Non-statutory plant

Statutory plant (pressure and lifting equipment, etc.)

Non-statutory plant

Fig. A1.1   Overall scope of the CSWIP plant inspector scheme

Note the following points:

- A mature candidate route is available at Level 1.
- Levels 2 and 3 are preceded by stringent experience assessment and formal prequalification examinations.
- Certification for statutory inspections cannot be achieved until Level 2 (plus endorsements) have been achieved.
- The CSWIP scheme involves technical training examination and certification only; CSWIP cannot award academic certificates (HND, BSc, BTEC, etc.) or substitutes for them.

Level 1 personnel

An inspector certified to Plant Inspector Level 1 has a wide-ranging basic knowledge of technical issues relating to works (vendor) and in-service inspection of mechanical plant. The inspector will be able to perform straightforward unsupervised inspections in accordance with an inspection and test plan (ITP) for new equipment or a written scheme of examination (WSE) for in-service equipment which is not subject to statutory requirements (Pressure Equipment Directive (PED), PSSRs, LOLER, etc.). The inspector will be able to choose the inspection method, liaise with technicians performing the test, and provide clear reports, with recommendations, for final assessment by others. The inspector will be capable of making basic subjective judgements on non-

Fig. A1.2   CSWIP plant inspector scheme structure

controversial inspection results but will require assistance when dealing with more technically complex situations.

Level 2 personnel

An inspector certified to Plant Inspector Level 2 (basic) is able to carry out all duties for which a Level 1 inspector is qualified but is able to add a deeper level of technical understanding and interpretation. The inspector also has a wider scope of knowledge relating to pressure and lifting equipment and their manufacturing and test procedure. A Level 2 inspector can make more detailed comparison of inspection results with code/defect acceptance criteria and justify conclusions using simple code calculations. In cases where corrosion and other in-service defects are found, the inspector is able to allocate severity levels and contribute input to a risk-based-inspection (RBI) scheme, if applicable.

An inspector certified to Plant Inspector Level 2A/B/C (i.e. with additional endorsements 2A, 2B, and 2C) is able to perform all duties for which a Level 2 (basic) inspector is qualified and, in addition has training in the technical aspects of various inspection-related duties likely to be required as part of an organization acting as 'Competent Person' under the provision of relevant legislation such as the PSSRs and LOLER. In the case of the PSSRs, the inspector's role is limited to the category of pressure system for which the endorsement is held, i.e.:

- Endorsement 2A: Minor systems (PSSRs ACoP 105a).
- Endorsement 2B: Intermediate systems (PSSRs ACoP 105b).

Level 3 personnel

An inspector certified to Plant Inspector Level 3 (basic) would be capable of the following: assuming full responsibility as 'technical authority' for an inspection operation and staff; establishing techniques and procedures; interpreting codes, standards, specifications, and procedures; and designating the particular test methods, techniques, and procedures to be used. The inspector shall be deemed to have the competence to interpret and evaluate results in accordance with existing codes, standards, and specifications and have sufficient practical technical background to select methods and establish techniques and to assist in establishing acceptance criteria where none are otherwise available. The inspector will also be able to assume responsibility for a 'user inspector' operation and demonstrate a clear and justified view of the subject of risk.

An inspector certified to Plant Inspector Level 3A/B/C/D/E/F/G/H/I (i.e. with additional endorsements) is able to perform all duties for which a Level 3 (basic) inspector is qualified and, in addition, perform the duties related to the specific endorsements held i.e.:

- Endorsement 3A: Inspection of 'major systems' (PSSRs ACoP 105c).
- Endorsement 3B: Inspection of rotating equipment.
- Endorsement 3C: Statutory inspection of lifting equipment (LOLER).
- Endorsement 3D: Inspection of pressure vessels: API 510.
- Endorsement 3E: Inspection of storage tanks: API 653/EEMUA 159.
- Endorsement 3F: Inspection of offshore installations.
- Endorsement 3G: Advanced painting and linings inspection.
- Endorsement 3H: Failure investigation and the expert witness role.
- Endorsement 3I: RBI management.

An inspector who has passed endorsement 3A is able to perform various inspection-related duties as part of an organization acting as 'Competent Person' under the PSSRs dealing with 'major systems' (PSSRs ACoP 105c). In addition the inspector is able to draw up and/or certify written schemes of examination for minor, intermediate, and major systems.

Important note. The CSWIP scheme involves technical training and certification only. CSWIP cannot award certificates that guarantee compliance with the competency requirements of UK in-service inspection legislation or non-statutory documents such as UKAS RG2. There is no such system for doing this in the UK — competence can only be proved or disproved in the courts.

CSWIP plant inspector certification scheme training syllabus

Certification	Modules
Plant Inspector Level 1 **(2 week duration)** **Level 1 week 1**	• The plant inspector: roles and duties (1). • QA/inspection in context. • Inspection safety. • Inspection: background skills. • Legislation, rules, and regulations. • Inspection and test plans (ITPs). • Inspecting materials. • Visual examination skills. • Visual examination of welds. • Inspection and basic non-destructive testing (NDT). • Introduction to corrosion.
Level 1 week 2	• The plant inspector: roles and duties (2). • Using codes and standards. • Introduction to pressure equipment.

Fig. A1.3   CSWIP plant inspector certification scheme training syllabus

	• Inspection of pressure vessels. • Inspection of pipework systems (API 570 introduction). • Inspection of storage tanks (API 653 introduction). • Inspecting painting/rubber glass-reinforced plastic (GRP) linings. • An introduction to RBI. • Inspection reporting.
Level 2 Plant Inspector (basic) (Basic course: 2 weeks duration) **Basic course Level 2 week 1**	• Safety briefing: Pressure tests. • Inspection of cranes (works inspection). • Inspection of cranes (site inspection/introduction to LOLER). • Plant inspector competency/ skills assessment. • Inspecting pressure relief valves (PRVs). • API 510 pressure vessel inspection code. • API 510 assessment exercise. • Pressure Equipment Directive; introduction.
Basic course Level 2 week 2	• L2W2 entry questionnaire. • Plant inspector and ISO 9000: 2000/EN 45004. • Case study: HSE investigation. • On-site calculations. • Pressure testing. • Inspecting heat exchangers. • High-temperature/boiler corrosion
Level 2 Endorsement 2A: 2 days **Level 2 Endorsement 2B: 2 days**	• Statutory PSSRs inspection of 'minor systems' (ACoP 105a). • Statutory PSSRs inspection of 'intermediate systems' (ACoP 105b).

Fig. A1.3   (Continued)

Level 2 Endorsement 2C: 2 days	• Statutory (LOLER) inspection of lifting equipment.
Level 3 Plant Inspector (basic) (Basic course: 2 week duration) Basic course Level 3 week 1	• Role of the plant inspector (3): investigations, disputes, and claims. • Plant remnant life assessment. • Fitness-for-purpose assessments (1). • Advanced corrosion and criticality reporting.
Basic course Level 3 week 2	• Fitness-for-purpose assessments (2) (BS 7910/API 579). • RBI implementation. • Advanced inspection reporting. • Inspection economics and management.
Level 3 Endorsement 3A: 2 days	• Statutory PSSRs inspection of 'major systems' and steam plant (ACoP 105c).
Level 3 Endorsement 3B: 2 days	• Inspection of rotating equipment.
Level 3 Endorsement 3C: 2 days	• Statutory (LOLER) inspection of lifting equipment.
Level 3 Endorsement 3D: 2 days	• Inspection of pressure vessels: API 510.
Level 3 Endorsement 3E: 2 days	• Inspection of storage tanks: API 653/EEMUA 159
Level 3 Endorsement 3F: 2 days	• Inspection of offshore installations.
Level 3 Endorsement 3G: 2 days	• Advanced painting/linings inspection.
Level 3 Endorsement 3H: 2 days	• Failure investigation and the expert witness role.
Level 3 Endorsement 3I: 2 days	• RBI management.

Fig. A1.3   (Continued)

CSWIP is managed by the Certification Management Board, which acts as the Governing Board for Certification, in keeping with the requirements of the industries served by the scheme. The Certification Management Board, in turn, appoints specialist Management Committees to oversee parts of the scheme. All CSWIP Boards and Committees are comprised of member representatives from relevant industrial sectors and other interests. The training is provided by TWI Training and Qualification Services Limited.

Contact for the CSWIP plant inspector certification scheme

For further general information contact:
Mark Allen: Training, TWI Training and Qualification Services Limited, Granta Park, Great Abington, Cambridge CB1 6AL, UK.
Phone: +44 (0) 1223 891162. Telefax: +44 (0) 1223 891630.
Email:trainexam@twi.co.uk
Website: www.twi.co.uk

Appendix A2

Websites – quick reference

Organizations and associations

Table A2.1 Organizations and associations

Organization	URL
American Bureau of Shipping	www.eagle.org
American Consulting Engineers Council	www.acec.org
American Institute of Engineers	www.members-aie.org
American Institute of Steel Construction Inc.	www.aisc.org
American Iron and Steel Institute	www.steel.org
American National Standards Institute	www.ansi.org
American Nuclear Society	www.ans.org
American Petroleum Institute	www.api.org
American Society for Non-Destructive Testing	www.asnt.org
American Society for Testing of Materials	www.ansi.org
American Society of Heating, Refrigeration and Air Conditioning Engineers	www.ashrae.org
American Society of Mechanical Engineers	www.asme.org
American Water Works Association Inc.	www.awwa.org
American Welding Society	www.awweld.org
APAVE Ltd	www.apave-uk.com
Association of Iron and Steel Engineers (USA)	www.aise.org
British Inspecting Engineers	www.bie-international.com

Table A2.1   *(Continued)*

Organization	URL
British Institute of Non-Destructive Testing	www.bindt.org
British Standards Institution	www.bsi.org.uk
British Valve and Actuator Manufacturers Association	www.bvama.org.uk
Det Norske Veritas	www.dnv.com
Dti STRD 5:	www.dti.gov.uk/strd
Engineering Integrity Society	www.demon.co.uk/e-i-s
European Committee for Standardization	www.cenorm.be www. newapproach.org
Factory Mutual Global (USA)	www.fmglobal.com
Fluid Controls Institute Inc. (USA)	www.fluidcontrolsinstitute.org
Hartford Steam Boiler (USA)	www.hsb.com
Heat Transfer Research Inc. (USA)	www.htrinet.com
Her Majesty's Stationery Office	www.hmso.gov.uk/legis.htm www.hmso.gov.uk/si www.legislation.hmso.gov.uk
HSB Inspection Quality Ltd.	www.hsbiql.co.uk
HSE (Home page)	www.hse.gov.uk/hsehome.htm
International Standards Organization	www.iso.ch
Lloyds Register	www.lrqa.com
Manufacturers Standardization Society of the Valve and Fittings Industry (USA)	www.mss-hq.com
National Association of Corrosion Engineers (USA)	www.nace.org
National Board of Boiler and Pressure Vessel Inspectors (USA)	www.nationalboard.org
National Fire Protection Association (USA)	www.nfpa.org
National Fluid Power Association (USA)	www.nfpa.com
National Institute of Standards and Technology (USA)	www.nist.gov
Pipe Fabrication Institute (USA)	www.pfi-institute.org
Plant Safety Ltd.	www.plantsafety.co.uk
Registrar Accreditation Board (USA)	www.rabnet.com
Royal Sun Alliance Certification Services Ltd	www.royal-and-sunalliance.com/
Safety Assessment Federation	www.safed.co.uk

Table A2.1   *(Continued)*

Organization	URL
SGS(UK) Ltd	www.sgs.com
The Aluminum Association Inc. (USA)	www.aluminum.org
The Engineering Council	www.engc.org.uk
The Institute of Corrosion	www.icorr.demon.co.uk
The Institute of Energy	www.instenergy.org.uk
The Institute of Materials	www.instmat.co.uk
The Institute of Quality Assurance	www.iqa.org
The Institution of Mechanical Engineers (IMechE) Pressure Systems Group	www.imeche.org.uk www.imeche.org.uk/pressure/ index.htm
The Institution of Plant Engineers	www.iplante.org.uk
The United Kingdom Accreditation Service	www.ukas.com
The Welding and Joining Society	www.twi.co.uk/members. wjsinfo.html
Tubular Exchanger Manufacturers Association Inc. (USA)	www.tema.org
TUV (UK) Ltd	www.tuv-uk.com
TWI Certification Ltd	www.twi.co.uk
Zurich Engineering Ltd	www.zuricheng.co.uk

General technical information

Table A2.2   *General technical information*

Subject area	URL
Technical standards	www.icrank.com/Specsearch.htm www.nssn.org
Pressure vessel design	www.birdsoft.demon.co.uk/englib/pvesse www.normas.com/ASME/BPVC/guide.html www.pretex.com/Glossary.html www.nationalboard.org/Codes/asme-x.html
Boilers fire tube	www.kewaneeintl.com/scotch/scotivOb.htm www.hotbot.com/books/vganapathy/firewat.html
Pressure vessel software	www.chempute.com/pressure_vessel.htm www.eperc.jrc.nl www.mecheng.asme.org/ www.coade.com/pcodec/c.htm www.codeware.com/

Table A2.2 *(Continued)*

Subject area	URL
Transportable pressure receptables (gas cylinders)	www.ohmtech.no/ www.hse.gov.uk/spd/spdtpr.htm www.iso.ch/cate/2302030.html www.hmso.gov.uk/sr/sr1998/19980438.htm www.pp.okstate.edu/ehs/links/gas.htm www.healthandsafety.co.uk/E00800.html www.hmso.gov.uk/si/si1998/19982885.htm
Valves	www.yahoo.com/Business_and_Economy/ Companies/Industrial/Valves_and_Control www.highpressure.com www.flowbiz.com/valve_manual.htm www.valvesinternational.co.za/further.htm
Heat exchangers	www.britannica.com/seo/h/heat-exchanger/ www.geapcs.com www.chem.eng.usyd.edu.au/pgrad/bruce/h www.firstworldwide.com/heatex.htm www.heat-exchangers.com/heat-exchanger
Simple pressure vessels	www.egadvies.nl/ce/scope/DrukvateEN.html www.dti.gov.uk/strd
Boilers and HRSGs	www.normas.com/ASME/BPVC/qx0010.html www.boiler-s.com/boiler-s pages.hotbot.com/books/vganapathy/boilers.html www8.abb.com/americas/usa/hrsg.htm www.hrsg.com/
Welding	www.welding.com/ www.cybcon.com/Nthelen/1weld.html www.amweld.org/ www.mech.uwa.edu.au/DANotes/welds/home.html
Non-destructive testing	www.ndt.net/ www.dynatup.com/apps/glossary/C.htm
Failure	www.sandia.gov/eqrc/e7rinfo.html www.netaccess.on.ca/~dbc/cic_hamilton/ krissol.kriss.re.kr/failure/STRUC/MATE www.clihouston.com/asmfailureanalysis.htm
Corrosion	www.cp.umist.ac.uk/ www.icorr.demon.co.uk/about.html www.cranfield.ac.uk/cils/library/subje

Table A2.2   *(Continued)*

Subject area	URL
Technical reference books	www.icrank.com/books.htm
	www.powells.com/psection/MechanicalEng
	www.lib.uwaterloo.ca/discipline/mechen
	www.engineering-software.com/
	www.fullnet.net/dbgnum/pressure.htm
	www.engineers4engineers.co.uk/0938-but
ISO 9000	www.isoeasy.org/
	www.startfm3.html
Forged flanges	www.kotis.net/~porls2/ForgedFL.htm
	www.englink.co.za/htmlfiles/nclro2/BS3.1.htm
	www.maintenanceresources.com/Bookstore
	polyhydron.com/hhpl/SAL.htm
Creep	www.men.bris.ac.uk/research/material/projects/ fad.htm/
Air receivers	www.manchestertank.com/hor400.htm
Materials	www.Matweb.com
	www.pump.net/otherdata/pdcarbonalloysteel.htm
General on-line reference websites	www.efunda.com/home.cfm
	www.flinthills.com/~ramsdale/EngZone/d
	www.eev1.ac.uk
Safety valves	www.taylorvalve.com/safetyreliefvalves.htm

Directives and legislation

- For a list of new EC Directives and standards on pressure equipment, go to www.nssn.org and search using keywords 'pressure vessel standards' and 'EC'.

- The entry website for the Health and Safety Executive (HSE) is:

  www.hse.gov.uk/hsehome.htm.

- Statutory Instrument (SI) documents are available from Her Majesty's Stationery Office (HMSO) at:

  www.hmso.gov.uk/legis.htm.
  www.hmso.gov.uk/si.
  www.legislation.hmso.gov.uk.

- The Department of Trade and Industry (DTI) entry website is:

www.dti.gov.uk/strd.

- Background information on mutual recognition agreements relating to European directives in general is available on:

  http://europa.eu.int/comm/enterprise/international/indexb1/htm.

- Reference to most pressure equipment related directives (and their interpretation guidelines) is available on:

  www.tukes.fi/english/pressure/directives_and_guidelines/index.htm.

- A good general introduction to the Pressure Equipment Directive (PED) is available on:

  www.ped.eurodyne.com/directive/directive.html.
  www.dti.gov.uk/strd/pressure.htm.

- Guides on the PED from the UK DTI are available for download on:

  www.dti.gov.uk/strd/strdpubs.htm.

- The European Commission Pressure Equipment Directive website has more detailed information on:

  europa.eu.int/comm/dg03/directs/dg3d/d2/presves/preseq.htm or
  europa.eu.int/comm/dg03/directs/dg3d/d2/presves/preseq1.htm.

- The CEN website provides details of all harmonized standards:

  www.newapproach.org/directivelist.asp.

- For details of harmonized standards published in the EU official journal go to:

  www.europa.eu.int/comm/enterprise/newapproach/standardization/
  harmstds/reflist.html.

- A listing of European Notified Bodies for the PED is available from:

  www.conformance.co.uk/CE_MARKING/ce_notified.html

- All the sections of the text of the UK Pressure Equipment Regulations are available on:

  http://www.hmso.gov.uk/si/si1999/19992001.htm.

- The UK Pressure Systems Safety Regulation (PSSRs) are available from:

  www.hmso.gov.uksi/si2000/20000128.htm.

Appendix A3

Summary of in-service inspection requirements worldwide

Table A3.1 shows the situation (latest information 1999) regarding the in-service inspection of pressure equipment in countries worldwide (see footnote).

Table A3.1   Summary of in-service inspection requirements worldwide

Country	Is plant 'registration' required?	Are there specified periodicities for inspection?	Inspection period (months) for boilers	Inspection period (months) for pressure vessels	Inspection by government (G) or private (P) agency	Is postponement allowed?	Is an RBI-based approach accepted?
Australia	No	Yes	12–48	24–144	G, P	Yes	Yes
Bulgaria	Yes	Yes	Yes	48–96	G	Yes	No
California	Yes	Yes	12–18	36–60	G	No	No
Czech Republic	Yes	Yes	12	60	G	No	No
Germany	Yes	Yes	NK	NK	P	Yes	Yes
Greece	No	No	None	None	P	Yes	Yes
Hong Kong	Yes	Yes	12–24	24	G	No	No
India	Yes	Yes	12–24	24	G	No	No
Japan	NK	Yes?	NK	NK	NK	No	No
Malaysia	Yes	Yes	12–36	12–36	G	Yes	Yes
New Zealand	No	Yes	12–48	24–144	G, P	Yes	Yes
Republic of China	Yes	Yes	NK	NK	G	No	No
Republic of Korea	Yes	Yes	NK	NK	G	No	No
Singapore	Yes	Yes	12	24	P	Yes	Yes
Slovak Republic	Yes	Yes	12	12	G	No	No
South Africa	Yes	Yes	36	36	G, P	Yes	Yes
Taiwan	Yes	Yes	12–24	NK	G	No	No
Texas	Yes	Yes	12	24	G	No	No
Thailand	Yes	Yes	12	NK	G	No	No
United Kingdom	No	No	NS	NS	P	Yes	Yes

The above information was compiled during a study by a UK consultancy in 1999. The objective of the exercise was to compare how the inspection regimes in the different countries compared in terms of inspection frequency, depth and types of inspection undertaken and the methodology of the inspection regime. The response from individual countries was varied but the above chart represented the situation in 1999. Note that any countries have a system whereby periodicities can be changed as the result of a formally approved RBI study or similar. Courtesy AEATechnology Ltd (now Veritec Ltd).

LNK, not known

NS, non-statutory (i.e. period not defined in statutory instrument)

Appendix A4

EEMUA publications for mechanical plant and equipment

Figure A4.1 shows EEMUA (Engineering Equipment and Materials Users Association) documents relevant to in-service inspection. Contact: EEMUA, 45 Beech Street, London. www.eemua.co.uk.

EEMUA No.	
107	*Recommendations for the Protection of Diesel Engines for Use in Zone 2 Hazardous Areas.*
143	*Recommendations for Tube End Welding: Tubular Heat Transfer Equipment – Part1 – Ferrous Materials.*
147	*Recommendations for the Design and Construction of Refrigerated Liquefied Gas Storage Tanks.*
151	*Liquid Ring Vacuum Pumps and Compressors.*
153	*EEMUA Supplement to ASME B31.3.*
154	*Guidance to Owners on Demolition of Vertical Cylindrical Steel Storage Tanks.*
159	*Users' Guide to the Maintenance and Inspection of Aboveground Vertical Cylindrical Steel Storage Tanks.*
162	*EEMUA Supplement to BS 5500: for the Oil, Gas and Chemical Industries.*
164	*Seal-less Centrifugal Pumps: Class 1.*
167	*EEMUA Specification for Quality Levels for Carbon Steel Valve Castings.*
168	*A Guide to the Pressure Testing of In-Service Pressurized Equipment.*
169	*Specification for High Frequency Electric Welded Line Pipe (Obsolescent).*

Fig. A4.1 EEMUA publications relevant to in-service inspection

EEMUA No.	
170	*Specification for Production Testing of Valves – Part 1 – Ball Valves.*
171	*Specification for Production Testing of Valves – Part 2 – Plug Valves.*
172	*Specification for Production Testing of Valves – Part 3 – Gate Valves.*
173	*Specification for Production Testing of Valves – Part 4 – Butterfly and Globe Valves.*
179	*A Working Guide for Carbon Steel Equipment in Wet H_2S Service.*
180	*Guide for Designers and Users on Frangible Roof Joints for Fixed Roof Storage Tanks.*
182	*Specification for Integral Block and Bleed Valve Manifolds for Direct Connection to Pipework (Incorporating Information Sheet No. 20: Application Guidelines).*
183	*Guide for the Prevention of Bottom Leakage fromVertical, Cylindrical, Steel Storage Tanks.*
184	*Guide to the Isolation of Pressure Relieving Devices.*
185	*Guide for Hot Tapping on Piping and other Equipment.*
188	*Guide for Establishing Operating Periods of Safety Valve.*
190	*Guide for the Design, Construction, and Use of Mounded Horizontal Cylindrical Bulk Storage Vessels for Pressurized LPG at Ambient Temperatures.*
192	*Guide for the Procurement of Valves for Low-Temperature (Non-Cryogenic) Service.*
196	*Valve Purchasers' Guide to the European Pressure Equipment Directive.*
199	*On-Line Leak Sealing of Piping – Guide to Safety Considerations.*
200	*Guide to the Specification, Installation, and Maintenance of Spring Supports for Piping.*

Offshore

144	*90/10 Copper Nickel Alloy Piping for Offshore Applications – Specification: Tubes Seamless and Welded.*
145	*90/10 Copper Nickel Piping for Offshore Applications – Specification: Flanges Composite and Solid.*
146	*90/10 Copper Nickel Alloy Piping for Offshore Applications – Specification: Fittings.*

Fig. A4.1   (Continued)

EEMUA No.	
158	*Construction Specification for Fixed Offshore Structures in the North Sea.*
166	*Specification for Line Pipe for Offshore Pipelines (Obsolescent).*
176	*Specification for Structural Castings for use Offshore.*
194	*Guidelines for Materials Selection and Corrosion Control for Subsea Oil and Gas Production Equipment.*
197	*Specification for Fabrication of Non-Primary Structural Steelwork for Offshore Installation.*
	Noise
104	*Noise: A Guide to Information Required From Equipment Vendors.*
140	*Noise Procedure Specification.*
141	*Guide to the Use of Noise Procedure Specification.*
161	*A Guide to the Selection and Assessment of Silencers and Acoustic Enclosures.*
	General
101	*Lifting Points – A Design Guide.*
105	*Factory Stairways, Ladders and Handrails (Including Access Platforms and Ramps).*
148	*Reliability Specification.*
149	*Code of Practice for the Identification and Checking of Materials of Construction Pressure Systems in Process Plants.*
193	*EEMUA Recommendations for the Training, Development, and Competency Assessment of Inspection Personnel.*
195	*Compendium of EEMUA Information Sheets on Topics Related to Pressure Containing Equipment.*

Fig. A4.1   (Continued)

Appendix A5

SAFeD publications and fact sheets

The list below shows SAFeD (Safety Assessment Federation) publications relevant to in-service inspection. Contact: SAFeD Limited, Nutmeg House, 60 Gainsford Street, Butlers Wharf, London SE1 2NY. www.safed.co.uk.

Pressure Systems: Guidelines on Periodicity of Examinations
Reference: PSG1: 1997 (ISBN 1-901212-10-6).

The information contained in this publication is based on the collective experience of SAFeD member companies and gives practical guidance on the recommended intervals between successive examinations of pressure systems. It also considers those areas that should be investigated when considering an extension of existing intervals.

Potential Hazards Created by Water Hammer in Steam Systems
Reference: FS1: 1997 (ISBN 1-901212-15-7).

Incidents involving water hammer can cause serious or even fatal accidents. This factsheet:

- Outlines the causes of water hammer.
- Strongly recommends the replacement of cast-iron stop valves on steam boilers.
- Presents a five-point action plan to minimize hazards.

The Use of Accredited Inspection Bodies
Reference: FS2: updated 1999 (ISBN 1-901212-203).

This fact sheet explains that inspection bodies accredited to BS EN 45004: 1995 have the technical competence to perform inspection activities, while also possessing the infrastructure to deliver a quality service. Definitions of commonly encountered key terms are included.

Guidelines on the Thorough Examination and Testing of Lifts
Reference: LG1: 1998 (ISBN 1-901212-35-1).

The aim of this publication is to achieve consistency of examinations, agreed methods of testing and harmonized reporting requirements. The publication also presents periodicities for examinations and tests. These guidelines have been prepared in consultation with the Health and Safety Executive and other interested parties in the lift industry, and they take into account current legislation:

- The Provision and Use of Work Equipment Regulations 1998.
- The Lifting Operations and Lifting Equipment Regulations 1998. This Publication replaces HSE Guidance Note PM7.

This is a two-volume set which includes Test Certificates and Declaration Forms of hard copy and on CD.

PUWER 98 Risk-Based Compliance
Reference: PUWER 98: 1999 (ISBN 1-901212-60-2).

This document introduces a new service for assessing and evaluating compliance with PUWER 98 that takes due account of modern risk assessment and 'goal setting' techniques. The SAFeD approach aims to provide 'reasonably practicable' recommendations for risk minimization measures.

PUWER 98 places new obligations on employers to identify, control, and record risks associated with the use of plant and equipment in the workplace. This document draws on the collective expertise of its members – both in relation to the assessment of risks and in relation to the appropriateness of risk improvement measures – to enable a consistent approach to PUWER 98 to be taken.

Shell Boilers: Guidelines for the Examination of Shell-to-Endplate and Furnace-to-Endplate Welded Joints
Reference: SBG1: 1997 (ISBN 1-901212-05).

This publication is written specifically for owners, users, and Competent Persons performing duties under the Pressure Systems Safety Regulations 2000. It describes those procedures considered to be good practice, and incorporates advice based on the authors' extensive experience in assessing defects in shell-to-endplate welds and furnace-to-endplate welds in shell boilers.

Shell Boilers: Guidelines for the Examination of Longitudinal Seams of Shell Boilers
Reference: SBG2: 1998 (ISBN 1-901212-30-0).

This publication is aimed at owners, users, and Competent Persons who have duties under the Pressure System Safety Regulations 2000. Based on methods of measurement developed by inspection organizations, it describes procedures which will reduce the risk of boiler cracking and subsequent explosions.

These guidelines also:

- Provide detailed guidance on ultrasonic testing and report provision.
- Supersede AOTC Guidance Note GN3, produced in the late-1980s.

Appendix A6

The European Pressure Equipment Directive (PED): summary

Introduction – what's the PED all about?

It is mainly about compliance of *new* pressure equipment. Unlike many types of engineering products, the design, manufacture, and operation of pressure equipment is heavily affected by a mass of legislation and regulations. Whereas in the past these were predominantly national requirements, the current trend to European integration has led to a situation whereby the objective is to make the requirements pan-European.

In practice, most of the pan-European requirements influence the inspection of new equipment, rather than the inspection of the equipment in-service during its working life. There are some exceptions however, where the requirements for new equipment impinge on its use, particularly when the equipment is subject to refurbishment, or perhaps re-rating to accommodate changes of process conditions. PED compliance can therefore become an issue in in-service inspections. Pressure equipment that is 'substantially refurbished' or replaced not on a 'like-for-like' basis may need to be PED-compliant if there is any chance that it could be placed on the market for sale. Remember that the situation in the USA is different and more tightly regulated.

It is useful background to understand the basics of European legislation such as the PED, in anticipation of the fact that, despite the concentration on new-construction inspection at the moment, the general philosophy of European legislation will probably spread to in-service inspection as well, as time progresses. It seems probable that, in the future, the UK PSSRs, LOLER, etc. will be replaced by European Directives covering the in-service inspection of pressure equipment and lifting equipment. It is also probable that such directives will be constructed along similar lines to the PED.

The driving forces behind the PED and similar European directives

There are two main driving forces:

- The objective of the European Union is to eliminate (or at least minimize) barriers to trade between EU member states. This means, essentially, that the requirements governing pressure equipment should be the same in all member states. The mechanism is to implement this by issuing European *Directives*.
- The UK (and all other EU member states) is bound by law to comply with European Directives. They do this by reflecting the requirements of the Directives in their own national legislation, in the form of *Statutory Instruments*. The requirements of these are, in turn, reflected in the content of various *Regulations*. In the UK the body charged with implementation is the Department of Trade and Industry (DTI) in the form of its Standards and Technical Regulations Directorate (STRD).

The EU 'new approaches'

Pressure equipment legislation has been influenced by two European philosophies designed, nominally, to encourage the free circulation of products within the EU. These are known under their titles of:

- The new approach to product regulation ('the new approach'). And
- The global approach to conformity assessment ('the global approach').

The idea of 'the new approach' is that products manufactured legally in one EU country should be able to be sold and used freely in the others, without undue restriction. One important factor in this is the idea of technical harmonization of important aspects of the products, known as 'essential requirements'. This means that there is a distinction between the essential requirements of a product and its full technical specification.

The idea of 'the global approach' is that procedures should be in place for reliable conformity assessment of products covered by Directives and that this should be carried out using the principles of 'confidence through competence and transparency' and putting in place a framework for conformity assessment. Some elements of this are:

- Dividing conformity assessment procedures into *modules* (these are specifically relevant to new construction of pressure equipment).
- The designation of bodies operating *conformity assessment procedures* (called 'Notified Bodies').
- The use of the *CE mark*.
- Implementing quality assurance and accreditation systems (such as the EN 45000 and EN ISO 9000 series).

Since 1987, various European Directives have gradually come into force, based on the philosophy of these two approaches. Several with direct relevance to pressure equipment have already been issued and others are in preparation.

The role of technical standards

Traditionally, most EU countries had (and in most cases still have) their own well-established product standards for all manner of manufactured products, including pressure equipment. Inevitably, these standards differ in their technical and administrative requirements, and often in the fundamental way that *compliance* of products with the standards is assured.

Harmonized standards

Harmonized standards are European standards produced (in consultation with member states) by the European standards organizations CEN/CENELEC. There is a Directive 98/34/EC which explains the formal status of these harmonized standards. Harmonized standards have to be *transposed* by each EU country; this means that they must be made available as national standards and that any conflicting standards have to be withdrawn within a given time period.

A key point about harmonized standards is that any product that complies with the standards is automatically assumed to conform to the 'Essential Requirements' of the 'New Approach' European Directive relevant to the particular product. This is known as the 'presumption of conformity'. Once a national standard is transposed from a harmonized standard, then the *presumption of conformity* is carried with it. Note that the following terms appear in various Directives, guidance notes, etc. They are all *exactly the same thing*:

- Essential Safety Requirements (ESRs);
- Essential Requirements;
- Essential Health and Safety Requirements (EHSRs);

Compliance with a harmonized standard is not compulsory; it is voluntary, but compliance with it does infer that a product meets the essential safety requirements (ESRs) of a relevant Directive and the product can then carry the 'CE mark'

National standards

EU countries are at liberty to keep their national product standards if they wish. Products manufactured to these do not, however, carry the 'presumption of conformity' with relevant Directives, hence the onus is on the manufacturer to prove compliance with the ESRs to a Notified Body (on a case-by-case basis). Once compliance has been demonstrated, then the product can carry the CE mark.

The situation for manufacture of new pressure equipment

Pressure equipment is 'a product' so is subject to the type of controls exercised over products as described above. The situation is still developing (and will continue to do so) but Fig. A6.1 broadly summarizes the current situation. Partly for historical reasons, as well as technical ones, vessels which qualify as being 'simple' have their own Simple Pressure Vessels (SPV) Directive published in 1987. Following the full implementation of the Pressure Equipment Directive in 2002, it is the intention that the SPV Directive will be revised or withdrawn. Note that all the requirements shown in Fig. A6.1 refer to the design and manufacture of pressure equipment – they have no direct relevance (yet) to the inspection of pressure equipment during its working life in the UK. This is addressed by the Pressure System Safety Regulations 2000 (PSSRs), which apply to the UK only (see Chapter 5 for details).

Vessel 'statutory' certification

Currently, as the Pressure Equipment Directive (PED) has only been mandatory since May 2002, most pressure equipment in service was constructed to the previous system of 'vessel certification' rather than the PED. This means that most in-service inspections will be of pressure equipment that was constructed to this old system, rather than under the requirements of the PED.

Subject area

The directive covers pressure equipment and assemblies with a maximum allowable pressure (PS) greater than 0.5 bar. Pressure equipment means vessels, piping, safety accessories, and pressure accessories. Assemblies means several pieces of pressure equipment assembled to form an integrated, functional whole.

Intention of legislation

To remove technical barriers to trade by harmonizing national laws of legislation regarding the design, manufacture, marking and conformity assessment of pressure equipment

Coverage

It covers a wide range of equipment such as reaction vessels, pressurized storage containers, heat exchangers, shell and water tube boilers, industrial pipework, safety devices, and pressure accessories. Such equipment is widely used in the chemical, petro-chemical, biochemical, food processing, refrigeration and energy industries, and for power generation.

Implementation

The Pressure Equipment Regulations 1999 (SI 1999/2001) were laid on 19 July 1999 and came into force on 29 November 1999.

The Commission's proposal was submitted to the Council of Ministers on 15 July 1993 and was adopted by the European Parliament and the Council on 29 May 1997. It came into force on 29 November 1999 but compliance with its requirements was optional until 29 May 2002.

The implementation date for the Directive in all member states was 29 November 1999.

During a transitional period up to 30 April 2002, member states could permit the placing on the market of any equipment which complies with the legislation in force in that state on 29 May 1997 (the date of adoption of the Directive).

Included

The design, manufacture and conformity assessment of: Pressure equipment and assemblies with a maximum allowable pressure greater than 0.5 bar including:

- Vessels, piping, safety accessories and pressure accessories.
- Components and sub assemblies.

Exclusions

- The assembly of pressure equipment on site and under the responsibility of the user.
- Items of equipment for military use.
- Equipment for use in transport.
- Equipment which presents a 'relatively low hazard from pressurization'.
- Some equipment which is within the scope of another CE mark Directive (e.g. the Machinery Directive).

Fig. A6.1   PED summary – 'quick reference'

This system of certification was based on four premises:

- The need for certification was imposed or inferred by statutory legislation in the country where the vessel was intended to be *installed and used.*
- The need for certification was imposed or inferred by statutory legislation in the country where the vessel was *manufactured.*
- The need for certification was imposed or inferred by the company that would provide an insurance policy for the vessel itself and second- and third-party liabilities when it was in use.
- The manufacturer, contractor, or end-user *chose* to obtain certification because they felt that:

  - it helped maintain a good standard of design and workmanship;
  - it provided evidence to help show that legal requirements for 'due-diligence' and 'duty-of-care' had been met.

Note that three of these reasons were as a result of certification requirement being imposed (or at least *inferred*) by an external player, the other being a voluntary decision by one or more of the directly involved parties. The main reason for this was that there were some countries in the world where statutory requirements were (and still are) unclear, and sometimes contradictory or non-existent. The more risk-averse vessel manufacturers and contractors often assumed that certification *would* be necessary, even if evidence of this requirement was difficult to find.

Certification was therefore an *attempt* to assure the integrity in a way that was generally accepted by external parties. It used accepted national vessel standards or codes as benchmarks of acceptability and good practice. There was little attempt at harmonization of these standards. Certification did address issues of vessel design, manufacture, and testing, but only insofar as these aspects were imposed explicitly by the relevant standard – not more. Certification was evidence therefore of *code-compliance.* Note that compliance with the ASME code was a special case – because of its statutory implications in the USA.

In order to obtain full certification, the organization that issued the certificate will have had to comply with the activities raised by the relevant code. These differed slightly between codes but the basic requirements were the need to:

- Perform a quite detailed design appraisal.
- Ensure the traceability of the materials of construction.
- Witness NDT activities and review the results.

- Witness the pressure test.
- Monitor the manufacturing process.
- Issue a certificate (e.g. BS PD 5500 'Form X' or its equivalent).

The CE mark – what is it?

CE probably stands for *Communitee Européen*, i.e. the French translation of 'European Community'. It could also represent *Conformité Européen*. Unfortunately, it is far from certain that whoever invented the mark (some bureaucrat in Brussels) had anything particular in mind other than to create a logo which would be universally recognized in the European Union. Given all the national prejudices about language in the different countries of the EU, any original national identity was probably conveniently forgotten by the time it became 'official'. Hence it is best thought of as simply a convenient logo, without any deeper meaning. In its current context, a product can only have the CE mark fixed to it if it complies with all European Directives pertaining to that type of product.

PED – its purpose

The purpose of the PED is to provide for a legal structure whereby pressure equipment can be manufactured and sold throughout the EU without having to go through a local design approval and inspection regime in every member state. Figure A6.1 shows a summary. The objective is also to ensure common standards of safety in all pressure equipment sold within the EU, i.e. manufacturers are able to meet the requirements for approval in any member state of the EU, and do not have to repeat the process when selling goods in any other state.

The general idea is that manufacturers will have their equipment approved in their home country. Manufacturers outside of the EU may also have approvals and test work undertaken at their own factory (in many cases this is obligatory) but responsibility for compliance with the requirements of the Directive will ultimately rest on the person responsible for selling the product. The PED became mandatory in May 2002.

PED – its scope

The PED applies to the design, manufacture, and conformity assessment of pressure equipment and assemblies with a maximum

Typical equipment covered	Typical equipment excluded
• Shell and water tube boilers.	• The assembly of pressure equipment on site and under the responsibility of the user.
• Heat exchangers.	• Items of equipment for military use.
• Plant vessels.	• Equipment for use in transport.
• Pressurized storage containers.	• Equipment which presents a 'relatively low hazard from pressurization'.
• Industrial pipework.	• Some equipment which is within the scope of another CE mark directive (e.g. the machinery directive.)
• Gas cylinders.	
• Certain compressed air equipment.	
• Safety accessories:	
– safety valves,	
– bursting disc safety devices,	
– buckling rods,	
– controlled safety pressure relief systems,	
– pressure switches,	
– temperature switches,	
– fluid level switches,	
(where these are used in safety-related applications)	

Fig. A6.2 The PED – what does it cover?

allowable pressure greater than 0.5 bar. Vessels, piping, safety accessories, and pressure accessories are all included (see Fig. A6.2).

PED – its structure

The Directive defines a number of classifications for pressure equipment, based on the hazard presented by their application. Hazard is determined on the basis of stored energy (pressure–volume product) and the nature of the contained fluid. Assessment and conformity procedures are different for each category, ranging from self-certifica-

Risk category	Applicable modules
I	A
II	A1, D1 or E1
III	B1 + D, B1 + F, B + E, B + C1, H
IV	B + D, B + F, G or H1

Module	Description
A	Internal production control
A1	Internal production control with monitoring of final assessment
B	EC-type examination
B1	EC design examination
C1	Conformity to type
D	Production quality assurance
D1	Production quality assurance
E	Product quality assurance
E1	Product quality assurance
F	Product verification
G	EC unit verification
H	Full quality assurance
H1	Full quality assurance with design examination and special surveillance of final assessment

Fig. A6.3   PED risk categories and modules

tion for the lowest (Category I) hazard up to full ISO 9001 quality management and/or Notified Body type examination for Category IV equipment. Figure A6.3 shows the risk categories and their corresponding 'modules'.

The assessment procedures are arranged in a modular structure and manufacturers have the choice of which modules to select (within predetermined combinations) in order to best suit their application and manufacturing procedures.

PED – Conformity assessment procedures

Figure A6.4 reproduces the product classification table and the nine classification charts given in the PED. Refer to the Regulations themselves or *Guidance Notes on the UK Regulations (URN 99/1147)* for details of the conformity assessment procedures and full statement of the requirements.

The PED essential safety requirements (ESRs)

The PED defines certain requirements about the design and performance of pressure equipment in a series of Essential Safety Requirements (ESRs), contained in Annex 1 of the Directive itself. These are compulsory. Two important clauses from the PED are:

- *Clause 3.* 'The manufacturer is under an obligation to analyse the hazards in order to identify those which apply to his equipment on account of pressure; he must then design and construct it taking account of his analysis.'
- *Clause 4.* 'The essential requirements are to be interpreted and applied in such a way as to take account of the state of the art and current practice at the time of design and manufacture as well as of technical and economic considerations which are consistent with a high degree of health and safety protection.'

The schedule of ESRs is quite long and covers a wide scope of technical issues relating to the safe design and construction of the equipment. Full details are available in *Engineers Guide to Pressure Equipment* by Clifford Matthews, 2001 (Professional Engineering Publications) (ISBN 8-6058-298-2).

Chart 1   Vessels for Group 1 gases

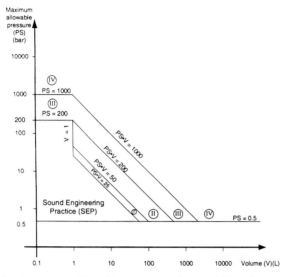

Exceptionally, vessels intended to contain an unstable gas and falling within categories I or II
on the basis of Chart 1 must be classified in category III

Chart 2   Vessels for Group 2 gases

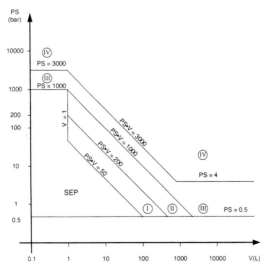

Exceptionally, portable extinguishers and bottles for breathing equipment must be classified
at least in category III

Fig. A6.4   PED – the nine classification charts

Chart 3 Vessels for Group 1 liquids

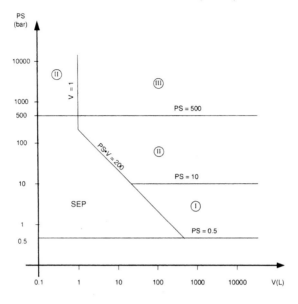

Chart 4 Vessels for Group 2 liquids

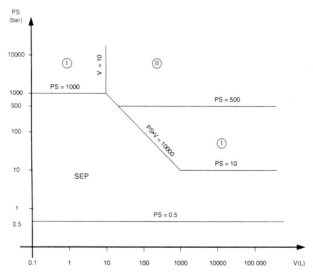

Exceptionally, assemblies intended for generating warm water at temperatures not greater than 110°C which are manually fed with solid fuels and have a product of pressure and volume greater than 50 bar litres, must be subject either to an EC design examination (Module B1) with respect to their conformity with Sections 2.10, 2.11, 3.4, 5(a) and 5(d) of the essential safety requirements, or to full quality assurance (Module H).

Fig. A6.4   (Continued)

Chart 5   Steam generators

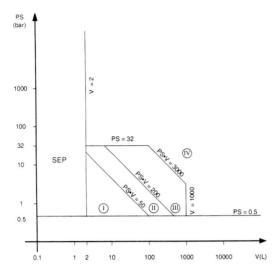

Exceptionally, the design of pressure cookers must be subject to a conformity assessment procedure equivalent to at least one of the category III modules.

Chart 6   Piping for Group 2 gases

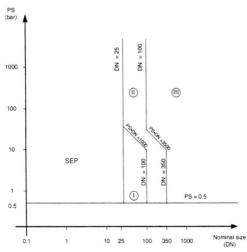

Exceptionally, piping intended for unstable gases and falling within categories I or II must be classified in category III.

Fig. A6.4   (Continued)

Chart 7 Piping for Group 2 gases

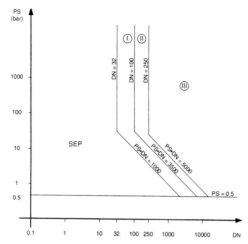

Exceptionally, all piping containing fluids at a temperature greater than 350°C and falling within category II must be classified in category III.

Chart 8 Piping for Group 1 liquids

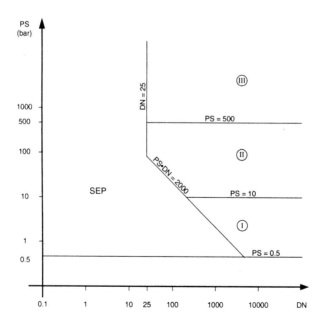

Fig. A6.4 (Continued)

Chart 9   Piping for Group 2 liquids

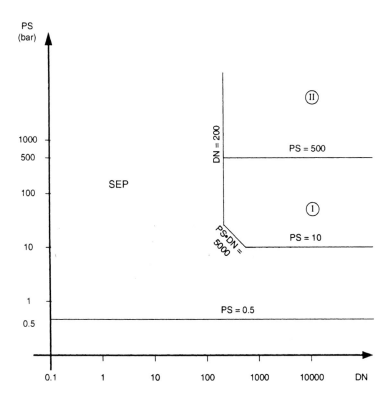

Fig. A6.4   (Continued)

Appendix A7

Degradation mechanisms (refining/ petrochemical applications) as defined in API RP 580

Table A7.1 Degradation mechanisms – thinning

Deterioration mechanism	Description	Behaviour	Key variables	Examples
Hydrochloric acid	Typically causes localized corrosion in carbon and low-alloy steel, particularly at initial condensation points (<400 °F). Austenitic stainless steels experience pitting and crevice corrosion. Nickel alloys can corrode under oxidizing conditions	Localized	pH, acid %, temperature, materials of construction	Crude unit atmospheric column overhead, hydrotreating effluent trains, catalytic reforming effluent, and regeneration systems
Galvanic corrosion	Occurs when two dissimilar metals are joined and exposed to an electrolyte	Localized	Joined materials of construction, distance in galvanic series	Seawater and some cooling-water services
Ammonia bisulphide	Highly localized metal loss due to erosion–corrosion in carbon steel and admiralty brass	Localized	Chemical concentrations, velocity, pH	Formed by thermal or catalytic cracking in hydrotreating, hydrocracking, coking, catalytic cracking, amine treating and sour water effluent and gas separation systems

Table A7.1 (Continued)

Deterioration mechanism	Description	Behaviour	Key variables	Examples
Carbon dioxide	Aqueous CO_2 corrosion of carbon and low-alloy steels is an electrochemical process involving the anodic dissolution of iron and the cathodic evolution of hydrogen. The reactions are often accompanied by the formation of films of $FeCO_3$ (and/or Fe_3O_4) that can be protective or non-protective depending on the conditions	Localized	Carbon dioxide concentration, process conditions	Refinery steam condensate system, hydrogen plant and the vapour recovery section of catalytic cracking unit
Sulphuric acid	Very strong acid that causes metal loss in various materials and depends on many factors	Localized	Acid %, pH, material of construction, temperature	Sulphuric acid alkylation units, demineralized water
Hydrofluoric acid	Very strong acid that causes metal loss in various materials	Localized	Acid %, pH, material of construction, temperature, velocity, oxidants	Hydrofluoric acid alkylation units, demineralized water
Phosphoric acid	Weak acid that causes metal loss. Generally added for biological corrosion inhibition in water treatment	Localized	Acid %, pH, material of construction, temperature	Water treatment plants

Table A7.1 (Continued)

Deterioration mechanism	Description	Behaviour	Key variables	Examples
Phenol (carbolic acid)	Weak organic acid causing corrosion and metal loss in various alloys	Localized	Acid %, pH, material of construction, temperature, amine type and concentration, material of construction, temperature, acid gas loading, velocity	Heavy oil and dewaxing plants
Atmospheric corrosion	The general corrosion process occurring under atmospheric conditions where carbon steel (Fe) is converted to iron oxide Fe_2O_3	General uniform corrosion	Presence of oxygen, temperature range and the availability of water/moisture	This process is readily apparent in high-temperature processes where carbon steels have been used without protective coatings (steam piping for example)
Corrosion under insulation (CUI)	CUI is a specific case of atmospheric corrosion where the temperatures and the concentrations of water/moisture can be higher. Often residual/trace corrosive elements can also be leached out of insulation material creating a more corrosive environment	General to highly localized	Presence of oxygen, temperature range, and the availability of water/moisture and corrosive constituents within the insulation	Insulated piping/vessels

Table A7.1 (Continued)

Deterioration mechanism	Description	Behaviour	Key variables	Examples
Soil corrosion	Metallic structures in contact with soil will corrode	General to localized	Material of construction, soil characteristics, type of coating	Tank bottoms, underground piping
High-temperature sulphide corrosion without H_2	A corrosive process similar to atmospheric corrosion in the presence of oxygen. In this case the carbon steel (Fe) is converted in the presence of sulphur to iron sulphide (FeS). Conversion rate (and therefore corrosion rate) is dependent on temperature of operation and sulphur concentration	General uniform corrosion	Sulphur concentration and temperature	All locations where there is sufficient temperature (450 °F minimum) and sulphur is present in quantities greater than 0.2%. Common locations are crude, and hydroprocessing units
High-temperature sulphidic corrosion with H_2	With the presence of hydrogen, a significantly more aggressive case of sulphidation (sulphidic corrosion) can exist	General uniform corrosion	Sulphur and hydrogen concentration and temperature	All locations where there is sufficient temperature (450 °F minimum) and sulphur is present in quantities greater than 0.2%. Areas of hydro-processing units–reactor feed downstream of the hydrogen mix point, the reactor, the reactor effluent, and the re-cycle hydrogen gas including the exchangers, heaters, separators, piping, etc.

Table A7.1 *(Continued)*

Deterioration mechanism	Description	Behaviour	Key variables	Examples
Naphthenic acid corrosion	Naphthenic acid corrosion is the attack of steel alloys by organic acids that condense in the temperature range 35 to 75 °F. The presence of potentially harmful amounts of naphthenic acids in crude may be signified by a neutralization number above 0.5	Localized corrosion	Naphthenic/organic acid concentration and temperature	Middle section of a vacuum column in a crude unit can also occur in atmospheric distillation units, furnaces and transfer lines
Oxidation	A high-temperature corrosion reaction where metal is converted to a metal oxide above specific temperatures (975 °F for carbon steel, 140 °F for 9 Cr–1 Mo)	General uniform corrosion	Temperature, presence of air, material of construction	Outside of furnace tubes, furnace tube hangers, and other internal furnace components exposed to combustion gases containing excess air

Table A7.2 Degradation mechanisms – stress corrosion cracking

Deterioration mechanism	Description	Behaviour	Key variables	Examples
Chloride cracking	Cracking that can initiate from the inner or outer diameter of austenitic stainless steel equipment, primarily due to fabrication or residual stresses. Some applied stresses can also cause cracking	Trans-granular cracking	Acid (chloride) concentration, pH, material of construction, temperature, fabrication, stresses approaching yield	Externally present in equipment with poor insulation and weatherproofing, downwind of cooling water spray and equipment exposed to fire water. Internally wherever chlorides can be present with water such as atmospheric column overheads of crude units and reactor effluent condensing streams
Caustic cracking	Cracking primarily initiated from the inner diameter of carbon steel equipment, primarily due to fabrication or residual stresses	Intergranular and transgranular cracking	Caustic concentration, pH, material of construction, temperature, stress	Caustic treating sections, caustic service, crude unit feed preheat desalting, sour water treatment, steam systems

Table A7.2 (Continued)

Deterioration mechanism	Description	Behaviour	Key variables	Examples
Polythionic acid cracking	Cracking of austenitic stainless steels in the sensitized condition (due to high-temperature exposure or welding) in the presence of polythionic acid in wet, ambient conditions. Polythionic acid is formed by a conversion of FeS in the presence of water and oxygen	Inter-granular cracking	Material of construction, sensitized microstructure, presence of water, polythionic acid	Generally occurs in austenitic stainless steel materials in catalytic cracking unit reactor and flue gas systems, desulphurizer furnaces and hydroprocessing units
Ammonia cracking	Cracking of carbon steel and admiralty brass	Inter-granular cracking	Material of construction, temperature, stress	Generally present in ammonia production and handling such as overhead condensation where ammonia is a neutralizer

Table A7.2   (Continued)

Deterioration mechanism	Description	Behaviour	Key variables	Examples
Hydrogen-induced cracking/stress orientated hydrogen-induced cracking	Occurs in carbon and low-alloy steel materials in the presence of water and H_2S. Deterioration of the material properties is caused by atomic hydrogen generated through corrosion diffusing into the material and reacting with other atomic hydrogen to form molecular hydrogen gas in inclusions of the steel. Deterioration can take the form of blisters in stress-relieved equipment and cracking in non-stress-relieved equipment	Planar cracks (blisters), trans-granular cracks as blisters progress toward welds	H_2S concentration, water, temperature, pH, material of construction	Anywhere that H_2S is present with water; such as crude units, catalytic cracking compression and gas recovery, hydroprocessing, sour water, catalytic reforming and coker units
Sulphide stress cracking	Occurs in carbon and low-alloy steel materials in the presence of water and H_2S	Trans-granular cracking, normally associated with fabrication, attachment, and repair welds	H_2S concentration, water, temperature, pH, material of construction, post-weld heat treatment condition, hardness	Anywhere that H_2S is present with water, such as crude units, catalytic cracking compression and gas recovery, hydroprocessing, sour water, catalytic reforming and coker units

Table A7.2 (Continued)

Deterioration mechanism	Description	Behaviour	Key variables	Examples
Hydrogen blistering	Occurs in carbon and low-alloy steel materials in the presence of water and H_2S. Deterioration of the material properties is caused by atomic hydrogen generated through corrosion diffusing into the material and reacting with other atomic hydrogen to form molecular hydrogen gas in inclusions of the steel. Deterioration can take the form of blisters in stress-relieved equipment and cracking in non-stress-relieved equipment.	Planar cracks (blisters)	H_2S concentration, water, temperature, pH, material of construction	Anywhere that H_2S is present with water such as crude units, catalytic cracking compression and gas recovery, hydroprocessing, sour water, catalytic reforming and coker units
Hydrogen cyanide	Presence of hydrogen cyanide can promote hydrogen deterioration by destabilizing the iron sulphide protective surface scale	Planar cracks (blisters) and trans-granular cracking	Presence of HCN, H_2S concentration, water, temperature, pH, material of construction	Anywhere that H_2S is present with water such as crude units, catalytic cracking compression and gas recovery, hydroprocessing, sour water, catalytic reforming and coker units

Table A7.3 Degradation mechansims – metallurgical and environmental failures

Deterioration mechanism	Description	Behaviour	Key variables	Examples
High temperature hydrogen attack	Occurs in carbon and low-alloy steel materials in the presence of high temperature and hydrogen, usually as a part of the hydrocarbon stream. At elevated temperatures ($>500\,°F$), deterioration of the material properties is caused by methane gas forming fissures along the grain boundaries. Molecular hydrogen, as a part of the process composition diffuses into the material and reacts with carbon from the steel, forming methane gas	Inter-granular fissure cracking	Material of construction, hydrogen partial pressure, temperature, time in service	Typically occurs in reaction sections of hydrocarbon processing units such as hydrodesulphurizers, hydrocrackers, hydroforming and hydrogen production units
Grain growth	Occurs when steels are heated above a certain temperature, beginning about $1100\,°F$ for low-carbon steel and most pronounced at $1350\,°F$. Austenitic stainless steels and high nickel–chromium alloys do not become subject to grain growth until heated to above $1650\,°F$	Localized	Maximum temperature reached, time at maximum temperature, material of construction	Furnace tubes failures

Table A7.3 *(Continued)*

Deterioration mechanism	Description	Behaviour	Key variables	Examples
Graphitiza-tion	Occurs when the normal pearlite grains in steels decompose into soft weak ferrite grains and graphite nodules usually due to long-term exposure in the 825–1400 °F range	Localized	Material of construction, temperature, and time of exposure	FCC reactor overhead
Sigma phase embrittlement	Occurs when austenitic and other stainless steels with more than 17% chromium are held in the range of 1000–1500 °F for extended time periods	General-ized	Material of construction, temperature, and time of exposure	Cast furnace tubes and components, regenerator cyclones in FCC units
885 °F embrittlement	Occurs after aging of ferrite-containing stainless steels at 650–1000 °F and produces a loss of ambient temperature ductility	General-ized	Material of construction, temperature	Cracking of wrought and cast steels during shutdowns
Temper embrittlement	Occurs when low-alloy steels are held for long periods of time in temperature range of 700–1050 °F. There is a loss of toughness that is not evident at operating temperature but shows up at ambient temperature and can result in brittle fracture	General-ized	Material of construction, temperature, and time of exposure	During shutdown and start-up conditions the problem may appear for equipment in older refinery units that have operated long enough for this condition to develop

Table A7.3   (Continued)

Deterioration mechanism	Description	Behaviour	Key variables	Examples
Liquid metal embrittlement	Form of catastrophic brittle failure of a normally ductile metal caused when it is in, or has been in, contact with a liquid metal and is stressed in tension	Localized	Material of construction, tension stress, presence of liquid metal	Mercury is found in some crude oils and subsequent refinery distillation can condense and concentrate it at low spots in equipment such as condenser shells
Carburization	Caused by carbon diffusion into the steel at elevated temperatures. The increased carbon content results in an increase in the hardening tendency of ferritic steels. When carburized steel is cooled a brittle structure can result	Localized	Material of construction, temperature and time of exposure	Furnace tubes having coke deposits are a good candidate for carburization
Decarburization	Loss of carbon from the surface of a ferrous alloy as a result of heating in a medium that reacts with carbon	Localized	Material of construction, temperature	Carbon steel furnace tubes. Result of excessive overheating (fire)
Metal dusting	Highly localized carburization of steels exposed to mixtures of hydrogen, methane, CO, CO_2, and light hydrocarbons in the temperature range of 900–1500 °F.	Localized	Temperature, process stream composition	Dehydrogenation units, fired heaters, coker heaters, cracking units, and gas turbines
Selective leaching	Preferential loss of one alloy phase in a multiphase alloy	Localized	Process stream flow conditions, material of construction	'Admiralty brass' tubes used in refinery cooling water systems

Table A7.4 Degradation mechanisms – mechanical failures

Deterioration mechanism	Description	Behaviour	Key variables	Examples
Incorrect or defective materials	Engineering failures can happen due to incorrect or defective materials being installed	N/A	Equipment design, operating procedures	A manufacturer may substitute what they consider to be an equivalent or better material than specified. A stainless steel fitting is not necessarily an improvement over a carbon steel fitting, especially as far as pitting corrosion or SCC is concerned
Mechanical fatigue	Failure of a component by cracking after the continued application of cyclic stress which exceeds the material's endurance limit	Localized	Cyclic stress level, material of construction	Reciprocating parts in pumps and compressors and the shafts of rotating machinery
Corrosion fatigue	Form of fatigue where a corrosion process, such as pitting, encourages the mechanical fatigue process	Localized	Cyclic stress, material of construction, pitting potential of the process fluid	Steam drum headers, boiler tubes
Cavitation	Caused by the rapid formation and collapse of vapour bubbles in liquid at a metal surface as a result of pressure variations	Localized	Pressure head value along the flow of process stream	Reverse faces of pump impellers, elbow, etc.

Table A7.4 (Continued)

Deterioration mechanism	Description	Behaviour	Key variables	Examples
Mechanical deterioration	Typical examples are the misuse of tools and equipment, wind deterioration, careless handling when equipment is moved or erected	N/A	Equipment design, operating procedures	Flange faces and other machined seating surfaces may be damaged when not protected
Overloading	Occurs when loads in excess of the maximum permitted by design are applied to equipment	N/A	Equipment design, operating procedures	Hydrostatic testing can overload supporting structures due to excess weight applied. Thermal expansion and contraction can cause overloading problems
Overpressur-ing	Application of pressure in excess of the maximum allowable working pressure (MAWP) of the equipment under consideration	N/A	Equipment design, operating procedures	Excess heat as a result of upset process condition can result in overpressuring, blocking off equipment which is not designed to handle full process pressure
Brittle fracture	Loss of ductility, i.e. the material has low notch toughness or poor impact strength	Localized	Material of construction, temperature	Can occur during equipment pressurization in absence of precautionary measures

Table A7.4 (Continued)

Deterioration mechanism	Description	Behaviour	Key variables	Examples
Creep	High-temperature mechanism in which continuous plastic deformation of a metal occurs while under stresses below the normal yield strength	Localized	Material of construction, temperature, applied stress	Furnace tubes and supports
Stress rupture	Time to failure for a metal at elevated temperatures under applied stress below its normal yield strength	Localized	Material of construction, temperature, applied stress, time of exposure	Furnace tubes
Thermal shock	Occurs when large and non-uniform thermal stresses develop over a relatively short time in a piece of equipment due to differential expansion or contraction	Localized	Equipment design, operating procedures	Associated with occasional, brief flow interruptions or during a fire
Thermal fatigue	Thermal fatigue differs from thermal shock in that the rate of temperature changes experienced is much greater and the magnitude of the temperature gradient is much less	Localized	Equipment design, operating procedures	Bypass valves and piping with heavy weld reinforcement in cyclic temperature service are typically prone to thermal fatigue

Appendix A8

European and American associations and organizations relevant to pressure equipment and inspection activities

Acronym	Organization	URL	Postal address	Telephone	Fax
ABMA	American Bearing Manufacturers Association	www.abma-dc.org	2025 M Street NW Suite 800 Washington, DC 20036	00-1-(202)-367-1155	00-1-(202)-367-2155
ABS	American Bureau of shipping (UK)	www.eagle.org	ABS House 1 Frying Pan Alley London E1 7HR	00-44-(207)-247-3255	00-44-(207)-377-2453
ACEC	American Consulting Engineers Council	www.acec.org	1015 15th Street NW #8O2 Washington DC 20005	00-1-(202)-347-7474	00-1-(202)-898-0068
AGMA	American Gear Manufacturers Association	www.agma.org	1500 King Street Suite 201 Alexandria, VA 22314	00-1-(703)-684-0211	00-1-(703)-684-0242
AIE	American Institute of Engineers	www.members-aie.org	1018 Aopian Way El Sobrante CA 94803	00-1-(510)-223-8911	00-1-(510)-223-8911
AISC	American Institute of Steel Construction Inc.	www.aisc.org	One, E Wacker Drive Suite 3100 Chicago IL 60601-2001	00-1-(312)-670-2400	00-1-(312)-670-5403
AISE	Association of Iron and Steel Engineers (USA)	www.aise.org	3 Gateway Center Suite 1900 Pittsburgh PA 15222-1004	00-1-(412)-281-6323	00-1-(412)-281-4657

Acronym	Organization	URL	Postal address	Telephone	Fax
AISI	American Iron and Steel Institute	www.steel.org	1101 17th SE NW Suite 1300 Washington DC 20036	00-1-(202)-452-7100	00-1-(202)-463-6573
ANS	American Nuclear Society	www.ans.org	555 North Kensington Avenue La Grange Park IL 60526	00-1-(708)-352-6611	00-1-(708)-352-0499
ANSI	American National Standards Institute	www.ansi.org	11 West 42nd Street New York NY 10036	00-1-(212)-642-4900	00-1-(212)-398-0023
AP	APAVE UK Ltd	www.apave-uk.com	Gothic House Barker Gate Nottingham NG1 1JU	000-44-(115)-955 1880	000-44-(115)-955 1881
API	American Petroleum Institute	www.api.org	1220 L Street NW Washington DC 20005	00-1-(202)-682-8000	00-1-(202)-682-8232
ASER-COM	Association of European Refrigeration Compressor Manufacturers	www.hvacmall.com	C/o Copeland GmbH Eichborndamm 141-175 D-1000 Berlin Germany	00-49-(30)-419-6352	00-49-(30)-419-6205

Acronym	Organization	URL	Postal address	Telephone	Fax
ASHRAE	American Society of Heating, Refrigeration, and Air Conditioning Engineers	www.ashrae.org	1791 Tullie Circle NE Atlanta GA 30329	00-1-(404)-636-8400	00-1-(404)-321-5478
ASME	American Society of Mechanical Engineers	www.asme.org	3 Park Avenue New York NY 10016-5990	00-1-(973)-882-1167	00-1-(973)-882-1717
ASNT	American Society for Non-Destructive Testing	www.asnt.org	1711 Arlington Lane Columbus OH 43228-0518	00-1-(614)274-6003	00-1-(614)-274-6899
ASTM	American Society for Testing of Materials	www.ansi.org	100 Barr Harbor Drive W. Conshohocken PA 19428-2959	00-1-(610)-832-9585	00-1-(610)-832-9555
AWS	American Welding Society	www.awweld.org	550 NW Le Jeune Road Miami FL 33126	00-1-(305)-443-9353	00-1-(305)-443-7559
AWWA	American Water Works Association Inc	www.awwa.org	6666 W Quincy Avenue Denver CO 80235	00-1-(303)-794-7711	00-1-(303)-794-3951
BCAS	British Compressed Air Society	www.britishcompressedair.co.uk	33-34 Devonshire Street London W1G 6YP	00-44-(207)-935-2464	00-44-(207)-935-2464

Acronym	Organization	URL	Postal address	Telephone	Fax
BCEMA	British Combustion Equipment Manufacturers Association	www.bcema.co.uk	The Fernery Market Place Midhurst West Sussex GU29 9DP	00-44-(1730)-812782	00-44-(1730)-813366
BFPA	British Fluid Power Association	www.bfpa.co.uk	Cheriton House Cromwell Business Park Chipping Norton Oxon OX7 5SR	00-44-(1608)-647900	00-44-(1608)-647919
BGA	British Gear Association	www.bga.org.uk	Suite 43 Inmex Business Park Shobnall Road Burton on Trent Staffordshire DE14 2AU	00-44-(1283)-515521	00-44-(1283)-515841
BIE	British Inspecting Engineers	www.bie-international.com	Chatsworth Technology Park Dunston Road Chesterfield D41 8XA	000-44-(1246)-260260	000-44-(1246)-260919
B.Inst. NDT	British Institute of Non Destructive Testing	www.bindt.org	1 Spencer Parade Northampton NN1 5AA	000-44-(1604)-259056	000-44-(1604)-231489

Acronym	Organization	URL	Postal address	Telephone	Fax
BPMA	British Pump Manufacturers Association	www.bpma.org.uk	The McLaren Building 35 Dale End Birmingham B4 7LN	00-44-(121)-200-1299	00-44-(121)-200-1306
BSI	British Standards Institution	www.bsi.org.uk	Marylands Avenue Hemel Hempstead Herts HP2 4SQ	00-44-(1442)-230442	000-44-(1442)-231442
BVAMA	British Valve and Actuator Manufacturers association	www.bvama.org.uk	The MacLaren Building 35 Dale End Birmingham B4 7LN	000-44-(121)-200-1297	000-44-(121)-200-1308
CEN	European Committee for Standardisation (Belgium)	www.cenorm.be www. newapproach.org	36, rue de Stassart B-1050 Brussels Belgium	00-(32)-2-550-08-11	00-(32)-2-550-08-19
DNV	Det Norske Veritas (UK)	www.dnv.com	Palace House 3 Cathedral Street London SE1 9DE	00-440-(207)-6080	00-440(207)-6048

Acronym	Organization	URL	Postal address	Telephone	Fax
DTI	DTI STRD 5 (UK)	www.dti.gov.uk/strd	Peter Rutter, STRD5, Department of Trade and Industry, Room 326, 151 Buckingham Palace Road, London SW1W 9SS	00-44-(207)-2151437	
DTI	DTI Publications Orderline (UK)			00-44-(870)-15025500	00-44-(870)-1502333
EC	The Engineering Council (UK)	www.engc.org.uk	10 Maltravers Street, London WC2R 3ER	00-440-(207)-240 7891	00-440-(207)-240 7517
EIS	Engineering Integrity Society (UK)	http://www.demon.co.uk/e-i-s	5 Wentworth Avenue Sheffield S11 9QX	00-44-(114)-262-1155	00-44-(114)-262-1120
EMA	Engine Manufacturers Association (USA)	www.engine-manufacturers.org	2 N.LaSalle Street, Suite 2200 Chicago IL 606	00-1-(312)-827-8700	00-1-(312)-827-8737
FCI	Fluid Controls Institute Inc (USA)	www.fluidcontrols institite.org	PO Box 1485 Pompano Beach FL 3306	00-1-(216)-241-7333	00-1-(216)-241-0105

Acronym	Organization	URL	Postal address	Telephone	Fax
FMG	Factory Mutual Global (USA)	www.fmglobal.com	Westwood Executive Center 100 Lowder Brook Drive Suite 1100 Westwood, MA 02090-1190	00-1-(781)-326-5500	00-1-(781)-326-6632
HI	Hydraulic Institute (USA)	www.pumps.org	9 Sylvian Way Parsippany NJ 07054	00-1-(973)-267-9700	00-1-(973)-267-9055
HMSO	Her Majesty's Stationery Office	www.hmso.gov.uk/legis.htm www.hmso.gov.uk/si www.legislation.hmso. gov.uk			
HSB	HSB Inspection Quality Ltd (UK)	www.hsbiql.co.uk	11 Seymour Court Tudor Road Manor Park Runcorn Cheshire WA7 1SY	00-44-(1928)-579595	00-44-(1928)-579623
HSE	HSEs InfoLine (fax enquiries)	www.hse.gov.uk/ hsehome.htm	HSE Information Centre, Broad Lane, Sheffield S3 7HQ		00-44-(114)-2892333
HSE	HSE Books (UK)	www.hse.gov.uk/ hsehome.htm	PO Box 1999, Sudbury, Suffolk CO10 6FS	00-44-(1787)-881165	00-44-(1787)-313995

Acronym	Organization	URL	Postal address	Telephone	Fax
HTRI	Heat Transfer Research Inc. (USA)	www.htrinet.com	1500 Research Parkway Suite 100 College Station TX 77845	00-1-(409)-260-6200	00-1-(409)-260-6249
IGTI	International Gas Turbine Institute(ASME)	www.asme.org/igti	5775 -B Glenridge Drive #370 Atlanta, G 30328	00-1-(404)-847-0072	00-1-(404)-847-0151
IMechE	The Institution of Mechanical Engineers (UK)	www.imeche.org.uk	1 Birdcage Walk London, SW1H 9JJ	00-44-(207)-222-7899	00-44-(207)-222-4557
IoC	The Institute of Corrosion (UK)	www.icorr.demon.co.uk	4 Leck Street Leighton Buzzard Bedfordshire LU7 9TQ	00-44-(1525)-851-771	00-44-(1525)-376-690
IoE	The Institute of Energy (UK)	http://www.instenergy.org.uk	18 Devonshire Street London W1N 2AU	00-44-(207)-580-7124	00-44-(207)-580-4420
IoM	The Institute of Materials (UK)	http://www.instmat.co.uk	1 Carlton House Terrace London SW1Y 5DB	00-44-(207)-451-7300	00-44-(207)-839-1702
IPLantE	The Institution of Plant Engineers (UK)	http://www.iplante.org.uk	77 Great Peter Street Westminster London SW1P 2EZ	00-44-(207)-233-2855	00-44-(207)-233-2604

Acronym	Organization	URL	Postal address	Telephone	Fax
IQA	The Institute of Quality Assurance (UK)	http://www.iqa.org	12 Grosvenor Crescent London SW1X 7EE	00-44-(207)-245-6722	00-44-(207)-245-6755
ISO	Intermational Standards Organization (Switzerland)	www.iso.ch	PO Box 56 CH-1211 Geneva Switzerland	00-22-749-011	00-22-733-3430
LR	Lloyd's Register (UK)	www.lrqa.com	71 Fenchurch Street London EC3M 4BS	00-44-(207)-7099166	00-44-(207)-488-4796
MSS	Manufacturers Standardization Society of the Valve and Fittings Industry (USA)	www.mss-hq.com	127 Park Street NE Vienna VA 22180-4602	00-1-(703)-281-6613	00-1-(703)-281-6671
NACE	National Association of Corrosion Engineers (USA)	www.nace.org	1440 South Creek Drive Houston TX 7708	00-1-(281)-228-6200	00-1-(281)-228-6300
NBBPVI	National Board of Boiler and Pressure Vessel Inspectors (USA)	www.nationalboard.org	1155 North High Street Columbus OH 43201	00-1-(614)-888-8320	00-1-(614)-888-0750
NFP	National Fire Protection Association (USA)	www.nfpa.org	1 Batterymarch Park PO Box 9101 Quincy MA 02269-9101	00-1-(617)-770-3000	00-1-(617)-770-0700

Acronym	Organization	URL	Postal address	Telephone	Fax
NFPA	National Fluid Power Association (USA)	www.nfpa.com	3333 N Mayfair Road Milwaukee WI 53222-3219	00-1-(414)-778-3344	00-1-(414)-778-3361
NIST	National Institute of Standards and Technology (USA)	www.nist.gov	100 Bureau Drive Gaithersburg MD 20899-0001	00-1-(301)-975-8205	00-1-(301)-926-1630
PDA	Pump Distributors Association	www.pda-uk.com	5 Chapelfield Orford Woodbridge Suffolk IP12 2HW	00-44-(1394)-450181	00-44-(1394)-450181
PFI	Pipe Fabrication Institute (USA)	www.pfi-institute.org	655 32nd Avenue, Suite 201 Lachine Quebec, Canada HT8 3G6	00-1-(514)-634-3434	00-1-(514)-634-9736
PS	Plant Safety Ltd (UK)	www.plantsafety.co.uk	Parklands Wilmslow Road Didsbury Manchester M20 2RE	00-44-(161)-4464600	00-44-(161)-4462506
RAB	Registrar Accreditation Board (USA)	www.rabnet.com	PO Box 3003 Milwaukee WI 53201-3005	00-1-(888)-722-2440	00-1-(414)-765-8661
RSA	Royal Sun Alliance Certification Services Ltd (UK)	www.royal-and-sunalliance.com/	17 York Street Manchester M2 3RS	00-44-(161)-2353375	00-44-(161)-2353702

Acronym	Organization	URL	Postal address	Telephone	Fax
SAE	Society of Automotive Engineers	www.sae.org	400 Commonwealth Drive Warrendale PA 105096-001	00-1-(724)-776-4841	00-1-(724)-776-5760
SAFeD	Safety Assessment Federation (UK)	www.safed.co.uk	Nutmeg House 60 Gainsford Street Butlers Wharf London SE1 2NY	00-44-(207)-403-0987	00-44-(207)-403-0137
SGS	SGS (UK) Ltd	www.sgs.com	SGS(UK) SGS House Johns Lane Tividale, Oldbury West Midlands B69 3HX	00-44-(121)-5206454	00-44-(121)-5223532
TEMA	Tubular Exchanger Manufacturers Association Inc.	www.tema.org	25 N Broadway Tarrytown New York, NY 10591	00-1-(914)-332-0040	00-1-(914)-332-1541
TUV	TUV (UK) Ltd	www.tuv-uk.com	TUV(UK) Ltd Surrey House Surrey Street Croydon CR9 1XZ	00-44-(208)-6807711	00-44-(208)-6804035
TWI	TWI Certification Ltd (UK)	www.twi.co.uk	Granta Park Great Abington Cambridge CB1 6AL	00-44-(1223)-891162	00-44-(1223)-894219

Acronym	Organization	URL	Postal address	Telephone	Fax
UKAS	The United Kingdom Accreditation Service	www.ukas.com	21-47 High Street Feltham Middlesex TW13 4UN	00-44-(208)-917-8554	00-44-(208)-917-8500
VGT	Verenigning Gas Turbine (Dutch Gas Turbine Association)	www.vgt.org/vgt	Burgemeester Verderiaan 13 3544 AD Utrecht PO Box 261 3454 ZM De Meern Netherlands	00-31-(30)-669-1966	00-31-(30)-669-1969
WJS	The Welding and Joining Society (UK)	http://www.twi.co.uk/ members.wjsinfo.html	Granta Park Great Abington Cambridge CB1 6AL	00-44-(1223)-891162	00-44-(1223)-894219
ZURICH	Zurich Engineering Ltd (UK)	www.zuricheng.co.uk	54 Hagley Rd Edgbaston Birmingham B16 8QP	00-44-(121) 4561311	00-44-(121) 4561754

Index